ASHEVILLE-BUNCOMBE TECHNICAL INSTITUTE

NORTH CAROLINA
STATE BOARD OF EDUCATION
DEPT. OF COMMUNITY COLLEGES
LIBRARIES

Asheville-Buncombe Technical Institute
LIBRARY
340 Victoria Road
Asheville, North Carolina 28801

DISCARDED

DEC - 6 2024

DAM GEOLOGY

To
CARLO SEMENZA

He has shown us the safety of a cupola dam, in Engineering; and the danger of submerging a natural valley, in Geology.

DAM GEOLOGY

R. C. S. WALTERS
B.Sc., A.K.C., F.I.C.E.

*Past President of the Institution of Water Engineers;
Past President of the British Section of the Société
des Ingénieurs Civils de France; Sometime
Member of Council of the Geological
Society*

LONDON
BUTTERWORTHS

THE BUTTERWORTH GROUP

ENGLAND:
BUTTERWORTH & CO (PUBLISHERS) LTD.
LONDON: 88 Kingsway, WC2B 6AB

AUSTRALIA:
BUTTERWORTH & CO (AUSTRALIA) LTD.
SYDNEY: 586 Pacific Highway, Chatswood NSW 2067
MELBOURNE: 343 Little Collins Street, 3000
BRISBANE: 240 Queen Street, 4000

CANADA:
BUTTERWORTH & CO (CANADA) LTD.
TORONTO: 14 Curity Avenue, 374

NEW ZEALAND:
BUTTERWORTH & CO (NEW ZEALAND) LTD.
WELLINGTON: 26-28 Waring Taylor Street, 1
AUCKLAND: 35 High Street, 1

SOUTH AFRICA:
BUTTERWORTH & CO (SOUTH AFRICA) (PTY) LTD.
DURBAN: 152-154 Gale Street.

First published 1962

Second edition 1971

©

Butterworth & Co (Publishers) Ltd. 1962, 1971

ISBN 0 408 70191 9

Printed and bound in Great Britain by R. J. Acford, Ltd., Industrial Estate, Chichester, Sussex.

PREFACE TO SECOND EDITION

This book is intended to be a kind of bridge between geology and engineering. The introductory chapters are followed by notes on those sites of geological interest with which I have been connected and those which I have visited (with few exceptions). I hope that the factual data presented, both geological and engineering, will be of some interest to others.

In preparing the second edition, I find that geological strata cannot be updated but their behaviour, after putting dams upon them, may be altered as we now have records of behaviour of some of the dams described in the first edition.

Scarcely an engineering paper has been written during the last ten years without considerable space being devoted to geology, whereas, formerly, the subject was often dismissed with the information that the foundation rock was base rock, which probably described it in both senses of the word! In view of the great interest in this subject a new chapter on Reservoir Geology has been added in this edition.

In order to deal adequately, and briefly, with recent dam constructions and their foundations, I have in mind those who may need to be acquainted with many facets of engineering which are dependent on geology. In this edition the introductory chapters which appeared in the previous edition on the influences of geological strata, hardness, slopes of valleys, slips and subsidences, coal mining, glacial moraines, earthquakes, grouting, have been retained.

In Part 2 the sections on Austrian and Swiss dams have been greatly extended and numerous additions have been made to other chapters notably those dealing with British, Italian and American dams. A description of the Kariba dam has also been included through the kindness of Mr. T. A. L. Paton and Dr. John Knill. In addition, new chapters have been added covering the dams of Australia, Canada, Pakistan and Turkey.

Records of movement and deformation, particularly in the Swiss dams are of great interest and I have also included through the kindness of Mr. A. A. Fulton and Mr. L. H. Dickerson, a British example of deformation on the Monah dam in Scotland, as well as some interesting data on costs of Scottish dams, one of which does not 'behave' at all well owing to the costs involved due to bad foundations.

The Appendices for the new geological dam sciences, particularly on soil mechanics, rock mechanics and seismic tests have been extended with the help of Dr. John Knill.

In the preparation of this edition, I acknowledge the assistance of many friends, engineers and geologists who have kindly read my notes on the various dams. There are so many that I have thought it best to refer to them at the end of each section. I would, however, specially like to mention the following:

PREFACE

Dr. John Knill and Mr. Michael Kennard for reading my notes on their paper on the Cow Green scheme.

Mr. G. P. Palo of the Tennessee Valley authority and his successor, Mr. G. H. Kimmons, for reading my notes from the papers on behaviour of pore pressures and how they were dealt with on the dams (described in the first edition) which I visited in 1958.

For the Swiss dams, I am especially indebted to the Swiss National Committee and particularly to Mr. Eduard Gruner who has read my notes and given permission for publication.

Mr. C. M. Roberts for his notes on the Monar dam and his partner, Mr. N. J. Cochrane, for those on the Clywedog dam.

Professor Edoardo Semenza, I thank especially for sending me his synthesis and maps of the geology of the reservoir area of the Vaiont dam.

I must again thank Mr. G. M. Binnie and his partners for verifying the notes on the Mangla and Dokan dams, the latter which was included in the first edition and suffered certain troubles on filling.

Mr. W. G. N. Geddes has kindly verified the notes on the Backwater dam, comparable with the Sylvenstein dam described in the first edition. Mr. A. B. Baldwin has very kindly given me permission on behalf of the Sheffield Corporation for my notes on the Waldershelf slip.

Acknowledgements are due to the following organisations for permission to reproduce certain articles and papers. The Institution of Civil Engineers, the Institution of Water Engineers, The Geological Society, the National Committees of I.C.O.L.D., particularly those of Switzerland and Austria.

Finally, I must acknowledge the help given to me by Mr. Saim Kale (formerly my assistant but who has now returned to Turkey) on Turkish dams and for putting me in touch with the Turkish Committee on large dams.

I thank Miss Joan Tucker for adapting many diagrams which have been used in this book, Mrs. Richmond for typing the material and my wife for reading the manuscript.

R. C. S. WALTERS

CONTENTS

PART 1: GENERAL GEOLOGICAL PROBLEMS

1. Types of Rocks as they Affect Dam Construction — 3
2. Surface Features of Valleys which Determine the Types of Dam — 8
3. The Topographical and Geological Conditions for Different Types of Dams — 15
4. Hazards of Preliminary Site Investigation — 19
5. Dams, Reservoirs and Lakes in Glaciated Valleys — 23
6. Leakage — 27
7. Early Development of Grouting of Dam Foundations — 30
8. Modern Grouting — 32
9. Settlement, Slips and Subsidences — 37
10. Mining and Dams — 40
11. Suspicious Old Dams — 44
12. Earthquakes and Dams — 50
13. Major Dam Disasters — 56
14. Reservoir Geology — 63

PART 2: TYPICAL GEOLOGICAL PROBLEMS

15. British Dams — 75
16. Austrian Dams — 142
17. Belgian Dams — 162
18. French Dams — 166
19. German Dams — 225
20. Holland — 234
21. Italian Dams — 236
22. Jugoslavian Dams — 278
23. Luxembourg — 280
24. Portuguese Dams — 284
25. Spanish Dams — 286
26. Swedish Dams — 289
27. Swiss Dams — 291
28. Turkish Dams — 307

CONTENTS

29.	Algerian Dams	311
30.	Irak	347
31.	Rhodesia	351
32.	Pakistan	354
33.	Japanese Dams	358
34.	U.S.A. Dams	363
35.	Canadian Dams	413
36.	Australian Dams	419
	Appendices	425
	Index	455

PART 1
GENERAL GEOLOGICAL PROBLEMS

1

TYPES OF ROCKS AS THEY AFFECT DAM CONSTRUCTION

Granite

There is no reason, in principle, why any dam should not be constructed on granites. These rocks are normally considered to be sound for bearing great pressures and they are generally watertight. Thus, the Marèges dam was constructed on granite with hardly any surface preparation. In the broad foundation and cut-off trench of the Argal dam in Cornwall were several patches of disintegrated granite, but these were reasonably hard and were capable of being grouted. The Sarrans dam, too, which had a broad foundation of some 11,000 m² of much decomposed granite, consumed 685 tons of cement in 81 bore holes which had an aggregate length of 2,800 m or 240 kg per m.

Other dams on granite consumed the following quantities: Seeuferegg (Grimsel, Switzerland), 252 kg per m; Spitallamm (Grimsel, Switzerland), 122 kg per m; and Couesque (Central Massif, France), 220 kg per m.

In certain granites, however, caution must be exercised, particularly in Cornwall where large masses of china clay appear. Thus it was originally intended that the De Lank dam, near Bodmin, being on apparently sound granite, should be constructed in concrete. China clay, however, was subsequently proved at the site and the design was altered to that of an earthen embankment (this dam has not yet been constructed).

Granite, therefore, must be subjected to investigation for fissures, disintegration, and particularly for china clay, even deep seated under sound granite, for it would hardly be feasible to anchor pre-stressed cables in china clay!

A dam was once suggested on granite-gneiss at Trerise, near The Lizard in Cornwall, and although this rock is generally sound, more fissures may be expected within it. This was confirmed by a bore hole which gave a sufficient quantity of water for a rural water supply; the construction of the dam was therefore abandoned.

Gabbros, Andesites, Dolerite and Basalt

Generally, these rocks are considered sound for supporting ordinary structures, other than water-retaining structures, but they are by no means to be trusted for dams and reservoirs. Serpentine also must be treated with suspicion. The Zerbine dam in Italy was constructed in 1925, but as the result of a heavy storm in 1935 which these rocks were not able to withstand it collapsed, drowning 100 people.

Dams on porphyritic rocks, lavas, trachytes, andesites and basalts, although they may be sound in themselves, are always more or less fissured and necessitate careful grouting. Gignoux and Barbier referred to the Rieutord dam on a tributary of the Loire which necessitated a considerable amount of grouting in basalt. On the other hand the Tirso multiple arch barrage in Sardinia is founded on trachytes and volcanic tuffs with little grout. Other examples, in America, of important dams on basalt are the Pitt River multiple arch dam, the Bull Run dam (Oregon), and the Black Canyon dam (Idaho). The Hoover dam* on andesite tuffs 726 ft. high when first filled in 1927 suffered considerable leakage which increased, necessitating grouting to a depth of up to 430 ft. between 1938 and 1947.

Amphibolites

Gneiss, mica schists and associated rocks may generally be considered to be satisfactory for sustaining bearing pressure and for water-tightness; for example, the Éguzon dam (Indre) having a height of 61 m consumed comparatively a small amount of cement. Gneiss and mica schist, particularly the latter, are less favourable owing to the mica which may facilitate slipping.

Where these rocks are associated, such as at the Bort dam, a very weak zone of disintegrated rock may be found between the junction of the gneiss and mica schists. The Forks dam, California (150 m), founded on gneiss and mica schists, had to be abandoned in 1929 because of bad foundation which occurred at the junction of these two rocks which, in themselves, were quite sound.

An early example of grouting is recorded for the Brüx dam, Bohemia, in 1914, which has a height of 53 m and is founded on gneiss. The leakage is reported to have been reduced from 132 to 28 l. per second.

The Chambon dam, however, when the surface of the gneiss *was* ascertained (and which was buried deeply under glacial deposits and was very uneven), proved to be sound!

A dam in Portugal, Castello-do-Bode, on which the micaschists were bedded, dipping upstream steeply, and more or less perpendicular to the direction of the flow of river, proved particularly favourable; similarly, the Swiss dam, Dixence, on gneiss, 87 m in height, proved quite satisfactory.

Metamorphic Rocks in general

Somewhat uncertain are the metamorphic rocks and intrusive igneous rocks, but many satisfactory dams have recently been constructed on them in Scotland (for example, Sloy, Pitlochry, Errochty, Shira and others), but grouting the foundations is generally essential. The usual types of dams

*Formerly Boulder dam; renamed after President Hoover.

adopted are gravity, buttress (in one case a pre-stressed buttress), and rockfill. In the upper part of the Tennessee Valley, also on Pre-Cambrian and Ordovician metamorphic rocks, gravity, buttress and rockfill dams have been constructed since World War II. In many of these cases hard rock may be covered with deposits of drift and alluvial material which if excessive may cause schemes to be abandoned.

Weathering at the surface may also have to be investigated as broken formations may prove exceptionally troublesome when the work is opened out.

An instance of a dam on granulite which proved to be much altered at the surface down to a depth of 20 m as well as broken and fissured was the Lavaud-Gelade dam in the Central Massif, Creuse, France; this site necessitated an extensive grouting injection with cement, clay and bentonite.

Limestone (Cambrian, Ordovician, Silurian, Carboniferous, Jurassic, Cretaceous and others)

There are no dams in Great Britain constructed wholly on limestone, although there are some on limestones associated with clay. A thick seam of Cornbrash Jurassic limestone was encountered at Sutton Bingham, Yeovil, and thin seams were cut off in the trench in Lower Lias Jurassic limestone at Eyebrook, Uppingham; both these are earthen embankments.

Concrete dams in France which have successfully been constructed on Jurassic limestone are those at Castillon, where difficulties owing to slips were surmounted and leakage overcome by an extensive grout scheme; the Chaudanne dam below Castillon, the Sautet dam on the Drac, and the recent St. Pierre-Cognet dam. The Génissiat dam on the Rhône was built on karstic cavernous Cretaceous limestone, the caverns being filled for the most part by glacial clay.

Limestone occasionally produces deposits of calcareous tufa which may necessitate some adjustment of a site owing to its weakness, such as at Beni Bahdel in the Jurassic. Dolomite limestones may give rise to large fissures approaching grottos and sites and designs also have to be modified, for example, the Fodda dam in Algeria, the original site of which had to be moved to an inferior one upstream.

The Camarasa dam (1933) in Spain founded on dolomite limestone, at the outset, had considerable leakage which necessitated very expensive grouting (186,000 tons of cement), and more recently (1960) the Dokan dam, Irak, also in dolomite, consumed 77,666 tons in the grout screen.

The dams in Northern Italy in the Trias and Jurassic Dolomites, do not suffer very much from karsticity although they have other difficulties. The Val Gallina dam, 105·2 m in height, an arch dam, founded on Upper Trias Dolomite limestone, necessitated an exceptionally large programme of grouting. The Waggital dam in Switzerland is noted for having a gorge 40–50 m deep in limestone filled up with alluvium; the limestone was grouted from several horizontal lines of tunnel. The Chickamauga is an interesting case in the Tennessee Valley of America which dealt with the permeability of limestone by injecting 59,500 tons. An interesting recent earthen embankment on soft Oligocene and Miocene limestone is that at Sarno, Algeria.

Grits

Many dams in the English Pennines are constructed on rocks often consisting of alternating seams of grit (sandstone) of the Carboniferous, often some 40–50 ft. thick, alternating with clay or shale of about the same thickness, and most of them are earthen embankments; some of the largest are Langsett (for Sheffield), the Longdendale reservoirs (for Manchester), Scar House (for Bradford) and Ladybower on the Sabden shales in the Derwent valley.

In the Scammonden valley near Huddersfield, an earthen embankment is nearing completion (1971), 242 ft. in height, which impounds 168 ft. of water and carries the Pennine motorway (M62) founded on Upper Kinderscout Grit and Scotland Flags of the Millstone Grit formation.

Figure 1. Lamaload dam. Example of a buttress dam on alternating shales and grits. A grit seam is seen above the round-headed buttress foundations under construction April 1961

(By courtesy of Mr. J. H. Dossett, Macclesfield Corporation, and Mr. J. Shaw, Macclesfield Water Board)

One or two concrete dams have been built recently on these alternating grits and shales, for example, a concrete gravity dam at Baitings for Wakefield, and at Lamaload, a buttress dam, under construction (1961) for Macclesfield (*Figure 1*). In addition, there are the older masonry dams of Derwent and Howden in the Derwent Valley.

Clay

Although clay formations may occasionally be thick and massive they are often associated with thin seams of grit, sandstone or limestone and such formations necessitate either earth or rockfill types of dams. The embankments of the Staines, Chingford, and other reservoirs in the Thames and Lee valleys may be cited as reservoirs wholly in London Clay and the Cheddar reservoir near Bristol in Keuper Marl.

Gravel, Sands and Boulder Clay

Several small dams of an agricultural nature and forming ornamental ponds have hitherto been built on gravel, sands and boulder clay, but naturally, unless there are special circumstances, such formations have not been considered satisfactory on which to build important dams. Perhaps the Trentabank reservoir (Macclesfield) may be cited as an earlier example (1925) of 'special circumstance', as it is built mainly on sand, and the Selset reservoir on 'Boulder Clay' is another example.

At Bough Beech, near Sevenoaks, a pumped storage scheme has just been completed (1971) on the Weald Clay where the embankment (curved downstream to give 1900 m.g.) is constructed with Weald Clay with seven horizontal filters of Lower Green sandstone formation; it impounds 70 ft. of water.

Similarly, abroad, great depths of soft materials, permeable or unpermeable, have been and are being built upon; to cite two examples, the Sylvenstein dam, Bavaria, in a dolomite gorge filled with several hundred feet of debris and the Serre Ponçon dam (under construction 1957), on the R. Durance, France, in a Lower Lias limestone valley filled with 100 m of soft material.

The first comparable British dam was opened by H.M. Queen Elizabeth II in December 1969 in the Backwater Valley near Balmoral. The embankment is founded on gravel, sands and Boulder Clay, 165 ft. in depth, resting on metamorphic schistose grits and phyllites.

2

SURFACE FEATURES OF VALLEYS WHICH DETERMINE THE TYPES OF DAM

The shape of a valley and the rock with which it is formed not only profoundly affect the type of dam, but also its dimensions.

The main types of valleys are as follows:

(a) Gorges . . . . Chord-height ratio under 3
(b) Narrow valleys . . . . Chord-height ratio 3-6
(c) Wide valleys . . . . . . Chord-height ratio above 6 or 7
(d) Flat country: plains

DAMS IN GORGES

Cupola or Dome
Where a dam can be fitted into a gorge having a crest chord-height ratio of under 3, and the rock in the gorge is capable of withstanding high pressures, and is not likely to move bodily, that is, fail by shearing, thin arch or thin cupola (or dome) dams have recently been successfully built.

The arch dam is curved in the horizontal plane while the dome dam is curved in both horizontal and vertical planes; the section of the dome dam is even thinner than an arch dam and has maximum strength comparable to the shell of an egg. On account of its thinness the pressure transmitted to the abutments is enormous, unless the abutments are thickened.

From the crest of a cupola dam it is a curious experience to lean over the parapet on the downstream side in order to see the downstream toe, and, similarly, if the reservoir is empty, to lean over the upstream parapet and not be able to see the upstream 'heel'!

The Marèges dam was one of the first (1935) thin cupola dams, having a crest of 247 m and a height of 90 m; the chord-height ratio being under 3. It is constructed on sound granite and the design broke away from the traditional gravity dam usually adopted in France before that time.

The Vaiont dam in Northern Italy would have been considered the finest example of a cupola dam for it has a maximum height of 875 ft. It is founded on Middle Jurassic limestone, dolomitised, in a gorge where the chord/height ratio at the crest level is 0·7, the height of the dam being about $1\frac{1}{2}$ times the width of the valley at the crest. The rock was expected to withstand the thrust of the arch abutments of 60 kg/cm² (approximately 60 ton/ft²) with one or two metres of water going over the overflow weir.

At the time of the disaster, on October 9th, 1963, a wave 200 m of water on the right bank and 100 m on the left bank above the crest flowed over the abutments, as shown by the level of destroyed vegetation, with hardly any damage at all to the dam.

Other modern cupola dams in gorges are the Val Gallina dam (1950) in the Upper Trias dolomite, a few miles south of Vaiont, which has a crest chord-height ratio of 2·7, and the Pontesei dam, constructed in 1956–57, in another Triassic limestone valley (the Maè) with a chord-height ratio of 1·66.

The Barcis cupola dam (1952–53), with a chord-height ratio of 1·25, was the subject of several geological reports owing to the unfavourable dip of the Cretaceous limestone, its unsoundness, and the threat of abnormal leakage from the reservoir.

At the site of the Sautet dam in France (1925), where the chord-height ratio is approximately two-thirds, a curved gravity dam was built on horizontally bedded Lower Lias hard limestones with thin soft clays interbedded. Undoubtedly, nowadays, a thin arch dam (if not a cupola) would have been built in this valley, and the gravity section would not have been adopted; at any rate it would have been reduced in thickness in such a gorge. This particular dam gave rise to considerable leakage from the reservoir upstream, as would any dam impounding up to 126 m of water in this particular valley.

In the same valley as the Sautet dam but completed (1957), the St. Pierre-Cognet dam, built on Upper Lias limestone with a chord-height ratio of about 1·4, is a most attractive thin arch dam.

In the valley of the Durance, also with a chord-height ratio of about 2, the thin arch Castillon dam has been constructed in extremely fissured Upper Jurassic limestone, for which major works were necessary, including pre-stressing the geology, or, expressed perhaps more scientifically, using pre-stressed concrete buttresses to hold the foundations together.

A double curvature arch dam has been built in a gorge below Loch Monar, 30 m west of Inverness on pre-Cambrian granulite, one of the few sites in Britain found suitable for a cupola or dome dam. The chord/height ratio is 3. The estimated cost was 10% lower than a gravity dam.

Thin arch dams can therefore be economically fitted into gorges where the chord-height ratio is under 3 and when the rock is or can be made sufficiently sound. Soundness of the foundation is of paramount importance for all arch and dome dams.

DAMS IN NARROW VALLEYS

Narrow valleys (as distinct from gorges) may be defined where the chord-height ratio of the dam is between 3 and 6.

The Chambon dam, constructed in 1921 on the river Romanche in France, is a curved gravity structure. It was founded in a valley filled with a good deal of permeable glacial material on gneiss, Trias limestones and Lias schists, which necessitated increasing the depth of the foundation, giving a chord-height ratio of about 3. Probably, even nowadays, any reduction in section would not be contemplated owing to the heterogenous nature of the foundation rocks at that site.

The French Bort dam on the Dordogne is a gravity arch structure in a valley where the chord-height ratio is approximately 3, and this dam, partly on weak clay between two faults of sound gneiss and micaschist, would not have warranted any form of thin arch or reduction in the gravity section; and, even for the gravity section, during construction, the French reported that the situation was at times 'agonizing'!

The Tignes dam (1947–52), one of the most noteworthy dams in France, has a height of 165 m and a chord-height ratio of about 2; it is founded in Lower Triassic quartzite, which is very hard; the existence of a pre-glacial valley, however, limited the depth of water, otherwise a greater height might have been attained.

In Italy the chord-height ratio of the Piave di Cadore dam was 5·5, and some form of gravity dam resting on a plug in a subsidiary gorge on the right bank was first thought of, but the final suggestion, which was adopted, was a thick arch dam; that is to say, the thickness of the dam was less than a gravity dam (65 per cent), and more than a thin arch dam.

It is probable that an increasing number of thick arch dams with a thickness of less than the gravity section will be constructed in the future as confidence is gained in (*a*) the reliability of models to confirm and even supplant the tedious mathematical analyses, some of which have to be founded on doubtful assumptions; and (*b*) the experience of strengthening weak foundations to carry the heavier unit pressures which are to be sustained compared with the standard gravity section.

It seems that the thinning and the curving of the straight gravity section will be brought about only by intelligent boldness and by examining the behaviour (for example, settlement) and durability (for example, weathering of materials) of such structures as the Piave di Cadore dam.

Something of this kind is being done at the Alto Rabagão dam in Portugal, part gravity and part double curved on weaker strata than 'that usually considered acceptable'.

In addition, those dams like the Castillon, the Chaudanne and the Bort which had foundation difficulties during construction, should they stand satisfactorily with the passage of years, would give great confidence in the efficacy of what may be called extempore foundation remedial works.

DAMS IN WIDE VALLEYS

In a wide valley practically every type of dam can be constructed except a single (thick or thin) arch. Indeed, it may be said that in such a valley the type of dam (for example, gravity, buttress, multiple arch, earth, rockfill) is governed primarily by the geology of the site and the proximity of materials from which the dam is to be made. A thick arch dam has been built, as stated above, for a chord-height ratio of 5·5 but a higher ratio than this so far as present knowledge goes means a gravity dam.

Gravity Dams

Examples of masonry or mass concrete gravity dams, or both, particularly where the rock is near the surface (averaging within say 30 ft.) are many.

The earliest large example in Great Britain is the Vyrnwy dam (masonry) for the water supply of Liverpool. The chord-height ratio of the dam is 7.

Other early examples of British masonry concrete gravity dams are to be found in the Derwent Valley, namely the Howden dam (chord-height 9); the Derwent dam (chord-height 10); and in Mid-Wales, the Caban Côch dam (chord-height 5); the Pen-y-Gareg dam (chord-height 4) and the Craig Côch dam (chord-height 4), and the more recent concrete dam, Claerwen (chord-height 5).

As all these dams are on sedimentary rocks of alternating shales and grits, it is doubtful, notwithstanding some of the chord-height ratios, if alternatives to gravity dams would have been adopted on such strata had they been built at a later date.

The present Aswan dam, gravity for a length of 500 m and buttress (pierced by sluices) for 1,400 m is another example of concrete and masonry in a wide valley of granite with soft igneous dykes, the total length of which is 6,300 ft., maximum height, 150 ft. and the length-height ratio, 42.

The gravity dam is generally adopted where rocks such as the sedimentary sandstones and limestones of moderate bearing strength have to withstand only about 8–10 tons per ft.^{2}.

For rocks of higher crushing strength in wide valleys, a pre-stressed concrete dam is a modern development for replacing a long gravity or buttress dam and for economizing in cost. Such a dam at Allt-na-Lairige is on the igneous rocks where they are capable of sustaining a pressure of 20 tons per ft.^{2}, and a pull of 1,176 tons per ft.^{2} from steel cables embedded in concrete in the rocks.

Buttress Dams

With suitable foundations capable of withstanding direct pressures and resistance to sliding, the buttress dam can usefully be adopted, as it is economical in material (for heights over 40 ft.) compared with a gravity dam. There are many examples; the recently constructed dams in Scotland are of especial interest as they are situated in wide valleys—Sloy and Lednoch each with chord-height ratios of 7, Errochty 10 and Shira 15.

There is an interesting long buttress dam (Fedaia) in Italy, which takes advantage of an outcrop of rock by having reverse curves and being joined to a gravity section where the height of the dam is less than 40 ft. The total length of the dam is about 2,000 ft.

Multiple Arch Dams

There are two interesting long multiple arch buttress dams in Algeria. One of them, the Beni Bahdel dam, is on alternating hard and soft strata which dip generally but across the axis of the dam; this cross dip necessitated adding struts between the buttresses to prevent lateral movement; the buttresses are supported on the harder sandstones. Owing to the shape of the valley on plan, the main overflow spillway is not at the Beni Bahdel dam, but at the Digue de la Route dam, where there is a bend on the river and a subsidiary valley which has been converted into an overflow channel.

The Digue de la Route dam incorporates a fantastic overflow spillway known as 'duckbills', a series of spillways at right angles to the arches.

The multiple arch variety of buttress dam is sensitive to 'dishonest' foundations, for the lateral thrust of the arches is taken by the buttresses and the arch is a monolith, whereas in the pure buttress type each half of the arch is a cantilever with an expansion joint on the crown of the arch.

Some differential settlement between the buttresses of a buttress dam may be permitted which would not be tolerated for a multiple arch.

Rockfill Dams

The wide valley is suitable for all forms of rockfill dams, particularly those valleys where there are nondescript strata. The Breaclaich dam near the south-west end of Loch Tay, Scotland, under construction (in 1959) on metamorphic and other complex rocks, is an interesting example. Loch Quoich is another example, 126 ft. in height and 1,050 ft. in length.

In the United States of America over 100 rockfill or partly rockfill, that is, composite, dams have been constructed; recently there have been two dams in the Tennessee Valley at Watauga on steeply dipping metamorphic Cambrian, and at South Holston on very wrinkled and variable metamorphic Ordovician.

The first rockfill dam having a sloping core was constructed in 1942 on metamorphic Cambrian rocks at Nantahala, Alcoa; the rock for the rockfill was obtained from the excavation of the overflow channel. The clay for the sloping core, some 30 ft. in thickness, was dug in the vicinity from the metamorphic schists, and the suitable filter sands for this was drawn from the sandy and gravelly material in the vicinity.

A similar rockfill dam, Lewis Smith (Sipsey Fork, Alabama), in the Pennsylvania Series in the Coal Measures (comparable with the Millstone Grits below our productive Coal Measures) was under construction (1959) in Alabama. Here the strata and shape of the valley are utilized for obtaining from a very long spillway diversion channel, constructed away from the dam, rock for the rockfill, clay for the impermeable core from the shales sandwiched between the grit-stone, and sand for the core filter.

Other examples are the Ghrib and Bou Hanifia rockfill dams where bends in the valley and favourable topography enable overflow channels to be constructed with the twofold object of keeping flood water away from the dam and providing rock for the embankments.

The Aswan 'high' dam under construction has been described by M. Taher Abou Wafa in communications *I.C.O.L.D.* II, 191 (1961) and III, 615 (1964).

The site of the dam is 7 km upstream of the existing dam and is situated in a valley flanked by fine grained granite and gneiss but having a great depth (to 200 m) of permeable sands and gravels.

A rockfill dam with a clay core and a long horizontal clay blanket and a grout curtain is envisaged. The maximum height of the dam will be 111 m and the length of the crest 3,500 m. Tests on the efficacy of grouting reduced the permeability from 2.5×10^{-2} to 2.3×10^{-4} cm/s in the coarse sand and from 6.1×10^{-3} to 3.6×10^{-4} cm/s in the fine sand.

Hydraulic Fill Dams

Such wide valleys are also suitable for the hydraulic fill dams where the materials are transported and laid down under water.

Earthen Embankments

More often, however, particularly in England, with so many wide valleys, the earthen embankment has been and is still the most suitable on many types of soft variable sedimentary strata, such as at Powdermill and Weirwood, on the Ashdown Sand of the Weald; Sutton Bingham on the Cornbrash and Forest Marble of the Jurassic; Eyebrook (Corby) on the clays and limestones of the Lower Lias; Langsett (Sheffield) on the Millstone grits and shales of the Carboniferous; Ladybower (Derwent Valley) on the Sabden shales also of the Carboniferous; Chew, near Bristol, on the Triassic marls and sandstones, and several hundred others.

Barrages

Although there are many so-called 'barrages' in narrow valleys, and in many both narrow and wide valleys there are barrages of earth, the writer rather restricts the term 'barrage' for such concrete structures as those along the river Tummel at Pitlochry, Dunalastair, Tummel and Gaur; on the Rhône at the several river diversion- and power-dams on canals; on the Danube at Jochenstein and the Tennessee valley and Mississippi barrages.

The largest recent development (1959) is that of the St. Lawrence valley constructed jointly by the United States of America and Canada. Here, to utilize sections of the St. Lawrence over a distance of approximately 183 miles, large barrages have recently been constructed across the river, flooding an immense amount of land and necessitating the removal of much property. The three major works are the Iroquois dam to regulate the flow of the river, the Robert Moses and Robert Saunders power dam (Barnhart Island dam), and the Beauharnois power dam wholly in Canada, to generate power. The ancillary works comprise many locks and subsidiary dams, for example, Long Sault Barrage.

In general, most sites of barrages across large rivers arise from some upheaval of hard rock over which the rivers flow swiftly, which causes sediment and other soft material to be washed away.

An excellent example is the Rance estuary barrage 700 m in length which sits on gneiss and bands of dolerite with little accumulation of alluvium owing to this fast flowing river.

DAMS ON PLAINS

Normally, dams are associated with valleys and are not built on level ground in the middle of sandy plains.

Examples of dams on plains are to be found on the new Rhône diversion canals, the Rhône being diverted, by means of gate-control barrages, into canals. These canals aggregate some 30 miles in length and lead the water from the Rhône to normal gravity section dams, built several miles away, on alluvial and permeable strata. Although the work of building such dams without having to keep in being a flowing river during construction sounds fairly simple, the problems arising in negotiating underground water difficulties, both during the construction of the dams and canals and subsequent maintenance, are enormous.

In the case of the Rhône, the underlying principles of these diversions are as follows.

(1) To gain hydro-electric power. As the flow of the natural Rhône is rapid and in places the river bottom has a steep slope, it is replaced and the water is diverted by a barrage across the Rhône into a canal having a very flat slope; hence, after a distance of the order of 10 miles, about 80 ft. head can be gained. At the end of the canal, a dam, a lock and a power-house are constructed. The tail-race water is led into a low-level canal which flows for another 10 miles or so delivering the water back to the old Rhône some 20 miles downstream of the diversion barrage.

(2) To improve navigation which is not possible, or at any rate is difficult, in the old Rhône. The new scheme enables craft to travel comfortably through the canals negotiating the dams by passing through locks.

(3) To improve irrigation in the plains through which the canals pass. As these canals are comparable with the size of the Suez Canal, irrigation facilities along the plains on which they are situated are important. Their correct alignment entails much geological investigation.

(4) By diverting the river water right away from the old river bed through undeveloped country, interference with the many towns, villages, railways and roads that exist along the Rhône valley is avoided.

Detracting from the great advantages of such works, the cost of which runs into several million pounds, are the great geological difficulties, not only in finding suitable sites for the dams but also in the construction of the canals through permeable materials, thus interfering with the natural underground water levels. Similar schemes have been and are being constructed along sections of the Rhine.

Other types of dams, on flat country or what may be considered to be dams, are the embankments of the large reservoirs of the Metropolitan Water Board, and the large reservoirs at Cheddar, Bristol.

Embankments along our coasts to keep out the sea are numerous. The largest example which may be cited to keep out the sea and fresh water in is that of the Netherlands Zuider Zee, an earthen embankment $18\frac{1}{2}$ miles in length.

The dams for reservoirs constructed in Canada and the United States of America for storing the greater part of the water which flows at night over Niagara, and diverted to power-stations during the day, are also remarkable dams. There are also many miles of ship canal with earthen embankments in the St. Lawrence scheme.

Thus, there are many instances of what may be considered to be dams the embankments of which are measurable in terms of miles in length and retained water well above ground.

3

THE TOPOGRAPHICAL AND GEOLOGICAL CONDITIONS FOR DIFFERENT TYPES OF DAMS

When the size of dam has been determined, the type of dam envisaged requires certain geological and topographical conditions which, for the main types of dams, may be stated as follows.

Gravity Dams

The gravity or mass concrete dam requires a site where there is 'hard' rock at or near the surface; the depth of soft material above the rock should not exceed 20–30 ft. thereby avoiding excavation. The rock should be capable of sustaining about 8–10 tons per ft.^{2} As such dams are nowadays constructed with concrete rather than masonry, materials for the aggregate, stone, and sand should be within a reasonable distance of the site, say, not exceeding 5–10 miles and reasonably accessible.

Gravity dams are particularly well suited where the length of the crest of the dam is five times or more than the maximum height. The chord at the crest is measured straight across the valley and the height from the crest to the rock on which the dam is founded.

Buttress Dams

The buttress concrete dam is particularly suitable where the rock is capable of sustaining 20–30 tons per ft.^{2} There is a substantial saving in concrete compared with the gravity dam, the buttress dam consuming between one-half and two-thirds the amount of concrete required for a gravity section.

Some additional skilled labour, however, may be considered necessary for form-work. Rather more special attention must be paid to the threat of any deterioration of concrete from the impounded water than is necessary with the thick gravity section.

The great advantage of this type of structure, in addition to economy of materials, is the elimination of a good deal of uplift pressure, the pressure resulting from water in the reservoir and possibly of water from the hillside rocks gaining access through or under any grout curtain and exerting pressure upwards underneath the mass concrete dam.

Buttress dams in many cases become more economical than gravity dams when the heights are over approximately 40 ft.

Multiple Arch Dams

The multiple arch concrete dam is a variety of the buttress dam and it may be considered desirable to adopt it where economy in materials is necessary, and more skilled labour for placing form-work is available. The chief geological criterion is that the rock must be absolutely reliable to withstand some 20–30 tons per ft.^{2} or more without any appreciable settlement (appreciable may be defined as one-third of an inch). There is some saving in concrete compared with the buttress dam, but in respect of uplift, corrosion, and economy over the gravity section the two types are similar.

Thick Arch Dams

The thick arch concrete dam can be built where the crest chord-height ratio is between 3 and 5 and where the rock of the valley is capable of sustaining or can be made to sustain something of the order of 30 tons per ft.^{2} A substantial saving in material compared with that of a gravity dam should be effected. Such dams are difficult to design on paper but exact dimensions appear to be quite well determined from trials on models.

Thin Arch Dams

Thin arch dams require valleys to have a crest chord-height ratio of under 3, with a radius of under 500 ft.

The pressure is considerable on the valley sides which must sustain some 50–70 tons per ft.^{2}

The cupola or dome type, that is, where there is a vertical radius of curvature in addition to a horizontal radius of curvature is most attractive, aesthetically. A French pioneer of this type of dam, M. Coyne, likened the Marèges dam to a racehorse full of sinews and muscles without superfluous weight like a gravity dam. (To English ears likening a dam to a racehorse does not appear to be a very good simile!) Where cement is expensive and labour is cheap, Signor Semenza has proved both thin and cupola dams to be entirely suitable; in limestone districts acid water troubles causing corrosion to the structure do not arise.

Rockfill Dams

Rockfill dams can be built where the following conditions exist.

(1) Uncertain or variable foundation unreliable for sustaining the pressure necessary for any form of concrete dam.

(2) Suitable rock in the vicinity. This rock can often be obtained from quarries close at hand or from the excavation for the overflow channel, or both. It must be hard and stand up reasonably well to variations of weather, especially frost, and must not disintegrate in air.

(3) An adequate amount of clay in the vicinity which may be inserted in the dam either as a vertical core or as a sloping core, usually the latter.

(4) Accessibility of the site and the width of the valley is suitable for the manipulation of heavy earth-moving machinery, caterpillar scrapers, sheepfoot rollers and large bulldozers.

Hydraulic Fill Dams

Hydraulic fill dams are suitable in valleys of soft materials and are constructed by pumping soft material duly consolidated up to moderate heights up to about 100 ft.

Earthen Embankments

The earthen embankment has been a common type of dam construction from early times to the present day where it has to be built of and on soft materials. The quantity of suitable earth available, it stability, permeability, depth and cost of foundation all have to be considered on their own merits for each site. Many earthen embankments have been built on hard rocks also suitable for concrete gravity dams; for example, at Falmouth there is an old earthen embankment and a new concrete dam both constructed on granite in the same valley.

In the Pennines there are some 200–300 earthen embankments constructed in valleys composed of alternating shales and grits; these shales and grits having crumpled and buckled and slumped down the valley sides, coupled in many instances with a fault or bulge, or both, in the bottom of the valley. Most of these valleys probably would never have been suitable for any form of concrete dam which would compete in cost with the earthen embankment, as too much would have had to be spent on the foundations.

The earthen embankment has proved and is still proving to be suitable for most of the relatively small gathering grounds composed of sedimentary strata in England, not only in the Pennines of the North, but in the Midlands and South; small gathering grounds do not warrant impounding a depth of water greater than 100 ft. (more often 50 ft.), for which the earthen embankment is eminently suitable.

Another factor is that the sedimentary rocks usually give rise to wide gathering grounds and there is no particular limitation in length; the perimeters of the earthen embankments of the reservoirs of the Metropolitan Water Board, for example, aggregate many miles.

Owing to the variable geology in English valleys, earthen embankments are usually more difficult to construct than the various forms of concrete dams on hard strata, and even quite low embankments which impound a depth of water of only 50 ft. may necessitate a trench approaching a depth of up to 100 ft. across a valley.

The Sheepstor dam necessitated 105 ft. depth of trench for a depth of 22 ft. of water impounded; it is true that this was in the days before grouting, but it is doubtful if grouting would have been successful in a mass of china clay in the granite.

Near the site there must be (a) clay to fill the trench, and (b) embanking material capable of standing safely, without slipping, to hold up a clay core.

Puddled clay may be replaced by concrete in the trench, particularly where the trench passes through seams of rock which are water-bearing, and in the embankment a concrete core may be substituted for a puddled clay core where clay is scarce in the vicinity of the dam.

Another advantage of the earthen embankment is that troubles due to deterioration of the structure by peaty waters of low pH do not arise.

Composite Dams

Not only can different types of dam be built in the same valley but the same dam can be of different types owing to the varying geological and topographical features of the dam site.

For instance, in the Tennessee Valley there is the Boone dam which consists of a gravity concrete structure, a rockfill dam, and an earthen embankment.

The Alwen (Birkenhead) dam is another example of a concrete and earthen dam. Many buttress dams also join up with gravity mass concrete dams at their haunches at the sides of the valley, and again at the centre have a mass concrete gravity dam to form a suitable overflow or spillway.

4

HAZARDS OF PRELIMINARY SITE INVESTIGATION

Although the Code of Practice for Site Investigation of the Institution of Civil Engineers recommends that up to 7 per cent of the cost of the structure is permissible for site investigation, one hesitates to spend, say, £35,000 on preliminary investigation for a £500,000 dam. Nevertheless, even such an expenditure in some cases may not be enough to investigate suitable borrow pits for (*a*) puddle and (*b*) embanking material as well as (*c*) bore holes and pits for ascertaining trench conditions, and (*d*) broad foundation bearing pressures for the embankment.

Much information may be missed in wide valleys, even with bore holes 100 ft. apart. Recently (1960), a trench at the Drift dam near Penzance revealed a width of 70 ft. of china clay which was missed by bore holes in the bottom of a granite valley (*Figure 2a and b*). The bore holes did discover, however, an old tin-mining drift which necessitated moving the site of the dam 50 yds. upstream. A third hazard encountered was the existence of enormous granite boulders which had to be removed in order that a satisfactory foundation might be provided (*Figure 2c*).

It is almost impossible to discover pockets of alluvial sloppy clay under the broad foundation of a large earthen embankment, such as occurred both at Weir Wood in the Ashdown Sand or at Sutton Bingham in the Forest Marble. The latter hazard necessitated the removal of 36,000 yds.³ of very soft alluvium and replacement with something more substantial at a cost of £23,000 at Weir Wood.

Another common occurrence is the liability of misinterpreting the material from bore holes in sedimentary strata, particularly when they are dug by percussion. Generally, the strata prove to be much harder when opened out in the trench as indicated by the bore holes.

At Sutton Bingham, concrete was substituted for puddle in the Forest Marble (Marl) trench. Similarly, at Weir Wood dam the Ashdown sandy clays were much harder than anticipated, and concrete was substituted for puddle. At Powder-Mill dam part of the trench of puddle was replaced by concrete to prevent leakage through hard beds of Ashdown Sand, and at the Eyebrook dam water-bearing seams in the Lower Lias clays were lined out with concrete, the rest of the trench being filled with puddle.

It is impracticable to sink a number of trial holes down to approximately 100 ft. deep and Soil Mechanical Analysis does not quite indicate the condition in large trenches.

Occasionally, bad mistakes are made from bore hole information; a classic case was that at the Silent Valley dam for the water supply of Belfast when a granite boulder at a depth of 50 ft. below the surface was assumed to be rock which proved to be nearly 200 ft. below the surface. This mistake is so well known that it is not likely to occur again!

F. C. Uren describes an interesting case which occurred in 1892 in which fissures in limestone were clogged by clay (reminding us of the clogging of the fissures under the Génissiat dam) and tested with water (again reminding us of the water tests prior to the construction of the St. Pierre-Cognet dam).

Figure 2a. Drift dam, Cornwall. In the foundations unbared in 1960, a seam of china clay 70 ft. in width and of unlimited depth was encountered in the hard granite which was missed in the exploratory trial holes and bore holes. Loading tests are seen to be in progress. On an average the china clay consisted of 40 per cent silt (0·0–0·06 mm), 20 per cent sand (0·06–2·0 mm), 40 per cent gravel (2–20 mm). Settlement of 10 tests registered up to $\frac{1}{2}$ in. with a load of 5 tons on a 12 in. square plate. Shear tests showed 0·4 of the load (tan 22 degrees)

It appears that the borough water engineer of Newport, S.Wales, proceeded, after a number of borings, to construct the Llanvaches reservoir, 400 million gal. (now known as the Wentworth reservoir). A puddle trench was excavated to a depth of 100 or 120 ft. at a cost of £50,000. Badly fissured rock was encountered and both G. H. Hill and R. H. Tiddeman reported that owing to the porous nature of the beds in and below the trench the reservoir could never be made watertight.

R. Etheridge, however, reported in favour of continuing the construction, but again another opinion from G. H. Hill and T. Hawksley, reported against continuing the scheme and the contractor was paid off.

The water engineer then proceeded to fill the trench with water to a depth of 60 ft. and it was found to be watertight; the dam was then built in 1894. Mr. A. E. Guild reported (1960) that the reservoir is still in existence and is sound, and indeed it has been quite trouble-free since it was constructed.

(b)

(c)

Figure 2b and c. Drift dam, Cornwall. (b) A pocket of china clay in the Cornish granite necessitated spreading the load of the gravity dam more evenly (from 4 to $2\frac{1}{2}$ tons/ft.²) over the weaker strata. This was done by constructing the triangular wedge on the upstream face. On the left, grouting through the cut-off trench is in progress. The expected settlement is of the order of $\frac{1}{4}$ in. after 6 months although up to 4 in. is possible. The dam was completed in June, 1961. (c) In the foundations of granite and disintegrated granite many large granite boulders were embedded and removed

(By courtesy of Mr. J. H. Blight, engineer to the West Cornwall Water Board)

This then is an early instance of fissures in limestone which are associated with clay tending to become clogged like those at Génissiat.

Geophysical prospecting of dam sites is not without value but the following instance is instructive.

At Nanpantan, near Loughborough, it was intended to build a concrete dam in a valley of the pre-Cambrian rocks. In this district the pre-Cambrian rocks at the bottom of the valley are overlain by the Keuper Marl of the Trias formation plus some boulder clay (*Figure 3*). A geophysical survey indicated 'rock' at a depth of about 30 ft. maximum, but trial bore holes

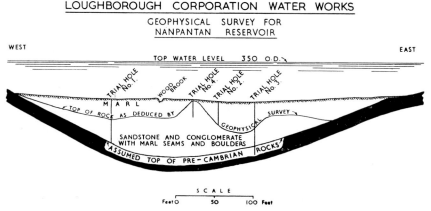

Figure 3. Nanpantan reservoir site investigation. A geophysical survey was made to ascertain the depth of the hard pre-Cambrian rock level with a view to constructing a concrete dam. The survey indicated that the broad foundation could be placed at a depth of over 25 ft. below the surface (with one exception), whereas the actual depth as proved by bore holes was over 75 ft. which would have necessitated an earth dam

indicated only tumbled boulders ('skerries') at this depth, the level of the true rock was as much as 60 ft. in depth. This would have necessitated substituting the intended concrete dam with an earthen or rockfill embankment.

There are many examples, however, where the accuracy of geophysical surveys are within 5 or 10 per cent of the correct depth to 'rock', but at the Sarrans dam site the overburden, owing to its varying permeability, affected the calculations of depth to the underlying granite.

BIBLIOGRAPHY

McIldowie, G. The construction of the Silent Valley Reservoir, Belfast Water Supply. *Trans. I.C.E.*, **239** 465 (1934–35).
Mitchell, P. B. Drift Dam (in discussion). *I.C.O.L.D.* III, 56 (1961).
Schlumberger, C. and Schlumberger, M. Application of electrical prospecting to the study of dam sites. *I.C.O.L.D.* IV 67 (1936).
Sunberg, K. Determination of Depth to Bedrock. *I.C.O.L.D.* IV 145 (1936).
Walters, R. C. S. Some geophysical experience in water supply. *J.I.W.E.*, 3, 436 (1949).

5

DAMS, RESERVOIRS AND LAKES IN GLACIATED VALLEYS

DAMS

Any valley down which a glacier has travelled is suspect.

Although the mechanics of the travel of glaciers are not understood entirely, it is known that they: (*a*) tend to travel in straight directions (on plan); (*b*) scrape up and transport the rocks over which they travel; (*c*) are capable of excavating deep holes (in section); and (*d*) on melting, deposit the material which they transport.

The presumptive evidence is that the pressure they exert, when on the move, is enormous, so much so that rather than follow the curves and meanders of the pre-glacial valley they carve out new valleys which are straight, or nearly straight.

Therefore, having once left an old valley, the straight newly made (epigenic) valley may be some miles from the old valley.

Depending on the meanderings of the old pre-glacial valley, therefore, the superimposed epigenic valley may coincide, nearly coincide, or be quite independent of the old valley and the deposition of morainic material may be in, near, or independent of the old valley. Hence at the dam site the presence or absence of this material will profoundly modify the cost of the dam.

Where an old valley and a new valley coincide all is well for the dam site. An interesting problem arose in connection with a suggested dam 150 ft. high, founded partly on granite and partly on metamorphic chiastolithic slates of the Skiddaw Ordovician series on the River Caldew, Cumberland (owing to the proximity of underlying granite, the slates have been altered and show little markings resembling the Greek letter χ). The valley is straight for a mile or so and as it is just below Skiddaw Mountain (3,000 ft.) a glacier had been at work. The left bank was obviously in sound rock which was visible at many places *in situ*, but the right bank was tamer, flatter, and in parts seemed to show seepage and was rather water-logged. Nevertheless, there was little surface indication of what was underneath. Four rows of bore holes, about 100 yds. apart, with 25 bore holes aggregating 2,000 ft. in length, were necessary to reveal the true significance of glacial interference. *Figure 4* shows a plan of the site, the surface of which ranged from about 1,050 O.D. to 1,250 O.D. and the contours of the underlying rock proved by

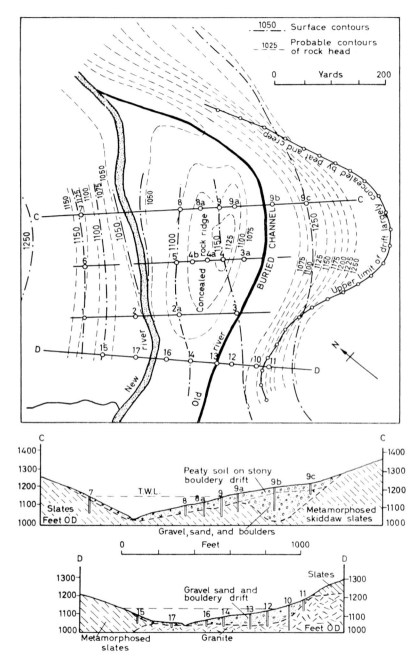

Figure 4. Drift-concealed pre-glacial valley revealed by borings at site of proposed Caldew Head reservoir

(Hollingworth, S. E. (1951). *J. Instn Wat. Engrs., Lond.* 5, 494)

the bore holes showed, on the left bank, only a few feet of overburden, while on the right, the morainic material reached a depth of over 150 ft.

What had happened was that the ancient valley had made a curve to the south at much the same bottom level (1,050 O.D.) as the new valley, and this old valley had been filled with permeable sand and gravel, as shown by the contours of the top of the rock and more strikingly by the sections.

There is a buried hump (Fr. verrou) of hard rock between the two valleys; the glacier preferred cutting a straighter course through the hard rock to the north of the old sharp bend in the river to the south.

The most economical site for the dam would be where the two valleys coincide.

Quite frequently excellent sites for reservoirs have been enclosed by two dams with a hard verrou between, owing to a glacier carving out a gorge not far distant from the old valley which is not filled (or not very much filled) with permeable soft moraine material. There is an excellent example of this in Switzerland, the Aussois dam, Savoy, and two others with two pairs of dams in the Kaprun valley scheme in Austria, where the new cleancut epigenic valley lends itself to cupola dams and the old weather-worn valley to gravity dams.

There is also a most interesting double valley, Glen Shira in Scotland, where there is a concrete gravity dam in a glacier-cut valley, and an earthen embankment in a pre-glacial valley associated with some morainic material. Higher up in this valley too, where there is little moraine, there is the remarkable Upper Shira straight buttress dam spanning the two valleys with the middle part of the dam low in height built on the verrou.

The site of the Tignes dam is another variation where the epigenic valley was much deeper than the pre-glacial valley, which is filled with rubbish. The author's authority for calling moraines boulder clay and drift 'rubbish' is John Tyndale's book on glaciers (1860). The level of the old valley determined the top water level in the Tignes reservoir.

RESERVOIRS

Among instances of reservoirs in glaciated valleys where dam sites are unaffected, but reservoirs may be, Génissiat reservoir in the permeable limestones of the Rhône may be cited. The site of the dam is in the gorge of the 'new' Rhône while the buried old Rhône which joins the new Rhône is several miles higher up the river. The level of the junction of the two Rhônes determined the top water level in the Génissiat reservoir. Not far distant there was also another ancient tributary, the Valserine; this was filled with rubbish and had to be grouted.

In the case of the River Drac, the new valley cut through the old river in several places, and the Sautet reservoir, near Grenoble, was built and was found to leak at the rate of some 50 million gal. per day through a moraine-filled valley about half-a-mile upstream of the dam. The water had submerged an unsuspected pre-glacial valley filled with permeable morainic debris. As this valley was some 100 m deep, and some 1,000 m wide at the top water level of the reservoir, it was too large and expensive in which to construct a cut off trench. Neither was a grout curtain considered feasible.

Lower down in the same valley the St. Pierre-Cognet dam was not built until the whereabouts and permeability of the material in a pre-glacial valley which would be submerged had been thoroughly investigated.

The Bort reservoir also submerged a moraine-filled valley, but as this valley did not reappear downstream of the dam for several miles it was assumed that leakage would not be of any consequence owing to the frictional resistance that would be incurred in so great a distance.

LAKES

A volume could be written on the formation of lakes by glaciers, but it is now recognized that most lakes have been cut by them out of the rocks, and hence these lakes are called rock-basins. Owing to the great weight of the glaciers which come down from great heights and their rate of travel, the ultimate effect, whether by ice or water pressure under the ice, is a hole. In 1857 the Mer de Glace travelled at the rate of up to 1 yd. in 24 hours. More recent observations on the flow of the Aletsch glacier have thrown further light on the subject, but as far as the Cumberland lakes are concerned, Marr suggested their formation in *Figure 5* (adapted).

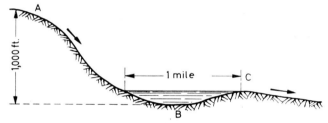

Figure 5. Lake formation. For many years it was assumed that lakes in glacial valleys were entirely formed by morainic deposits, but Marr (1916) is believed to be the first geologist to break away from this assumption and to say that the formation of a lake is due to the combination of processes of erosion and accumulation. The glacier coming from A carves a hole at B, which is the deepest part and in the middle of the lake; it then travels on to C where morainic material is deposited, thus giving rise to the wrong conclusion that the lake was caused by blockage owing to glacial deposits in the valley

As a matter of fact, the lakes of Thirlmere, Haweswater, Crummock and Ennerdale (for the water supplies of Manchester, Workington and Whitehaven) have all been deepened by dams at the ends, where a much less proportion of drift above the rock was revealed than is indicated by Marr's original diagram.

It seems, therefore, that the carving out of the lakes is analogous to a slip plane circle of Soil Mechanics.

In addition to active lakes there are extinct lakes; those completely filled in with moraine, perhaps forming a flat wide valley in which a post-glacial river flows on the moraine.

Such was the classic case of the siting of the Vyrnwy dam which was built between 1881–1892 on a buried Silurian limestone-shale rock-bar at the end of the site of a filled-in glacial lake.

6

LEAKAGE

LEAKAGE UNDER DAMS

Leakage under an earthen embankment is much more dangerous than that under a concrete dam, for the earthen embankment is usually built on soft material which is liable to be scoured out and it is itself also vulnerable to the inflow of water; whereas a concrete dam is usually built on rock which is not worn away so rapidly by the scouring action of water; and even a defective dam is not necessarily endangered by the passage of water through it or even under it.

The following examples show the significance of leakage.

(1) An earthen embankment at Falmouth was frequently giving trouble before it was repaired in 1942. It was built in the middle of the nineteenth century on granite, and was believed to have been overtopped, which was not unlikely, because it had at that time an overflow of only 9 ft. in length. The gathering ground was 1,770 acres, the capacity of the reservoir being 73 million gal. and the depth of water 20 ft. The length of the weir was increased from 9 to 30 ft. in 1930. Nevertheless, as the bank was constructed mostly with disintegrated granite it was then after 80 years or so rather shapeless, and there was leakage into the tunnel. Sometimes the leakage would become cloudy, and from time to time small subsidences like pot holes up to 18 in. deep would develop on the top of the bank near the valve tower. Thus, the junction with the valve tower and tunnel and puddle (if any!) in the bank came under suspicion.

Since the valve tower and tunnel were removed in 1942 and a new tower constructed with a proper junction with the clay which was found to exist, no further trouble has been experienced. Hence it may be inferred that with sound engineering design, a dam of rather inferior banking material of disintegrated granite (locally known as 'rabb') could be made to last over a century.

(2) A similar occurrence, though much more dangerous, occurred in the Ridgegate reservoir, Macclesfield, which impounded up to 41 ft. of water where the key between the puddled clay and the Millstone Grit at the ends and at the valve tower were found to be defective.

(3) An interesting case of leakage, presumably mainly under the dam, is that at Fedaia just below Marmolada, the highest mountain in Italy.

A considerable leakage occurs, which would be of much value for hydroelectric power at such a very high elevation. So far fluorescein has not shown where the leakage goes and English isotopes were to be used as tracers, but without result.

(4) The leakage in the Berlin Stettin canal is attributed by Terzaghi and Peck to capillary attraction. When the puddled clay core of this canal was 1 ft. above top water level the leakage was 450 gal. per min. (648,000 gal. per day) in a length of 12 miles. After the level of clay was raised to 2 ft. 4 in. above top water level the leakage was reduced to 100 gal. per min. (144,000 gal. per day).

(5) The assessment of unavoidable leakage by isotopes at the Rosshaupten dam near Füssen, Bavaria is another interesting incident.

LEAKAGE FROM RESERVOIRS

(1) When the Sautet dam, near Grenoble in France, was constructed in 1925 it was found that the leakage from the reservoir was about 50 million gal. per day, as stated above. After 35 years the leakage has fallen, naturally, to 35 million gal. per day through silting.

(2) In consequence of this loss, elaborate experiments were carried out to ascertain the probable leakage before the St. Pierre-Cognet dam, a few miles below the Sautet reservoir was built, and the forecast made by M. Ailleret of a leakage of $4\frac{1}{4}$ million gal. per day proved to be correct.

(3) In country over which glaciers have travelled anything may happen, as already mentioned, and it is not surprising that great precautions were taken in putting a grout curtain across a buried valley near Génissiat, and (4) the making of a detailed investigation to ascertain the site of a buried channel leading from the Bort reservoir, to avoid possible leakage.

(5) The top water level in the Tignes reservoir was largely determined by the level of an adjoining buried valley; the bottom of this valley, filled with permeable material, fortunately was not very deep.

(6) The site of a proposed reservoir at Caldew Head, Cumberland, revealed, after extensive investigation, a buried valley alongside the present River Caldew (see pages 23–25).

(7) The Yarrow reservoir for Liverpool is near the site of an ancient glacial gorge in the Millstone Grit filled up with glacial stones, gravel, clay, earth and sand to a depth of 100 ft. below the present river bed and was quite unsuspected from surface indications.

(8) A leakage of $6m^3$/sec. developed in the Dokan reservoir a mile away from the dam owing to a disturbance in the dolomite limestone.

(9) The filling of Broomhead reservoir was begun in May 1928. This produced 'runs of water' escaping from the earthen embankment in 1929, these 'runs' were reduced by inserting some 7,438 tons of cement in 50 boreholes aggregating 11,295 ft. and 6 relief holes aggregating 959 ft.

(10) Although serious leakage may be threatened in any area where glaciers have moved over, there are cases where leakage has been due purely to stratagraphical reasons, such as at the Dol-y-gaer dam built on a trough fault of Carboniferous Limestone between impermeable marls of the Old Red Sandstone.

Mr. Stanley R. Bate tells me that the reservoir had a capacity of 300 million gal. and had a leakage in 1912 of 11 million gal. per day which was reduced to $1\frac{1}{4}$ million gal. per day by grouting with lime.

The throw of this trough fault is about 2,000 ft. and is known as the 'Great Neath fault'. The Dol-y-gaer dam was submerged subsequently by the larger Pontsticill (Taf Fechan) reservoir constructed lower down the valley (*Figure 6*).

Figure 6. Dol-y-gaer (Pen-twyn) dam. In 1912 owing to the dam (impounding 300 million gal.) being built on a trough fault of Carboniferous Limestone, thrust down between the clays of the Old Red Sandstone formation, Mr. S. R. Bate reported that the leakage was some 11 million gal. per day. Subsequently grouting with lime reduced the leakage temporarily to $1\frac{1}{2}$ million gal. per day until the reservoir was submerged by the Pontsticill (Taf Fechan) reservoir built lower down the valley

The Carboniferous Limestone, being an excellent formation on which to build a dam, was highly permeable, although just above (and below) the dam site, the Old Red Marls were well suited to retain water but not to carry the weight of a concrete dam.

7

EARLY DEVELOPMENT OF GROUTING OF DAM FOUNDATIONS

One of the great advances of recent years for making a dam and reservoir watertight and safe is the practice of grouting.

In 1911, Lapworth[1] in his paper on the geology of dam trenches, said that grouting *might* be made use of as a last remedy, but he objected to the practice because it really amounted to guesswork. If a more certain method than grouting could be found he thought it would be preferable. In the discussion on this paper Mr. C. C. Smith said that grouting would be of value for stopping leakage through jointed and fissured rocks. He said fissured strata would sometimes silt up and Nature of her own accord would fill up crevices through which water passed, and therefore artificial grouting might materially improve the watertightness of a reservoir.

Gourley[2] reported in 1922 that a careful examination of the descriptions of recent water schemes in which extensive grouting has been carried out, chiefly in America, leads to the conclusion that our present methods might well be reconsidered with a view to reducing the cost of dam foundations and saving time without in any way sacrificing watertightness or stability.

In 1927, as President of the Institution of Water Engineers, Lapworth brought forward two papers[3,4] on the subject of grouting, and expressed the view that the process would seem to have a great future. Mr. H. P. Hill in discussion said that we were beginning to apply a new principle and that every successful application of the process would call for a great deal of close attention and experience, because the conditions varied in different sites and different strata. Consequently, the means of dealing with them had to be varied accordingly, and he then went on to describe an earlier, probably the first, incident of grouting in a reservoir near Holmfirth, where 4,000 or 5,000 tons of cement were injected into bore holes across the valley.

Attention must be called, therefore, in these early opinions of the efficacy of grouting, to the necessity of every case being treated on its own merits. A careful examination of the results of grouting is well worth while for future saving in the cost of dam foundations, saving time, securing watertightness and ensuring safety.

In the last 40 years the practice of grouting has developed enormously and great reliance has been put upon it; a number of reservoirs probably would never have been constructed but for the advance in the technique of grouting.

As it is still true that the practice very largely amounts to guesswork, it is therefore necessary to study the results of the guesses individually.

REFERENCES

[1] LAPWORTH, H. The geology of dam trenches. *Trans. I.W.E.* **16,** 25 (1911).
[2] GOURLEY, H. J. F. The use of grout in cut-off trenches. *Trans. I.W.E.,* **27,** 142 (1922).
[3] BARNES, A. A. Cementation of strata below reservoir embankments. *Trans. I.W.E.* **32,** 42 (1927).
[4] FOX, J. R. Pre-cementation of a dam trench. (Scout Dike) *Trans. I.W.E.,* **32,** 69 (1927).

8

MODERN GROUTING

There are several geological conditions which give rise to variations in grouting processes. Very broadly there are three general types of grouting: (1) *deep-seated grouting*, in some instances up to 300–400 ft. to prevent water leaking out of the reservoir and scouring out foundations underneath a dam of whatever materials it may be made; (2) *surface grouting of the foundations* of a dam (generally other than an earthen embankment) to secure adequate bearing pressure for the structure; and (3) *grouting in the structure itself*, particularly applicable to the various types of concrete dams.

Deep-seated Grouting

Some detailed descriptions of the process are given by way of examples in the accounts of the French dams at Castillon and Sautet. Very often holes of 1 or 2 in. in diameter are drilled at 5 or 10 ft. centres in stages of 10 ft., and cement, one part by weight to three parts of water, are pumped into the ground at a pressure of $1\frac{1}{2}$–2 times the maximum head of water when the reservoir is full. If the limestone proves to be karstic with large cavities grouting of this kind would not be applicable.

Where much cement is used or where an undue amount of cement is found to be used in one particular hole or one particular section of a hole, the cement-water ratio is increased (as at Sutton Bingham, Yeovil, in open limestone). This is common in limestones where fissures are much larger and more numerous than in other formations. Although many dams have been constructed on limestone in Northern Italy and in France, no dam has been constructed in England on wholly Chalk or Carboniferous Limestone, although there are instances in this country of beds of limestone of substantial thickness which have been successfully grouted.

In sandstone formations there are many examples in Great Britain of grouting, although generally such formations are associated with shale and clay (for example, the Pennine reservoirs of Yorkshire). In the majority of cases, perhaps in mixed strata consisting of permeable and impermeable formations, the practice of grouting is most widely applied. It seems easier to ensure watertightness when permeable beds are associated with impermeable beds rather than to rely on grouting to secure complete watertightness in permeable formations.

If the general dip of the strata is upstream, again, grouting is easier than when the general dip is downstream, particularly if permeable strata, which is submerged, crops out downstream.

There is still an element of uncertainty about the efficacy of grouting until the reservoir is built and filled, and the only criterion is the quantity of cement pumped into the ground, and this is generally expressed as the amount of cement used per foot of bore hole. More recently it has been expressed as the amount of cement used per square foot or per square metre of grout screen or curtain. The area of the grout curtain is that covered by the bore holes in section across the valley (there are numerous instances given with detailed figures for different formations).

It has been found impossible with ordinary cement to grout certain fine sandstones, particularly where the grains are smaller than that of cement. As water may percolate through such fine materials, chemicals (such as sodium silicate) are injected into the bore holes to lubricate the fissures sufficient to allow thin cement grout afterwards to percolate into them (Argal).

In semi-clay formations various forms of clay, for example, Bentonite a kind of Fuller's Earth, is very useful; in Great Britain, Fuller's Earth from Redhill is used, Bentonite being used mainly under French and Italian dams. The amounts injected are similarly expressed as the amount of chemical products per metre of bore hole or per square metre of grout screen.

Surface Grouting of the Foundations or Contact Grouting

Where it is important to strengthen the foundations in order to increase their resistance to bearing pressure, particularly where there are patches of weak rock for the support of concrete dams or abutments and, generally under a broad foundation for a gravity dam, shallow grouting, say, 10–20 ft. in depth of the surface rocks is now very general (for example, the new St. Pierre-Cognet dam on the Drac, France). The quantities of cement (and cement is generally used for this operation as it is not a question of watertightness but of strength) are expressed either in the amount used per foot or metre of bore hole or per square foot or square metre of dam foundation (measured on plan), or per cubic metre of strata grouted, measured according to the area and depth of the boreholes. After the operation has been completed it is possible to test the efficacy of the work by loading or seismic tests before the structure is built.

Grouting in the Structure Itself

Where a dam is built in sections and allowed to settle and cool, grouting may be used at the construction joints to convert the dam into a monolith; for an arch dam this is especially necessary. The amount of cement used is expressed in terms of per square foot or square metre of joint (for example, St. Pierre-Cognet dam).

Certain gravity dams may be made of weak concrete or hearting concrete for which grout may be applied or has to be applied if there is leakage.

Figure 7 (a). Glendevon Dam. Upstream face, left bank showing andesite and superficial debris. The dam had a concrete cut-off trench up to 55 ft. deep but no grout curtain. The total leakage was 500,000 gals./day. Inclined holes were drilled to make contact with the cut-off concrete in the trench. The leakage of the grout curtain was reduced from 500,000 gal./day to 1,500 gals./day or by over 99 per cent

Figure 7 (b). Glendevon dam. Upstream face, right bank.

COMMENTS AND CASE HISTORIES

An interesting paper of the National French Committee of I.C.O.L.D. on the use of cement for grouting limestone, quartzite, mica-schist, crystalline schist and gneiss confirms from experiments what others have found in practice :

(1) That pressures between 20 and 80 kg/cm are sufficient and higher pressures are not always necessary.

(2) That grout *improves* the strength of rock but does not make it *equal* to sound rock (*I.C.O.L.D.* I, 351 (1964)).

The pious suggestion expressed in the past that a *careful examination of the results of grouting is well worth while for the future saving of the cost of a dam,* the "future" has to some extent arrived! In a paper by F. F. Fergusson and P. F. F. Lancaster-Jones on testing of the efficiency of grouting (*I.C.O.L.D.* I, 125 (1964)) it is stated that a grout curtain in all formations *reduces* but does not *eliminate* permeability; and many relative efficiencies of curtains are given in a theoretical analysis. It is pointed out, however, that the surest proof of the success of a grout curtain is the elimination of a known leakage and actual examples are given by W. E. Perrott and P. F. F. Lancaster-Jones in Grouts and Drilling Muds; (Case Records of Cement Grouting' (Butterworths) page 56 (1963).

(*a*) GLENDEVON DAM, CUPAR, SCOTLAND (*Figures 7a* and *7b*). This concrete gravity dam for the water-supply of Cupar, Fife, founded on jointed andesite, 140 ft. in height with a crest length of 1,200 ft. and a cut-off trench, 55 ft. deep, leaked at about $\frac{1}{2}$ m.g.d. when filled in 1955. The water which went through certain joints in the dam, under the abutments and below, from the andesite, was clear. The author reported that a grout curtain was necessary but there should be no immediate danger.

In 1959 a sloping grout curtain was added at the upstream face (for which the reservoir had to be emptied) by inclined holes to go into the andesite below the cut-off trench to a depth of 70 ft. maximum. This was half the height of the dam. Grouting pressures were 2 lb/in²/per ft. of depth of water with holes at 5 to 20 ft. centres. Additional holes at 5 ft. centres were drilled and grouted to ensure contact with the base of the cut-off with the andesite. Curtain and contact grouting with 37,882 ft. of borehole consumed 510 tons of cement. Blanket grouting in addition took 10,833 ft. of drilling and 198 tons of cement, this with some additional 'stitch' grouting and minor leakages treated individually on the dam reduced the leakage to 1,500 gal./day or to 0·3% of the original leakage.

(*b*) THE DOKAN RESERVOIR (*q.v.* p. 347). This leaked some 6 m³/sec., two or three years after it had been filled. This was cured by an extension to the grout screen in a dolomite limestone unconformity involving a very deep grout screen, 150 to 160 m deep, 336 m in length.

(*c*) SHIRAWATA DAM, INDIA. This dam 6,600 ft. long and 127 ft. maximum height, was constructed between 1910 and 1916 with rubble masonry in lime mortar. Leakage in 1933 of 22 cusec.; 5,469 tons of cement in 200,000 ft. borehole were injected as a narrow grout curtain into the masonry and this reduced the leakage to 2 cusec.

In 1957 the leakage was again 4 cusec. and it was decided to regrout the dam and put a grout curtain at least 40 ft. into the foundation basalt, agglomerate and tuff of the Deccan Trap series which had never been grouted.

With 5 ft. spaced holes from the top of the dam and inclined holes from the downstream toe, curtains consuming 1,090 tons of cement injected into 180,000 ft. of hole reduced the leakage by 80%.

(d) CAMARASSA, SPAIN. Lugeon* in 1933 reported that a gravity arch dam 98 m in height constructed in 1926–31 (on the River Noguera Pallaresa, 400 m above the River Segke on dolomite limestone underlying shale) leaked at 220 m.g.d. (11·5 cumec.). A line of grout holes, at 8 m centres to the shale, reduced the leakage to 50 m.g.d. (2·5 cumec.) which remained constant for 30 years.

(e) DOL-Y-CAER (PEN-TWYN). An early example of grouting is shown in *Figure 6*, page 29.

These five examples seem to be in line with the previous section wherein sentiments were expressed on the future of grouting.

* LUGEON, M., Barrares et Géolorie, Dunod, Paris (1933)

9

SETTLEMENT, SLIPS AND SUBSIDENCES

SETTLEMENT

A study of the settlement of dams shows that most of it takes place in the first 2–3 years after construction, thereafter it is small. Settlement need not be considered as a symptom of slipping or subsidence as it is anticipated and is usually provided for by allowing $\frac{3}{8}-\frac{1}{2}$ in. per ft. of height of the embankment.

The Watauga rockfill dam (318 ft. in height) of Cambrian quartzite and shales with a camber of 9·2 ft. settled 2·6 ft. after 8 years, and the South Holston rockfill dam (285 ft. in height) of Ordovician sandstones with a camber of 10·1 ft. settled 2·85 ft. after $6\frac{1}{2}$ years. Little settlement took place after these periods.

SLIPS

Slipping takes place very often during construction or immediately after completion and is generally attributable to: (1) the *nature of material* and the inadequate design of the slopes of the bank in relation to the material; (2) the *speed* at which the bank is built; and (3) the *climatic conditions* during construction.

Nature of the Material

The nature of the material and the design of the slopes are very largely determined by the shear stress calculations (*see* Appendices).

Speed of Modern Machinery

Modern machinery, for example, bulldozers, caterpillar scrapers and sheep-foot rollers are so heavy and rapid in movement that inherent water in clay is squeezed and the clay becomes unstable and slides owing to water pressure within it. For this reason the Muirfoot dam constructed with Boulder Clay had to be reduced in height, and work on the Usk dam had to be suspended for several months. Such a condition is known as 'pore water pressure' and to control this there is apparatus available for measuring it.

Climatic Conditions

Bad weather conditions are responsible for much trouble during construction owing to the difficulty of properly consolidating the embankment material. Prolonged rain tends to settle on a half-formed bank as at Chingford (Metropolitan Water Board) and produces slips.

In constructing earthen embankments in England it was the practice to bring up the bank by inclining the surface towards the centre of the dam. After a period of prolonged rain on the half-constructed embankment, water accumulated which was found to cause slipping.

Slips in Valleys

The flooding of a valley can cause a slip to take place in the natural valley sides and if any of these take place near the dam or overflow works it becomes, to say the least, embarrassing.

A slip in the Ewden valley occurred soon after the construction of the Broomhead earthen dam for Sheffield; the natural grits and shales of the Millstone Grit threatened to block the overflow weir (*Figure 8a*).

Figure 8a. Broomhead earthen dam. A slip in alternating grits and shales occurred threatening to block the overflow weir

The Ghrib dam in Algeria was similarly threatened and this necessitated very heavy construction for the overflow channel.

The Pontesei (Maè) reservoir valley was threatened by a serious slip, fortunately away from the dam but unpleasantly near the draw-off valve tower. Higher up this valley, 500 m from the dam, there was a large fall of rock into the reservoir causing a sort of tidal wave which the cupola dam successfully withstood.

SUBSIDENCES

If an embankment has been placed on an unstable foundation and the camber has not been able to maintain the crest at its correct level, subsidence rather than normal settlement is envisaged. *Figure 8b* shows settlement which occurred on the Dowdeswell dam near Cheltenham.

Similarly, subsidence rather than settlement may be an attribute of very old embankments which have become decrepit, chiefly through neglect, and some are so shapeless that they could not possibly have been built to the

Figure 8b. Dowdeswell dam. After many years the embankment was found to have become dished, thereby giving insufficient freeboard and is now seen to be in the process of being raised

form in which they now exist. Vegetation and trees often mask their bizarre shapes, but fortunately more notice is taken of them now than in the past owing to the Reservoirs (Safety Provisions) Act, 1930.

10

MINING AND DAMS

From time to time the question of mining under a reservoir and its dam arises.

In Wales there was once some doubt as to whether it was safe for coal-miners to be below the reservoir when they mined, or after they had mined to live in their houses on the surface below the dam.

It is well known that buildings may be totally destroyed and that small service reservoirs may become a shambles. At Ilkeston in Derbyshire, the Shipley reservoir was totally destroyed, and the Marlpool, a 2 million gal. reservoir constructed of reinforced concrete and steel columns supporting a thin reinforced concrete roof, just managed to survive several feet of subsidence.

Mining coal under an impounding reservoir, containing many million gallons of water might be disastrous, as the reservoir would be on sedimentary strata (Carboniferous in England), and the dam, particularly if an earthen embankment, might be washed away in a matter of minutes.

The King's Mill dam at Sutton-in-Ashfield, near Mansfield, Nottingham, is an example of an earthen structure impounding 18 ft. of water and 90 million gal. constructed on Coal Measures and under the threat of coal-mining. Thus, a stream flowing towards the dam at 1,000 ft. from it, started flowing upstream away from the dam, owing to subsidence.

The Lluest Wen reservoir belonging to the Pontypridd and Rhondda Joint Water Board* is above five principal coal seams (eleven in all), the top one being 1,000 feet and the bottom one, the '9 ft.' seam, being 1,400 ft. under the dam (*Figure 9a and b*).

The problem arises which is the more valuable—the value of the coal (if sterilized), or the value of the main water supply of the Rhondda Valley? Although this old reservoir containing 242 million gal. of water may not be in itself very valuable as a structure, it is the water that it can collect and store reliably and safely throughout dry periods that is of value.

Hence, how near is it possible to mine the five seams without disturbing the dam?

The only available information about the reservoir is that the capacity is 242 million gal., the maximum depth is 65 ft., the length of the crest is

---

*Mr. G. Ewart Roberts who has kindly read this chapter confirms that his Board, the Pontypridd and Rhondda Joint Water Board, has readily given permission for the use of information regarding the Lluest Wen reservoir.

(a)

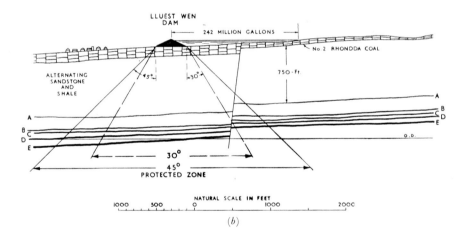

(b)

Figure 9. (a) and (b) Lluest Wen dam. Some consideration has been given to the question of mining 4 or 5 coal seams aggregating some 20 ft. in thickness under this dam. It impounds 242 million gals. with a maximum depth of 65 ft. and a crest of 660 ft. It is situated in the populous Rhondda Valley. A pillar of coal, at an angle somewhere between 30–45 degrees is suggested under the dam for protection against subsidence

    A denotes Gorllwyn seam   C denotes 4 ft. seam
    B   ,,   2 ft. 9 in. seam   D   ,,   6 ft. seam
                                            E   ,,   9 ft. seam

660 ft., and it has a 15 in. draw-off pipe and a 12 in. washout. No information is available on the cut-off trench, if any, under the embankment, nor how the embankment is keyed into the ground or into the sides of the valley, or whether there is any drainage system, and how the draw-off pipes are taken through the bank.

On December 23rd, 1969 a 'large hole' was recorded on top of the dam, large enough 'for a horse and cart to fall into' and from reports it would seem that the cause could have been a broken draw-off pipe under the embankment comparable to that shown on *Figure 10a* on page 45.

In April 1957, Mr. A. Young[1] (Chief Engineer of the National Coal Board), Mr. H. A. Longden (Director General of Production), and Mr. B. L. Metcalf (Chief Engineer) stated that it was necessary to protect a pit-head by sterilizing a zone of coal within an angle of 30 or 45 degrees to the vertical down to the deepest seam that will be worked.

The 1·8–1 ratio (29 degrees) was also suggested many years ago in a case of coal-mining under a reservoir for the water supply of Sydney (Wade[2], and quoted in full by Leggett[3]), as the only case concerning coal-mining under a large impounding reservoir.

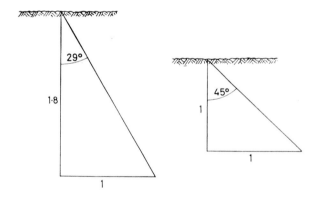

The dam concerned, the Cataract dam, Australia, is 150 ft. in height, concrete gravity, retaining 21,000 million gal. of water. It is built on the Hawksbury Sandstone overlying Coal Measures, and the 'Bulli' coal seam is 1,600 ft. below the base of the dam. The Water Board suggested reserving a pillar of 440 yd. 1,320 ft. (35 degrees, 1–1·4). The distance was finally settled by a joint report of Mr. Pittman (Government geologist) and Mr. Atkinson (Chief Inspector of Coal and Shale Mines) as follows.

(1) The coal could safely be worked under the *reservoir* owing to the intervening shales, and the resulting subsidence would not endanger the water supply by undue leakage.

(2) To ensure the stability of the Cataract *dam*, a solid pillar of coal should be left to extend 300 yd. beyond the base of the wall in all directions.

For the depth of 1,600 ft., the 900 ft. width is approximately 30 degrees or 1·8 vertical to 1 horizontal, closely coinciding with the recent view for guaranteeing stability for pit-heads.

These opinions obviously have been well considered and suggest that to protect such a vulnerable structure as a dam, mining should not be done within an angle of 45 degrees; certainly not less than 30 degrees from it.

REFERENCES

[1] YOUNG, A., LANGDON, H. A., and METCALF, B. L. *J.I.C.E.*, **8,** 671 (1957).
[2] WADE, L. A. B. *Min. Proc. I.C.E.*, **178,** 1 (1909).
[3] LEGGETT, R. F. *Geology and Engineering*. New York and London; McGraw-Hill, (1939).

11

SUSPICIOUS OLD DAMS

The Reservoir (Safety Provisions) Act 1930 has resulted, after 40 years, in many dams being suspect. The Act provides that not only should all dams be designed and constructed by an engineer on a panel of engineers created by the Act for the purpose, but also that all existing dams which retain more than 5 million gal. above the surface of the ground, must be inspected at least every 10 years.

Many British dams have been in existence for 100 years or more; up to 1930 some had been neglected, while others had been well-maintained. For some there are detailed plans of the foundations still in existence, and for others there are no plans whatever in existence. Hence, the examination of such reservoirs entails very careful investigation of what may appear to be trivial matters but which prove to be highly dangerous.

Many of the old dams are masked by trees and vegetation on earthen embankments, and in Great Britain nearly all the dams concerned are constructed of earth of unknown consistency. It is unknown how much clay was used to ensure watertightness and how far the clay cut-off trench (concrete was not used) was carried below the level of the river bed. Because an earthen embankment has stood for 50–100 years it is not to be assumed that it will safely stand for another similar period.

It has been found as a result of investigations, that any form of dampness in dry weather at the bottom of the embankment is suspicious even if there were no slipping of the embankment itself.

For example, the Bottoms dam (30 ft.) at Macclesfield, after a slip on the embankment had been opened out, was proved to have two 12 in. draw-off pipes buried in the bank, a circumstance which might have led in the course of time to failure of the dam (*Figure 10a*).

Where cloudy water appears, for instance, in tunnels, or where leakage increases without apparent cause and particularly where it becomes cloudy or where subsidences are noticed on the embankment, or both, such occurrences must be investigated immediately.

Another example of decrepitude occurred with the Blatherwycke dam, Northamptonshire, when suspicious dampness occurred where a supply pipe had been laid on the embankment and a brick culvert had subsided about 18 in. (*Figure 10b*).

A factor which may be a contributory cause of collapse is the lowering of the water, impounded by an earthen embankment, particularly during

Figure 10a. The 12 in. outlet pipes from Bottoms dam which were broken and drawn 2 in. apart by the weight of the earthen embankment. This was exposed in 1929
(Reproduced by courtesy of Mr. J. H. Dossett, Macclesfield Corporation)

Figure 10b. Blatherwycke dam. This earthen embankment is founded on soft Estuarine and Lacustrine deposits of the Lias of Northamptonshire. Some wetness occurred on the embankment, the brick tunnel had subsided by 18in. and a large pipe had been laid in the embankment. The outlet works were reconstructed and the pipe removed at a cost of £30,000 (1959). Part of the overflow weir is on the right of the sloping wall
(By courtesy of Charles Oliver, and Stewarts and Lloyds)

Figure 11 (a)

Figure 11 (b)

(c)

Figure 11. Island Barn reservoir. (a) *This reservoir was planted with trees to satisfy a landowner's objection to the reservoir interfering with amenities. The embankment settled and the cause was found to be the penetration of tree roots into the puddle. Owing to the reservoir bank having settled some 200 trees had to be taken up, the roots of which were up to 35 ft. in length.* (b) *Example of extensive area of roots uncovered for the foundations of a housing estate at Macclesfield. (From the collection of Mr. J. H. Dossett, engineer to the Macclesfield Corporation.)* (c) *Showing roots entering the puddled clay core*

(By kind permission of Mr. H. F. Cronin and Mr. W. M. Lloyd Roberts on behalf of the Metropolitan Water Board)

droughts lasting some months; the top of the embankment, including the clay puddle, may become baked and, on rapid filling, leakage may occur through the baked strata thus weakening the embankment.

Terzaghi has referred to the probability of water leaking out of a reservoir (when filled to the top water level) by capillary attraction over the top of the puddled clay core. As this clay is usually 6 in. below the top of the bank or 4–5 ft. above normal top water level, it is very rarely kept damp except by ordinary rainfall and is nearly always dry and more or less fissured. Hence, it is suspected that water by capillary attraction could quite well seep through the fissures of baked clay.

Mr. J. H. Blight, engineer and manager to the West Cornwall Water Board, noticed that certain leakage of a small reservoir at Boscathno (near Penzance) always stops when the water level is lowered a foot or two; this may well be due to capillary attraction because the water never rises above top water level as the reservoir is not subjected to floods, and hence any clay above this level would never become plastic and would always remain dry. In addition, the purer the clay the greater the contraction and hence the greater the fissures.

However, in not one of nearly thirty-five *old* English reservoirs which have had to be examined under the Reservoirs (Safety Provisions) Act has deterioration occurred as a direct result of any deep-seated geological trouble.

Trouble has nearly always developed from engineering defects such as the spillway being inadequate to cope with floods, the embankment having slumped, leaks occurring, particularly near a valve tower, defects developing in puddle on the upstream slope, the old practice of burying draw-off pipes directly in earth rather than putting them in a tunnel under the embankment, as well as physical deterioration of pitching or the growth of trees.

An instance reported by Cronin[1] occurred at the Island Barn reservoir embankment which had been planted with trees to satisfy a landowner's objection to the reservoir on the grounds of interfering with amenities. In 1950 the embankment had settled and investigations revealed that large tree roots had actually penetrated into the puddle. Some 200 trees were removed in a distance of 350 yd. with roots up to 35 ft. in length (*Figure 11a and c*). *Figure 11b* shows extensive tree roots unearthed at a housing estate at Macclesfield.

Mr. A. F. C. Verney, resident sub-agent through Mr. A. L. Bennett, Land Agent of the Buckinghamshire County Council, tells me he estimates that the roots of Beech trees become suspect after about eighty years when they deteriorate and water drains through them; whereas the roots of Oak trees are sound after a hundred years without rotting, when they deteriorate and just clog. The roots of Scots Pine remain sound generally for one hundred and fifty years and those of Elm stand up to wet conditions.

Although live roots may act for a time as a binding factor for stabilizing soil, some can also act, when dead or dying, as channels for water in a dam.

EMERGENCY EMPTYING OF RESERVOIRS

One of the questions often asked is how long does it take to empty a reservoir in an emergency. No general answer can be given but for hydro-electric concrete dams, power to operate turbines and gates is usually available, which enables the water-level to be lowered in a short time. As hydro-electric installations are generally in mountainous districts, flooding and damage downstream is not so much a problem but in any event it can, to some extent, be controlled by shutting down the turbines.

In water-supply dams, particularly earthen embankments, which are in remote situations lowering the water-level is effected, as a rule, by opening the scour and washouts on the supply pipe and this may take several days even in dry weather.

Any form of breaching the reservoir can be done only after safeguarding the inhabitants by their temporary evacuation, and or by pumping water out of the reservoir by fire-engines, sand-bagging, where practicable, to delay sudden collapse. Such methods were used successfully on the old Lluest Wen reservoir (see p. 41) on December 23rd, 1969, a very inconvenient date, during very wet weather.

BIBLIOGRAPHY AND REFERENCES

Cover, G. W. Repairs to an old earthen embankment at Macclesfield. *Trans. I.W.E.*, **36,** 282 (1931).

[1] Cronin, H. F. The planting of trees on reservoir banks. *J.I.W.E.*, **5,** 548 (1951).

12

EARTHQUAKES AND DAMS

Another geological impact on dams is that of earthquakes.

In China, in known earthquake country, the rockfill construction was adopted for the Shin Mung dam (for the water supply of Hong Kong).

The Cogoti dam in Chile is a rockfill dam which successfully withstood a severe earthquake without material damage.

An interesting example of damage by an earthquake to a concrete dam occurred in Algeria. On September 9, 1954, the town of Orléansville was subjected to a violent earthquake (Orléansville still looked like a much bombed town in June 1955) and the Pontéba dam distant 20 km from the town was much damaged.

The epicentre of the earthquake, determined by M. Rothe, was near the Sidi Djillali mosque which is 3·5 km from the Pontéba dam. The earthquake was classed as 9° or 10° on the international scale and caused 1,500 deaths and 5,000 injuries in the Orléansville district.

The dam, 18 m in maximum height, with a crest of 80 m in length is founded on Miocene blue clays and was constructed in 1870. The right bank rose 55 cm above the left bank and although the crest was unharmed, both ends were damaged (see Bibliography, M. Trevin).

Figure 12 shows that the right bank of the valley in which the dam is situated had risen relatively to the left bank, the dam being tilted and, as shown, a large flow of water went over the left end and none over the other and a good deal of cracking occurred as the result of the uplift.

It was difficult for the engineers to decide if and how the dam should be repaired, and for the geologists to decide if and when another earthquake was likely to occur!

Comrie, in Scotland, not far distant from the Pitlochry Tummel dams, has been known for some years to be an epicentre of earthquakes, but nothing untoward had occurred here or anywhere else in Great Britain until 1957 when the following extraordinary occurrence took place.

On February 11, 1957, Mr. Bates, the Borough Engineer of Loughborough, happened to be inside the filter-house at the foot of the Blackbrook dam 5 miles west of Loughborough. He reported a severe earth tremor accompanied by noise and considerable vibration, and on going outside to find the cause, he ascertained that the coping stones on top of the dam had been disturbed, and that many cracks had appeared in the parapet walls and several hair

(a)

(b)

Figure 12. Pontéba dam. Algeria. (a) Crest tilted by an earthquake as shown by unequal depth of water flowing over it. (b) Cracking of masonry on right bank is visible

cracks in the inspection galley inside the dam. The dam, which is over 50 years old, has a maximum height of 68 ft., is 482 ft. in length at the crest, and is constructed of masonry and concrete of gravity section.

The epicentre was placed at Diseworth 4 miles north of the dam by Dr. Dollar as a result of 1,500 replies to an appeal for information, and Mr. Tillotson, seismologist, reported that the tremor was Force No. 8 grade on the seismic scale which, as there are only 10 grades, must be considered high. It is thought that the tremor was in two pulses with an interval of 3 seconds.

The engineer's report gave the following further details. These notes have been read by Mr. John S. Bates who gave formal permission for publication.

Coping

An examination of the upstream and downstream parapet walls from the walkway on the top of the dam revealed that all the large coping stones, for the entire length on both parapet walls, had been lifted from their beds, that is, from their horizontal mortar joint (which had, of course, disintegrated) and the stones had dropped back on the mortar, taking up a slightly different position, so that in some cases it was possible to see daylight underneath the stone. Each stone is 4 ft. in length by 3 ft. in width by 1 ft. in depth, or 12 ft.[3] in all, having a weight of about 15 cwt.

In addition to the horizontal mortar joints being disintegrated, every vertical joint also was disturbed and broken. Some of the mortar and stone chippings had fallen on the walkway (*Figure 13a and b*).

Weir Arches

There were six vertical cracks detected, above each arch, above the overflow weir and under the centres of the coping stones.

An examination under the arches showed that the masonry facing in some places may have been parted from the brickwork; also, there were occasional cracks in the mortar joints of the masonry facing above the arches.

Downstream Face

Examination of the downstream face revealed cracks (seen with and without binoculars) extending to a depth of about 20 ft. below top water level. Thus, on the second bay from the end there was a very noticeable crack that extended vertically through three joints in the masonry (stone coping, string, and bullnosed courses), and zig-zagged down via the masonry joints of the facing to grass level.

Similarly, in the fourth and fifth bays from the right, there were other cracks, which extended to some 15 ft. below top water level and a considerable amount of calcareous tufa was disturbed on the face. Another crack was also observed in the seventh bay just below the arch.

On the left hand end, there was a crack on the downstream side in the fifth bay which could be traced from about 20 ft. below the top of the parapet.

Tufa

The amount of calcareous tufa on the downstream face suggested that a certain amount of seepage had occurred during the 50 years since the dam was built, and it may well be that the dam in the past had been disturbed by tremors of less severity which had passed unnoticed. It was striking to see how much of the surface calcareous tufa, which normally sticks very hard on to the masonry, could be removed easily by hand.

Gallery

At the entrance to the gallery, at both ends, at three or four brick courses below the manhole cover, lateral shifting had occurred in the horizontal bedding of the brickwork to the extent of half an inch.

An examination of the 5 ft. × 28 in. inspection gallery was made, which revealed many hair cracks of ½ mm or less, obviously newly formed, but in Mr.

(a)   (b)

Figure 13. Blackbrook dam, Loughborough. (a) Upstream side of the dam, shaken by an earthquake in February, 1957. Repairs are being carried out by cradles slung over the parapet. (b) Downstream side of the dam. Repairs were carried out by a cradle over the first and second arches

Bates' opinion they were tending to close since the earthquake. Nevertheless, they could be traced all round the gallery in many cases, particularly at construction joints and through the calcareous tufa, which had formed in the past from seepage at these joints.

Interesting and perhaps somewhat significant horizontal cracks were observed on the left bank side half way up the staircase.

Left End Abutment

An inspection was made on the left bank side of the dam where it enters the natural pre-Cambrian rock; here at the foot of the dam the rock was seen to be steeply dipping at some 45 degrees upstream, and there were certainly definite signs of recent falls of rock and newly formed cracks in the natural rock which

(a)

(b)

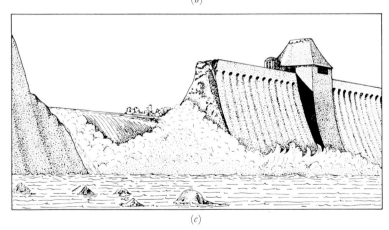

(c)

Figure 14a, b and c. The Möhne dam. The dam is curved gravity masonry structure, built 1909-13, and is used for water supply, flood control, and to generate a small amount of power. It is 132 ft. in height and impounds over $33,000 \times 10^6$ million gal. The dam was breached by a hole 250×42 ft. between the towers in 1942 for which some 2 tons of explosive would probably be necessary. The dam has been repaired and it is hardly possible to see the join between the old and the new work

would normally be considered of a minor nature, but significant in the present investigation.

No unequal settlement of the overflow weir or spillway was observed after a very close investigation.

There was some suspicion that some leakage had been caused, for before the earthquake certain springs (not attributed to leakage from the reservoir) usually showed a depth of $\frac{1}{2}$ in. (500 gal. per day) over a V-notch below the dam. On February 11, the flow increased to $5\frac{1}{2}$ in. (200,000 gal. per day). By March 6, however, this had dropped to $2\frac{3}{4}$ in. (36,000 gal. per day), and subsequently the flow reverted to normal conditions obtaining before the earthquake.

It is interesting to note that the epicentre was located at Diseworth on low lying ground of Keuper Marl, underlain unconformably by the pre-Cambrian rocks which outcrop at the surface at the dam in the high-lying Charnwood Forest. It was stated by Hunter and Keefe at the fifth Dam Congress, 1955, that epicentres are generally in low-lying ground near, but not in a mountain range, and hence dams located in a mountainous region are less liable to suffer than those in adjoining plains.

Comparable to earthquake shock, the bombing of the Möhne dam, Germany, may be cited (*Figure 14a, b and c*).

This dam, built in 1913, is of gravity section and is 132 ft. in height. Although it was severely damaged by British bombers in 1942 by a hole 42 ft. in depth and 250 ft. in length, neither the dam nor its foundations were otherwise disturbed. No doubt the curvature helped as well as the gravity section to localize the damage.

BIBLIOGRAPHY

TREVIN, M. Les effets du séisme de septembre, 1954, sur deux barrages de la région d'Orléansville Algerie. *I.C.O.L.D.* II, 214–217 (1964).

HUNTER, J. K. and KEEFE, H. G. Special problems relating to the construction of dams in active volcanic country. *I.C.O.L.D.* III, 511 (1955).

DOLLAR, A. T. J. The Midland earthquake of February 11th 1957. *Nature*, **179**, 507–510 (1957).

WALTERS, R. C. S. Damage by earthquake to Blackbrook dam, Loughborough. *I.C.O.L.D.* II, 1 (1964).

13

MAJOR DAM DISASTERS

It may truly be said that the causes of most of the major dam disasters are unknown; where a great volume of water is suddenly let loose in the valley by the bursting of a dam, the whole valley, so to speak, is washed clean and there are no traces whatever of the precise cause, and hence in the majority of cases the cause of a disaster can only be surmised.

One of the best authenticated reasons for a disaster is that of the St. Francis dam, California, situated 45 miles north of Los Angeles. This 200 ft. high dam, built for the purposes of water supply, was completed in 1926; it overflowed in March 1927.

The dam was of mass concrete, curved with a radius of 500 ft., of gravity section, and was built upon a fault between conglomerate on the right bank and schist on the left bank. It collapsed suddenly on March 12, 1928, drowning 400 people and causing £4,000,000 damage.

A block of 2,000 or 3,000 tons on the right bank, that is, on the conglomerate side, was carried half a mile downstream; a similar block on the left bank, on the schists, toppled over; the vertical planes of cleavage took place at the contraction joints of the dam (*Figure 15*).

After this great calamity, the conglomerate on the right bank was found to have a crushing strength of 50 lb. per in.^{2} or 3 tons per ft.^{2} which was not very good, and a laboratory test revealed that when a piece of it, about the size of an orange was placed in water 'a startling change takes place. Absorption proceeds rapidly, air bubbles are given off, flakes and particles begin to fall, and the water becomes turbid with suspended clay. After 15 minutes to an hour the rock has disintegrated into a deposit of loose sand and small fragments covered by muddy water'.

As the dam was taken down some 8–10 ft. below ground level, the hillside on the right bank disintegrated resulting in the right-hand end of the arched dam becoming unsupported and being carried downstream. On the left bank end, built on schists (dipping towards the thalweg) with slatey laminations, there was a tendency to sliding, so that when the right end collapsed, the immense volume of water scoured the schists on the left causing sliding which resulted in the downstream toe becoming unsupported, and causing the left end of the dam to topple over.

The central and highest part of the dam stood without much movement which is evidence that its design (a true gravity section) construction and

(a)

(b)

(c)

Figure 15. St. Francis dam, California. (a) *The dam before the failure; the outline of the portion which is still standing is indicated on the face of the dam.* (b) *After the failure: note the change in profile of the right slope at the dam from that shown in* (a); *note also the distinct demarcation of the plane between the conglomerate rock and schist as exposed along the right side of the valley. All this was caused by a block of 3,000 tons having been torn away from the right bank and carried for half a mile downstream.* (c) *Left abutment looking upstream. Rock face against which dam abutted completely gone or covered by landslides. Block on schists toppled over, in foreground*

(Reproduced by courtesy of the Editor of *Discovery*)

foundations were sound, the excavation for this part being carried down to 30 ft. below the surface.

One of the major disasters in France was that of the Bouzey dam which drowned 86 people. It failed because a soft sandstone zone present in the Lower Trias was scoured out under the dam although the water was only 15 m deep. Grouting the foundations was unknown in 1895.

The Les Cheurfas dam, Algeria, is another case where positive evidence is forthcoming, namely, the weak conglomerate of a Quaternary deposit was mistaken for a sound older Miocene conglomerate.

The foregoing dam disasters are due to water undermining the foundations, and there are some 100 dams in America of between 1864 and 1876 whose failure is attributed to this cause.

A similar cause of failure in 1900, through excessive leakage through open joints in the foundations, was that of the Austin dam, Texas, built on alternating limestones, shales and clays of the Cretaceous. The failure was attributed to water getting into the joints of the limestones during a period of the great flood, and it was thought that the clays and shales were lubricated as approximately a 500 ft. length of the dam slid some 60 ft. downstream.

The causes of failures of English dams have never been satisfactorily proved. Thus, the collapse in 1850 of the first Woodhead embankment built on alternating grits and clays was probably due to percolation through the grits, and the Bilberry dam (Holmfirth) on a similar formation, failed in 1852 through water forcing its way through fissures in the rock and through upward pressure rapidly washing away the earthen embankment.

The precise cause of the Dale Dyke (Bradfield, near Sheffield) disaster of 1864 is unknown. It was built in the Pennines in similar rocks of alternating grits and shales coupled with somewhat poor construction of the earthen embankment. Although many embankments have slipped when the greatest precautions have been taken in selecting materials and placing them, it is more probable that the real cause of the Dale Dyke disaster was the undermining of the foundations caused by water percolating the gritstone under the embankment owing to the cut-off trench not being deep enough, and, again, the absence of any grouting, a procedure which was unknown in those days. The embankment would become saturated with water and naturally slip.

Following the disaster, the coroner was so keen to get evidence which would attach the blame to some particular person (the engineer) that all the evidence was centred with this end in view, and he seemed highly prejudiced and biased rather than anxious to hold an impartial inquiry undertaken to ascertain the facts and causes of the disaster on a scientific basis.

The last disaster which occurred in Great Britain was that at Dolgarrog in North Wales in 1926 in which 16 people were drowned. The precise cause of the collapse was described in a report of 15,000 words which may be summarized to the effect that the dam, although sited on glacial clay and peat, with a maximum height of only 20 ft., the cut-off wall, believed to be only 6 ft. deep, was not carried down to sufficient depth below the surface. The reservoir was constructed 16 years before the burst took place and it was built to raise the level of the lake Eigiau, the dam being three-quarters of a mile in length. The initial rupture was believed to have occurred at the outlet of the reservoir in the culvert going under the dam.

Two major dam disasters occurred in 1959.

On January 9, the Vega de Tera dam on Lake Sanabria burst and caused considerable damage to the town of Rivadelago, with loss of life. The dam, 70 ft. in height, was under construction and was owned by Hidroélectrica Monocabril who own other dams in the province of Zamara. Although the dam is in a granite area, a possible reason for the disaster is that differential

Figure 16(a). The Malpasset dam, France. View shows the dam site and the valley of the Reyran after the catastrophic failure of the dam. The motor road which entailed blasting is shown on the top left-hand corner some 300 m from the left abutment. Some geological weakness, not entirely understood, below the foundation half way up the left abutment, was attributed by the commission to the dam failure

(Reproduced by courtesy of the Editor of *The Engineer*, and of *Associated Newspapers Limited*)

settlement of buttresses, due to variation of strata under each, caused excessive strains between one buttress and another.

On December 2, the left foundation of the Malpasset dam, 5 miles northeast of Fréjus in southern France, gave way (*Figures 16(a) and (b)*). The dam was a cupola of 66·5 m in height, built in 1952, crest 1·5 m in thickness, base thickness 6·91 m, radius 105 m, chord at crest level 180 m, chord-height ratio about 3, and its collapse was due entirely to geological causes. The capacity of the reservoir was 25×10^6 m³.

A geological map shows that the dam was founded in an area of Gneiss and Permo-Carboniferous rocks, and M. Dargeau stated that the valley is in a synclinal carboniferous zone enclosed by the metamorphic horizons of the base Massif of Esterel.

The Commission of Inquiry, appointed by the French Ministry of Agriculture stated that perhaps certain events in the life of the dam encouraged infiltration of water which softened the left abutment causing it to be swept off the left bank.

Some of these events during the six years of the life of the dam are set out in the table below.

Year	Month	O.D. of level in reservoir, m	Remarks
1953	December	—	
1954	April 30	51·30	
1954	May 24	60·60	Reservoir filling.
1955	September	79·75	
1956	July	83·85	
1957	November	—	When, at the left end of the dam, the reservoir keeper felt explosions of blasting in connection with the motor road construction distant 300 m.
1958	Feb 7	—	Conference of engineers determined that blasting with charge 300 kg tolamite 300 m from the dam was harmless.
1958	July	87·30	
1959	July	94·10	
1959	September	93·5	
1959	Nov 15	95·2	Reservoir keeper reported water oozing 20 m downstream of right bank. Heavy rain.
1959	Nov 27		Heavy rain.
1959	Nov 28		Heavy rain.
1959	Nov 29	95·75	'Seepage' on right bank assumed to be due to heavy rain.
1959	Nov 30 18.00 hrs.	97·0	Heavy rain.
1959	Dec 1 midnight	99.0	Heavy rain rendered communication precarious.
1959	Dec 2 18.00 hrs.	100·0 100·12	A site conference in the afternoon instructed the reservoir keeper to open sluices, which he did successfully. There was no vibration.
1959	19.30 hrs.	100·09	Reservoir keeper left site. Water level lowered by 3 cm.
1959	21.10 hrs.		In his house, 1500 m downstream of dam, he heard explosions; a violent blast of air opened doors and windows. There was a flash of light which went out. The dam was destroyed instantly.

The Commission based their findings on many detailed reports of eminent specialists on various matters which could be the cause of failure, i.e. sabotage, aerial attack, meteorites, valve control, workmanship of construction, quality of cement, aggregate and concrete, efficiency of grouting, (curtain, contact and structural), design and abutment (the thrust of which is stated at 4,850 tonnes for a flood level of 102 m O.D. for the critical upper 16 m of the dam).

As the design abutment thrust was 13,100 tonnes, and similarly all other normal reasons and many others were ruled out by the Commission, one is left with the cause of the disaster being geological.

The Commission summarises the Reports of Prof. Goguel, F.G.S. of Marseilles as follows.

'The foundation rock of the dam is a gneiss impregnated with numerous veins of pegmatite which generally appears sound and resistant. (From which we understand a hard crystalline rock of quartz-mico-felspar, with veins of felspar crystals embedded in a quartz matrix.)

'This gneiss, like most gneisses, has been subjected to a later primitive crystallization and all the specimens showed traces of deformation and sometimes recrystallization which, in their restoration, calcite and sericite has appeared, either in fissures, seen even in cracks in quartz, or within crystals themselves. The strength of the gneiss depends on the arrangement of its crystals and the presence or absence of water in its pores; although plentiful and mechanical resistance is thereby little affected.'

'Where gneiss has deposits of sericite, the joints between the layers are covered with fine silky grains and sliding is facilitated due to the grains, especially when the rock is saturated with water. Sericite was found in specimens taken from the 53 m. O.D. level below the foundation of the broken dam.' (One can imagine that sericite schist being an unctuous substance, derived from orthoclase, potash-felspar, might tend to cause slipping.)

'The attitude of the gneiss varies a good deal from place to place; joints break into the rock at more or less regular intervals and are readily noticeable at the surface; below they exist *in situ* almost everywhere but they do not seem to follow any form of mechanical discontinuity.'

'The site of the Malpasset dam appears exceptional owing to many faults, particularly in areas where gneiss is injected by pegmatite, there faults appear owing to the scouring of the valley sides at the disaster, and were not in their natural position covered with vegetation.'

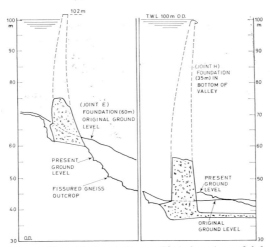

Figure 16b. Malpasset Dam. Typical sections of left embankment after the disaster. (Ministère de l'Agriculture, Commission d'Enquête du barrage de Malpasset 1960).

The Commission's conclusions were:—

(1) The design and customary calculations for the work were correct.

(2) The construction of the work was very good, particularly the concrete and its junction with the foundation rock.

(3) The cause of rupture must be looked for only *below* foundation level.

(4) The most probable cause of the disaster may have been a weak plane or fault upstream nearly parallel for some distance and a little below the foundation of the arch in the upper part of the left bank. The large distortion already in the foundation has been increased locally by this plane of weakness.

(5) The absence of a deep grout curtain has had no bearing on the disaster nor has the type of arch dam adopted.'

The lesson which seems to be learnt from this disaster, for the records of which thanks are due to the Commission, is:

(1) That even a strong rock, such as gneiss, must be critically examined.

(2) That the high part of abutments, especially if the valleys are steep, should be investigated with a view to prolonging the concrete abutments into the valley sides.

(3) That additional curtain grouting in steep sided valleys should be prolonged into the hillside, if necessary constructed from an adit or by inclined holes.

(4) Do we know enough about the effect of even small vibrations through strata?

The Dalgarrog disaster, coupled with previous disasters in Great Britain and with those abroad, gave rise to the Reservoirs (Safety Provisions) Act which was passed in 1930. In many countries there are certain government regulations set out to be followed in the design of dams owing to their vulnerability. Thus, in Italy, certain criteria have to be followed for the design of arch dams, and in the United States of America nearly every State has its own regulations, some of these differing slightly between one State and another.

Nevertheless, in all the 50 States, permission has to be given by a Department of Conservation of Water Resources, generally engineering or geological, or both, for the construction of all new dams over 10 ft. in height in some States and more often over 20 ft. in height in others. Alternatively, or perhaps in addition, if the dams impound above the surface of the ground more than 3 acre-feet in some States 15–20 acre-feet is the common quantity in many States and over 100 acre-feet in one or two others, permission must be sought. As an acre-foot is approximately 43,500 ft.^{3} = 326,000 U.S. gal. = 270,000 Imperial gal., our Reservoir Act specifying 5 million gal., therefore, seems a happy mean.

BIBLIOGRAPHY

DARGEAU, M. Malpasset Dam, *Travaux*, 1955.
EDITORIAL *Engineer, Lond.*, **208,** 774 (1959).
EDITORIAL Malpasset Dam, *Wat. & Wat. Engng*, **64,** 27 (1960).
GRUNER, EDWARD. Dam disasters. *I.C.E.*, **24,** 47 (1963).
MINISTÈRE DE l'AGRICULTURE. Commission d'Enquête du barrage de Malpasset (August, 1969).
WALTERS, R. C. S. The great Californian dam disaster. *Discovery*, **9,** 184 (1928); also *Engng News Rec.* (1928); and *Engineering, Lond.*, (1928).

14

RESERVOIR GEOLOGY

Unsuitability of Ashop valley for reservoirs

On page 93 it is stated that *dams* over a hundred feet in height in the Derwent valley are founded on Yoredale alternating sandstone and shales, much wrinkled and folded extending to 70 ft. *below the surface* with open joints 9 in. in width. It might have been said also that certain reservoir sites in the Pennines were rejected because of the instability of such strata *above the surface* in the *reservoir* area.

Thus in 1917 'a detailed examination of the Ashop valley and the sinking of trial pits showed that, owing to the existence of very large landslides, it would not be prudent to construct a reservoir so far up the Derwent valley as the site of the proposed reservoir'. Instead, the River Ashop was diverted by a tunnel to the Bamford filters (below the Lady Bower dam), and thence to the Derwent reservoir.

Waldershelf slip into Broomhead reservoir in the Ewden valley

A brief mention was made on page 38 of a reservoir for Sheffield in the Ewden valley where a slip occurred of natural grits and shales below the Rough Rock of the Millstone grit which threatened to block the overflow weir. This was the subject of a paper by J. K. Swales, C.B.E., published privately by the Institution of Water Engineers on May 19, 1932. A second paper by Lawrence Bendelow was also published privately by the Institution of Civil Engineers on October 30, 1943.

Anxiety was felt at the time for, in the same area, not far distant from the Ewden valley, as described on page 58, there are three reservoirs which had failed on similar alternating sandstones and shales. These are:

1850. The first Woodhead embankment.
1852. The Bilberny dam near Holmfirth.
1864. The Dale Dyke, Bradfield, near Sheffield.

The Ewden reservoirs were supervised by Mr. J. K. Swales, Mr. Walter Nicholson (of T. & C. Hawksley) and Dr. Herbert Lapworth (with Colin Clegg, Constructional Engineer up to 1929).

The Ewden valley reservoirs were constructed between 1913 and 1929 viz:

	Broomhead	More Hall
Capacity	1,120 m.g.	464 m.g.
Top water level	583 O.D.	480 O.D.
Depth of water above lowest draw off	91 ft.	71 ft.
Crest length	991 ft.	917 ft.
Gross yield	8,045,000 gallons per day	
Total cost by 1932	£2,017,395	

In 1924 a slip occurred which was thought to be due to excavating too deeply for the by-wash channel (*Figure 17*).

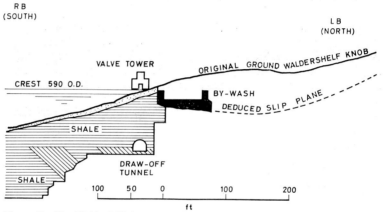

Figure 17. The Waldershelf slip. When the Broomhead reservoir was under construction in 1924 excavation for the bywash channel adjacent to the overflow weir may have started the slip. (The slip plane was deduced later.)

A deep trench filled with rubble for draining water from the area had been constructed parallel with the outer limits of the slip but by 1928 it had been pushed a distance up to 31 ft. (*Figure 18*).

In addition 400,000 yds.³ of slipped material had been removed for lightening the load.

In 1930 some further movement was detected, such as cracks in the by-wash wall, overflow weir and valve shaft. By this time Mr. W. J. E. Binnie had joined the advisers.

Seven exploratory holes to 90 ft. deep were dug and these pointed the way for exploring the high part of the hillside. A heading from trial hole No. 2 was, therefore, driven northwards to a little beyond borehole Y3, and two other boreholes were drilled (Y2 and Y1). These showed sandstone, Rough Rock, some 125 ft. in thickness *charged with water*, with a level of 752 O.D. when first found, which lowered to 700 ft. O.D. (*Figure 19*). This depth of 52 ft. of water representing the storage of a large volume of water, doubtless supplied the lubricant for the slip and was responsible for the pressure in the shales.

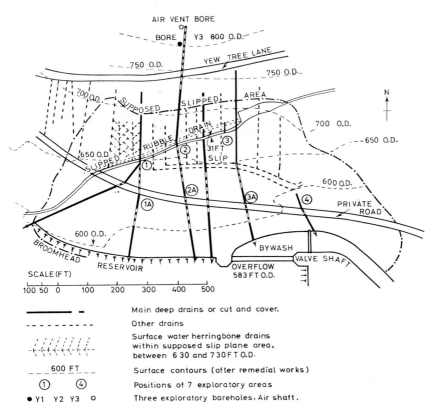

Figure 18. The Waldershelf slip. In the early stages of the slip (1924–30) a rubble drain had been constructed and was pushed up to a distance of 31 ft. Deep rubble drains were constructed running North to South in the direction of the slip-courses. Seven large exploratory holes were dug and several deep drains were constructed in cut and cover to discharge drainage water into the reservoir and a long heading was drilled northwards towards the thick water-bearing sandstone which had been revealed by three deep boreholes, Y3, Y2, Y1

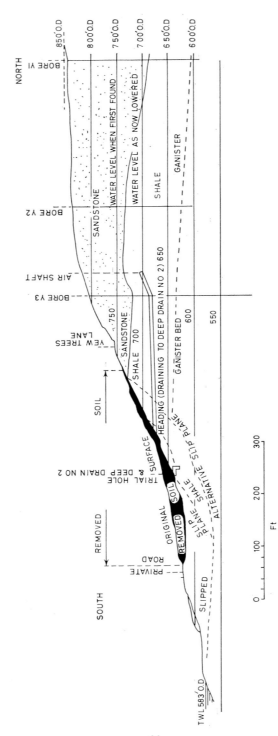

Figure 19. The Waldershelf slip. Deep boreholes revealed thick sandstone charged with water at a level of 752 O.D. which lowered to 700 ft. when drained by the heading; a major factor in stabilizing the slip

The rest of the work was to release the water from the shales within the slip by adding deep drains (4 in. to 12 in. with suitable 2 in. inlets) and removing surplus slipped material. Surface water was taken from the hill-slope by a system of herringbone rubble drains connected with the deep drains.

The works were completed in 1935 and no further movement has been recorded since Mr. Lawrence Bendelow gave his paper in Sheffield in 1943.

Total length of deep drains (av. 18 ft.)	1,900 yd.
Total length of drainage headings	385 yd.
Total length of surface drains	4,100 yd.
Amount of material removed for lightening, shale and rock (8%)	145,000 cu. yd.

The author is grateful to Mr. A. B. Baldwin, M.Eng. (General manager and engineer of the Sheffield Corporation Waterworks) for reading these notes and he would like to acknowledge the encouragement he has had from Mr. Gordon M. Swales, who with his father Mr. J. K. Swales was associated with the Waldershelf slip.

Slip at Scarborough

Slips of coastal cliffs (with the sea at their feet) may have some analogy with those in banks of reservoirs.

Between 1915 and 1921 the Royal Albert Drive and sea wall at the North Bay at Scarborough moved horizontally up to 7 ft. At an exceptionally low tide heavy gales had removed overlying sand exposing a line of fracture in the shales which were uplifted parallel with the coast, their dip into the coast being 70° to 80°.

This indicated a deep seated movement, a slip plane, of the underlying Estuarine clays of the Upper Lias as well as the obvious slipping of the semi-permeable Boulder clay above them.

The foregoing instance might suggest the desirability of emptying a reservoir, where slipping on its perimeter is suspected, to see if there is disturbance or rise of strata in the bottom.

Scammonden valley, Huddersfield

A new reservoir impounding up to 168 ft. of water submerges in part a depth of 50 ft. of Scotland Flags (grits and shales). These strata, on the left bank, dip about 1° into the valley and 3° or 4° parallel with the valley in the downstream direction. On the right bank the same dips are maintained, the dip being into the hillside. The general appearance of the right side is much less disturbed than the left side. Before filling has begun only surface slides of about 6 in. have been noted on the left side.

Hollins service reservoir, Macclesfield

As an illustration of behaviour of strata in a steep valley in the Pennines, *Figure 20* shows, on the left, gritstones dipping at about 5° into the hillside.

On the right the grits bending over, disintegrating and preparing to slide downhill.

When the work was first opened out, revealing the strata dipping into the valley and disintegrating, doubt was expressed by some as to whether the site was suitable.

Figure 20. Hollins Service Reservoir Foundation, Macclesfield. The foundation for this reservoir is on a steep escarpment on the Millstone Grit which appears to dip into the valley on the right whereas the true dip is into the hillside on the left

Sautet reservoir, France

The question of impounding water in valleys is touched upon on page 181 where there is a large leakage, 20 m.g.d. from the reservoir upstream of the dam through the valley side of glacial 'gravel' into another valley. It is on too large a scale to construct any form of grout curtain and any form of plastering the bank, to keep the water in, would collapse and cause a good deal of the bank to slip when the water is drawn from the reservoir.

Cow Green reservoir (A problem of reservoir geology)

This reservoir is on the River Tees about 10 miles north-west of Middleton-on-Teesdale. The top water level was decided, after an extensive investigation, to be 1,603 O.D. for a composite concrete gravity dam and embankment 300 yds. upstream of Cauldron Snout.

As the reservoir area in the Tees valley is some 300 ft. *higher* than the small valley of the Harwood Beck, distant about 2 miles to the east, and as the height of the dam is 73 ft., the head of water on the Harwood Beck, when the reservoir is full, could be of the order of 373 ft.

The intervening two miles between the Tees valley and lower Harwood Beck consists mainly of permeable Melmerby Scar limestone, as shown diagrammatically on *Figure 21*. It is a high-lying area, called a col, with an average elevation of 1,650 O.D., with Widdybank Fell—highest part 1,717 O.D. This area includes old mine workings (barytes and lead) and was found to be water-bearing. Hence a long investigation, in terms of

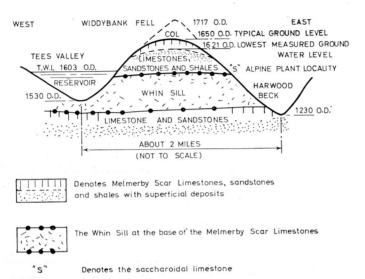

Figure 21. *Cow Green Reservoir. There was a large amount of research needed to ascertain the extent of impermeability of the whinstone and to find a reliable (minimum) water level between the River Tees and Harwood Beck. As the underground water level is higher (1,621 ft. O.D.) than the top water level in the reservoir (1,603 ft. O.D.), it is reasonable to suppose that the reservoir will be watertight. The Whin Sill at the base of the Melmerby limestone which was converted into a 'saccharoidal' limestone, weathering to a granular calcareous soil (S'). This soil produced a species of flora believed to be a relic of the cold period following the Ice Age*

years, was made to ascertain the probable minimum level of underground water in the limestones between the two valleys.

Mr. Michael Kennard and Dr. Knill, who have kindly read the foregoing notes, stated in their description of the Scheme that the assessment of the feasibility of the Cow Green Reservoir would not be straightforward owing to:

(1) The presence of limestone, associated with karstic conditions in the reservoir basin;
(2) The dip of the strata out of the reservoir basin, particularly in the region of the col;
(3) The presence of springs, seepages and streams at various localities and the permanence of mine water discharge;
(4) The possibilities of aggressive action by the reservoir waters on the limestone, and
(5) The influence of the extensive mine workings.

In 1956 the investigations were primarily devoted to the determination of the elevation of the top of the Whin Sill in the col area, with the object of determining whether the Sill would form a low permeability (e.g. 10^{-4} to 10^{-6} cm/sec.) barrier to seepage from the reservoir.

The 1964–65 exploration involved a very long investigation but, shortly, the lowest underground water level in the col between the Tees and Harwood Beck was estimated never to be lower than 1,621 O.D., which is well above the chosen top water level of the reservoir, 1,603 O.D.

The natural water in the strata could therefore be considered a suitable curtain for retaining water in the reservoir.

Arctic Flora

In the passage of the Bill through Parliament a good deal of comment concerned the maintenance of special Arctic flora believed to be a relic of the colder period which followed the Ice Age. The area of the Melmerby Scar limestone had been affected by the high temperature of the intrusion of the Whin Sill, which converted it into a 'saccharoidal' or 'sugar' limestone this weathered to a granular calcareous soil, which appeared to be suitable for the preservation of such plants.

Slides in Austria

In recent years there seems to have been an epidemic of slides in Austrian valleys, e.g. (*a*) into the Gepatch reservoir in the Inn valley extensive slides in the reservoir area, chiefly of talus on gneiss (described at length in *I.C.O.L.D.* **1** 37 & 669 (1967); (*b*) downstream of the Gmuend dam, also in the Inn valley, constructed on quartz schists and sandstones in the mountainous region at an elevation of 1,190 O.D. there were three rock slides.

These slips relative to the Gmuend dam, *Figure 22* were as follows:

(1) Nov. 24 1963. 40 m from left abutment of dam (2,000 m³).

(2) May 9 1964. 20 m from left abutment of dam. This slide was rather more extensive than (1). Slides (1) and (2) were on the flatter left bank where the rock dips into the valley at about 50°.

(3) June 9 1969. At the right abutment (250 m³). This slide was on the steep right bank where the rock dips into the hillside. The slip extended from the right abutments for about 30 m downstream.

The slides may have to be partly attributed to the stilling pool and flood discharge from the dam and supplementary grouting of this area may prove beneficial in reducing further slips. Although little damage was visible or other untoward effects observed by clinometer readings, and no repairs would appear to be necessary, the matter was treated seriously and a massive concrete block 14,000 m³ was built against the right abutment. Pre-stressed concrete arches 10,240 m³ on the left downstream slope were also built after the two rock slides.

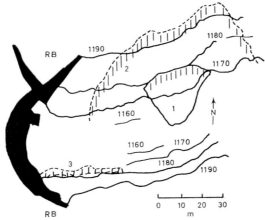

Figure 22. Gmuend or Gerlos Dam, Austria. As the left bank of the valley is flatter than the right bank a block of concrete was made to act as an abutment to enable a symetrical arch dam (39 m in height) to be made when the dam was constructed in 1943–45. After 20 years the strata downstream of the dam deteriorated and three slides occurred, (1) in 1963 (2) in 1964 both on the left bank and (3) in 1967 on the right bank which necessitated extensive precautionary strutting both abutments of the dam

Fall of rock into Pontesei (or Maè) reservoir 1959, Italy

The Maè is a tributary of the Piave and flows in Permian and Upper Trias limestones, these strike one generally as resembling those in England, something between Carboniferous and Jurassic limestones. They are jointed and fissured and blocks can become loose, but they can carry loads and are not cavernous. The late Dott. Ing. Carlo Semenza described in some detail (*see* page 257) how a 'huge landslide of 3×10^6 m³ fell into the reservoir and produced a wave of several metres over the crest with the disturbance of only one pendulum and no damage to the dam'. This was on March 22, 1959.

The great slide of Cretaceous strata into the Vaiont reservoir occurred at 22.39 on October 9, 1963, $4\frac{1}{2}$ years later, and when this occurred the late Mr. Sidney R. Raffety said the record of the Maè slide was prophetic.

The Vaiont disaster

It was soon after the Maè slide in early 1959 that Professor Edoardo Semenza (the son of Dott. Ing. Carlo Semenza) began a remarkable *analysis of the geological studies on the Vaiont landslide from 1959–1964*. This record, in Italian, describes in detail the first indications of movement of the Toc plain, the opening of cracks, the slow movements over the last three years before October 9, 1963 when 250 to 300×10^6 m³ (600 million tons, 200 m in height and 1,800 m in length) fell in a few seconds into the reservoir.

In 1961 at a garden party at the British Embassy in Rome the author gave Dott. Ing. Carlo Semenza a silver spoon from London inscribed 'Domine dirige nos'—he said rather emphatically *that is not a bad idea*. The late Mr. J. A. Banks said afterwards that Carlo Semenza was very worried, when we all met in Rome in 1961 (I.C.O.L.D. Meeting) about the slips that were taking place in the Vaiont valley.

From the above it seems that the common denominator of the cause of these troubles, slips, leakages, and disasters is *water*, and they show that, as greater depths and areas are impounded, more detailed examination of the *reservoir* geology will be of paramount importance just as it is now considered necessary for the *dam* geology.

BIBLIOGRAPHY

Ashop Valley
THOMPSON R. W. S. The Diversion of the River Ashop. *I.C.E.*, *C.C.V.I.*, 152 (1920).

Broomhead Reservoir
SWALES, John, K. Notes in respect of the Ewden Valley Scheme and a description of remedial works with the Broomhead Reservoir. *I.W.E.* (19th May 1932).
BENDELOW, Lawrence. Remedial Works in connection with the Waldershelf slip, Broomhead Reservoir. *I.C.E.* (30th October 1943).

Cow Green Reservoir
KENNARD, M. F. and KNILL, J. L. Reservoirs on limestone with particular reference to The Cow Green Scheme. *I.W.E.*, **23**, 87 (1969).

Slides in Austria
HARNINGER, G. and KROPATSCHEK, HANS. The rock slides downstream from Gmuend Dam, Austria. *I.C.O.L.D.*, III, 657 (1967).

Slip at Scarborough
WALTERS, R. C. S. Geology of the Scarborough District. *I.W.E.*, 1927.

PART 2
TYPICAL GEOLOGICAL PROBLEMS

15

BRITISH DAMS

Almost all dams in England and Wales are for water supply (*Figure 23*), whereas those in Scotland are mainly for hydro-electric purposes.

Most dams in England and in South Wales are of earth owing to the soft sedimentary Mesozoic strata on which they are built, whereas those in Scotland (except those in the Lowlands) and in Mid and North Wales, are mostly of concrete or masonry owing to the hardness of the Igneous, Metamorphic and Palaeozoic strata.

NOTES ON ENGLISH DAMS (see *Figure 23* for location)

There are two or three hundred English dams constructed on sedimentary strata which generally consist of alternating limestones, sandstones and clays.

Some of the valley sections of these were described by Lapworth[1] in 1911 showing how such strata give rise to slips which may occur rapidly, and surface creeps which take place slowly.

If water is brought into contact with slips, the sandstones above the clay slide even more rapidly. If water is brought into contact with strata which have moved laterally (creep) it would penetrate into the joints and fissures and wrinkles, and so undermine the structure.

Hollingworth, Taylor and Kellaway[2] in 1943, described in detail some other interesting cases of valley formations in the Northamptonshire Iron Field which illustrate the contacts between the Inferior Oolite (variable Estuarine beds, Northampton Sand and Ironstone) and the Lias clays.

They have given names and origins to the types of slip, creep and wrinkling as follows.

Cambers—Cambers are strata which have become inclined or bent in a curve towards a valley, due to slipping on wet clay.

Gulls—In their simplest form, gulls are vertical open cracks. There are many varieties due to different causes, particularly in cambered strata.

Bulges—Bulges are forms of wrinkling, or wrinkling of strata is associated with them. They are produced in the bottom of a valley by the weight of superincumbent strata above them.

Although there are often faults in a valley which may explain its origin, bulges are disconcerting, because if they include permeable seams it is

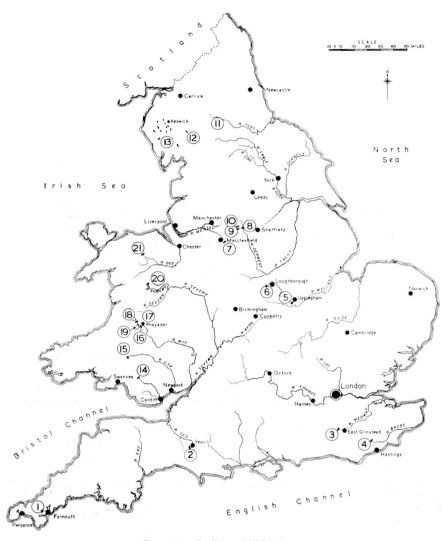

Figure 23. English and Welsh dams

(1) Argal, near Falmouth, Cornwall.
(2) Sutton Bingham, near Yeovil, Somerset.
(3) Weir Wood, near East Grinstead, Sussex.
(4) Powder-Mill, near Hastings, Sussex.
(5) Eye Brook, near Uppingham, Rutland.
(6) Blackbrook, near Loughborough, Leicestershire.
(7) Trentabank, near Macclesfield, Cheshire.
(8) Ladybower ⎫
(9) Derwent   ⎬ Derwent Valley, Derbyshire.
(10) Howden   ⎭
(11) Selset, Tees Valley, Yorkshire.
(12) Haweswater, Westmorland.
(13) The Lake District, Cumberland and Westmorland.
(14) Lluestwen, near Rhondda, South Wales.
(15) Usk, near Brecon, South Wales.
(16) Caban Côch   ⎫
(17) Pen-Y-Gareg ⎬ near Rhayader, Radnor, Mid-Wales.
(18) Craig Côch   ⎪
(19) Claerwen   ⎭
(20) Vyrnwy, Montgomery, North Wales.
(21) Alwen, Denbigh, North Wales.

difficult to decide how deep to take the cut-off trench and how much to rely on grouting.

It is found, however, that the steep curvature of a bulge may lessen with depth and that a satisfactory cut-off can be secured by a compromise between cut-off trench and cut-off curtain.

Similarly, there are many instances of intense wrinkling at the surface which also tends to die out with depth. Such movements, Hollingworth suggests, may be due to lateral forces acting independently from the two sides of the valley.

As glaciers have at one time been at work in the vicinity of some of these sites, one might be tempted to attribute such disturbances to redeposition of the strata, but there is evidence, he says, against this view.

In the south of England there are similar bulges without the necessity for calling in glaciers to account for their origin; where glaciers have passed over soft rocks it is, however, possible that the wrinkles and bulges in the bottom of a valley may have been enlarged owing to the additional weight having caused slumping.

That the strata of most of England are soft, capable of sustaining only small or variable bearing pressure, compared with the igneous or metamorphic rock foundations of Scotland, is the main reason why earth structures rather than concrete structures are adopted.

Other reasons are the availability of materials more suited to earthen embankments than to concrete dams. As labour in Great Britain lies somewhere between cheapness in Europe or Africa and costliness in America, the relative economy between materials and labour is not so accentuated in Great Britain as it is abroad.

In the London district there are long earthen embankments built on the flat ground of the valley gravels on London Clay which ensures water-tightness; the water for the supply of London is conserved by storing the water of the Rivers Thames and Lea by embankments 20–30 ft. above ground with trenches 20–30 ft. below ground sunk into the London Clay through the overlying gravels. These embankments have a perimeter of many miles and are in effect very fine dams.

There are similar embankments in the Bristol district at Cheddar on the flat lands of the valley of the Axe on the Keuper Marl of Somerset.

Earth dams have recently been built on the Ashdown Sand, a semi-permeable formation in Sussex, in which in the past many wells have been sunk for underground water. As a large part of the south of England is covered by Chalk, several hundred feet in thickness, no dams have been built there.

In south-west England there are several earth and concrete and masonry dams in the Trias, Culm-measures and granites of Devonshire and Cornwall. The wrinkling, bulges and cambers in the sedimentary strata and the unreliability of the granites in the China Clay districts of Cornwall, as well as tin mining, present special problems.

In the Midlands and in the Pennines of Cheshire, Lancashire and Yorkshire, which are largely covered by the alternating grits and sandstone of the Carboniferous, almost all of the several hundred dams are of earth constructed for the water supplies of Macclesfield, Stockport, Manchester,

Bury, Oldham, Nelson, Bradford, Leeds, Barnsley, Wakefield, Sheffield, Durham and Newcastle, to name only a few.

Concrete or masonry dams in this area include Scarhouse (Bradford), Ryburn and Baitings (Wakefield), Blackbrook (Loughborough), and those in the Derwent valley.

In the extreme north-west of England in the metamorphic much glaciated lake district, there are the masonry dam of Thirlmere and the concrete buttress dam of Haweswater, both for the water supply of Manchester.

Powder-Mill Dam (near Hastings, Sussex)*

The Powder-Mill dam, built between 1929 and 1933, is of interest as it was the first of any size on the Sussex Ashdown Sand which is a permeable formation, and in which some grouting was resorted to as well as a mixed concrete and puddle trench.

Area of gathering ground	1,120 acres
Area of water surface	51½ acres
Capacity of reservoir	188 million gal.
Top water level	78 ft. O.D.
Maximum depth of water	33 ft. 6 in.
Maximum depth of cut-off trench	50 ft.
Carrying capacity of bellmouth overflow (3 ft. depth)	1,200 cusec
Carrying capacity of bellmouth overflow (2 ft. depth)	625 cusec

Geology

The gathering ground is comprised of Wadhurst Clay, on the higher ground, resting on the upper part of the Ashdown Sand which is exposed in the bottom of the valley.

The junction of the Wadhurst Clay and Ashdown Sand marks the locality of the Sussex ironstone; indeed, the site at Powder-Mill was occupied by a small dam which created an ancient 'hammer' pond to forge iron, presumably for cannon balls, found during the construction of the reservoir.

Dam trench

The dam itself is wholly founded in Ashdown Sand which has only a slight general dip upstream; therefore it might be prone to leak.

The Ashdown Sand, however, is best described as consisting of alternating beds of sandy clay, clayey sand, sandy marl and marly sandstone—extremely heterogeneous; but as the trial bore holes and pits seemed to indicate clay material a puddled clay cut-off was specified. Another reason for adopting clay was the difficulty of obtaining stone for concrete aggregate.

In the course of the excavation of the trench, however, hard sandstone beds in a bulge, with much wrinkling of the strata, were proved, so that this part of the trench was filled with concrete rather than puddled clay. The junction between the puddle and the clay was made with an inclined groove in the concrete into which the puddle fitted (*Figures 24 and 25*).

* Acknowledgement is due to the Hastings Corporation through Mr. D. J. Walker for permission to include these notes.

Owing to the slight dip upstream it was necessary to cut two wing trenches on the right and left bank respectively through the top sandstones which would be in contact with the water; these were carried downstream for a distance of about 250 ft., thus increasing the resistance to potential leakage of water passing round the end of these trenches (referred to by Dr. Herbert Lapworth as 'wire-drawing').

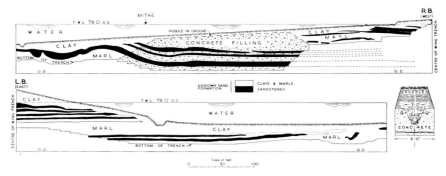

Figure 24. Powder-Mill dam, Sussex. This was the first large earthen embankment built on the soft sedimentary strata of the Weald. It was constructed on the site of a small bank for a Hammer Pond, for the old Sussex iron industry; old cannon balls were found on the site. It was originally intended to put puddled clay in the whole of the trench which was anticipated to be in soft strata, but some of these proved to be much harder than was anticipated and concrete was substituted (see top half) with a suitable junction for the puddle (see enlarged diagram on right)

Figure 25. Powder-Mill dam, Sussex. The trench, 1,200 ft. in length, was taken to 80 ft. below top water level or a maximum of 65 ft. below the surface in Ashdown Sand considered to be permeable. The view is taken from the left looking towards the right bank. The trench had to be close-boarded owing to the fineness of the strata. The embankment of Ashdown Sand material is seen on the right

Owing to the bulge or anticline, permeable sandstone became exposed in the bottom of the reservoir, so a grout screen, starting from the mitre and going at an angle of about 45 degrees in a north-east direction upstream in the reservoir area, was put for additional security.

Owing to the great depth of the impermeable marl which was revealed by trial holes, a mitre pointing upstream was adopted which enabled excavation to be reduced and hence showed some saving in cost (*Figure 26*).

The remainder of the main trench, and both the wing trenches, were filled with puddled Wadhurst Clay, and the embankment was constructed

Figure 26. Powder-Mill dam, Sussex. Showing the trench at the mitre under construction looking towards the left bank. As the dip of the marls in the Ashdown Sand is generally upstream, some depth in excavation was avoided by inclining the trench

of Ashdown Sand with a Wadhurst Clay core. No leakage from this reservoir has ever been traced.

Weir Wood Dam (Sussex)*

The Weir Wood reservoir was completed in 1950 for the water supply of the new town of Crawley and other places in East Sussex.

The following are the main dimensions of the scheme.

Rainfall—annual average	33 in.
Area of gathering ground	6,500 acres
Capacity of reservoir	1,270 million gal.
Gross yield	4 million gal. per day
Area of water surface	300 acres
Top water level	245 ft. O.D.
Maximum height of dam	41 ft.

*Acknowledgement is due to the Weir Wood Water Board through Mr. E. E. H. Cage for permission to include these notes.

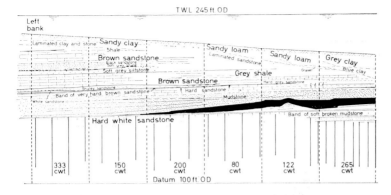
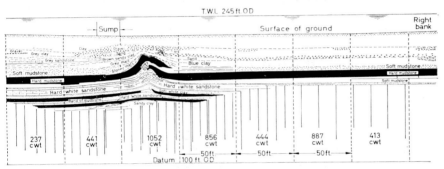

Figure 27. Weir Wood dam, Sussex. The dam, built of Ashdown Sand with a Wadhurst Clay core, is founded in a valley of Ashdown Sand having a 'bulge'. The section (looking downstream) shows the centre of the valley where the bulge is revealed in part of the trench (up to 70 ft. in depth and 650 ft. in length). In this section 274 tons of cement were absorbed in the grout curtain, equivalent to 1·33 cwt. per yd.^{2} of screen. It is noticed that the bulge flattened out between 60–70 ft. below the surface. On the right bank of the valley beyond the section for a distance of 800 ft. up the valley side the strata in the trench dipped at about 1 or 2 degrees, generally flatter than the surface whose slope also was towards the thalweg. This section, 800 ft. in length, consumed no less than 500 tons, or 2·25 cwt. per yd.^{2} of screen, whereas on the left where the strata dip into the hillside at a regular angle of 2 degrees for the distance proved 450 ft., consumed only 77 tons of cement, or 0·6 cwt. per yd.^{2}. Higher up the hill on the left, however, there is a capping of Wadhurst Clay which may have helped this lower consumption. The average grouting pressures were 40–50 lb/in^2.

Maximum length of dam	1,670 ft.
Length of overflow	106 ft.
Capacity of overflow at 3 ft. depth	1,500 cusec
Flood's Committee Upland flood	2,400 cusec
Maximum depth of trench	90 ft.
Volume of bank	350,000 yd.^{3}
Cost	£500,000

The reservoir and dam, like that at Powder-Mill, is built on the top horizon of the Ashdown Sand; the section of the trench is shown in *Figure 27*; this also shows a very good example of bulged strata which flattened out when the excavation was carried deeper.

The Ashdown Sand resembled mainly mudstones, and some half a million gallons of water per day were pumped continuously from the trench during construction when at its deepest. Although the many bore holes indicated that the strata would be sandstone and clay, when the trench was opened out much harder material was met with like that at Powder-Mill reservoir, and so the trench was filled with concrete.

The grout screen was carried down as shown over an area of 29,000 yd.^{2} into which some 800 tons of cement were injected.

As the embankment consisted of granular material, no pore-pressure readings were considered necessary, but some 20,000 yd.^{3} of weak clay under the embankment were excavated and replaced by sandstone.

Eye Brook Dam (Leicestershire, Rutland)*

The Eye Brook dam near Uppingham is an earthen embankment 1,650 ft. in length and 40 ft. in height with a trench in the Lower Lias which shows

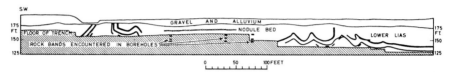

Figure 28. Eye Brook dam trench (part). Showing limestone rock bands in trench which necessitated substituting puddle by concrete in this part of the trench

wrinkling bulges and cambers (noticed by Hollingworth). The water-bearing limestones indicated the desirability of replacing puddle by concrete in parts of the trench (*Figure 28*). The water is used for Stewarts and Lloyds Works at Corby, Northamptonshire.

Sutton Bingham Dam (near Yeovil, Somerset)[3]

The Sutton Bingham reservoir was built (1952) for the water supply of the Yeovil district and it was important that leakage should be reduced to the minimum because the capacity of the reservoir was limited by the main London-Exeter railway line which could not be diverted.

The following data show that the gathering ground is large for the size of the reservoir, which is small, and from which the yield for water supply has to be guaranteed in times of drought.

Rainfall—average annual	38	in.
Area of gathering ground	7,470	acres
Capacity of reservoir	575	million gal.
Gross yield	$2\frac{3}{4}$	million gal. per day
Area of water surface	142	acres
Top water level	182	ft. O.D.
Maximum height of dam	44	ft.
Maximum length of dam	1,045	ft.
Length of overflow	180	ft.
Capacity of overflow at 3 ft. depth	2,500	cusec

*Acknowledgement is due to the Secretary of the Corby (Northants) and District Water Company through Mr. G. C. S. Oliver for permission to include these notes on the Eye Brook dam.

I.C.E. Upland flood	4,500	cusec
Maximum depth of trench	75	ft.
Volume of bank	160,000	yd.³
Cost (including road diversions)	£450,000	

Geology

The geological structure of the Jurassic formation was not apparent until the trench was opened out. Most of the reservoir area submerged impervious Forest Marble (clays and mudstone). At the dam site there were bulges in the bottom of the valley, a fault and an apparent anticline in the overlying very permeable Cornbrash limestone which was 35 ft. in thickness (*Figure 29*).

On the right bank there was a fault of nearly 100 ft. throwing down the overlying Oxford Clay, cutting out the Cornbrash, hence it was possible to

Figure 29. Sutton Bingham dam. Section of Cornbrash limestone 35 ft. in thickness on right bank tunnel excavation. Owing to the permeability of this rock special precautions had to be taken

extend the trench and grout curtain to seal off all possible leakage through the Cornbrash which dipped downstream (*Figure 31*). The Oxford Clay (as well as the underlying Cornbrash) consumed some 120 tons of cement in a distance of only 200 ft. which may be attributed to creep of these beds.

The central section of the trench in the 'bulge' area of the Forest Marble (marl or mudstone) which contained some limestone seams, consumed some 300 tons of cement under a pressure of 1 lb. per in.² for each ft. of cover; the maximum pressure was 50 lb. per in.² Little water was encountered in this trench. Owing to hardness of the Forest Marble and the great thickness of Cornbrash limestone, it was decided to fill the trench with (pumped) concrete.

As the excavation in the trench on the left bank was proceeded with, the full extent of the Cornbrash overlying the Forest Marble became visible, and since it is a very permeable limestone it was necessary to cut it out so

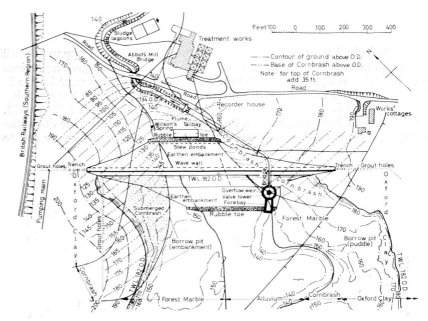

Figure 30. Sutton Bingham Dam. Plan of works showing (a) area of Cornbrash covered by water, (b) surface contours, (c) contours of base of Cornbrash (35 ft. thick). As the top of the Cornbrash would be somewhere near Wilson's Spring and at the point marked 134 O.D. in the river bed, leakage would occur at these points, so a grout screen, in part duplicated, was put near the left end of the dam, and extended upstream until the base of the Cornbrash rose to above top water level (182 O.D.)

(Reproduced by courtesy of the Secretary of The Institution of Civil Engineers)

far as reasonably practicable, and to rely on the impermeability of the underlying Forest Marble (*Figures 31 and 32a and b*).

A decision had to be made, however, as to how much overlying Oxford Clay and Cornbrash should be excavated, because although the Oxford Clay upstream of the trench was a useful cover, the outcrop of the 35 ft. of permeable Cornbrash would be submerged and in direct contact with the water.

It was decided, therefore, that a reasonable compromise would be to proceed as follows.

(1) Cut out the Cornbrash in the trench completely for a distance of 450 ft., that is, 100 ft. beyond the top water level, on the left bank.

(2) Investigate the Cornbrash downstream and plot contours of its base on the left bank.

(3) Insert a double screen of grouting, running upstream at right-angles to the trench.

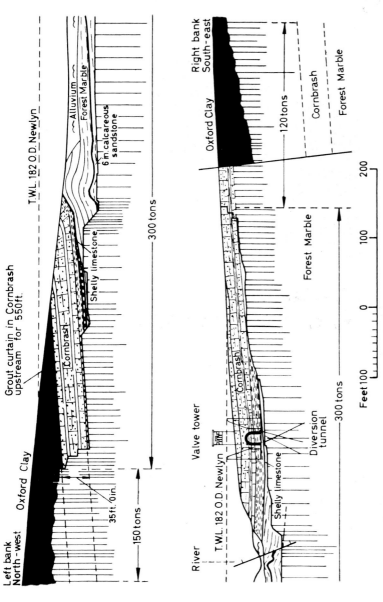

Figure 31. Sutton Bingham dam. Section of dam trench showing Oxford Clay faulted against Cornbrash on right bank which enabled leakage through the Cornbrash permeable rock to be completely cut off by the concrete filled trench and grout screen. Two bulges in the Forest Marble and the depth of the cut-off trench and grout curtain are shown in the central section. On the left bank the trench was taken some way under the Oxford Clay to a depth of 70 ft. below the surface. To secure an adequate cut-off, a wing grout curtain upstream was necessary (see Figure 30)

(Reproduced by courtesy of the Secretary of The Institution of Civil Engineers)

Figure 32. Sutton Bingham dam. (a) Typical view of timbered trench in Forest Marble Clay which extended to 70 ft. below the surface. (b) The trench was filled with concrete in which a groove was formed to receive the clay of the puddled clay core of the embankment

(Reproduced by courtesy of Holland, Hannen and Cubitts)

(4) Because some of the Cornbrash below top water level was to be used for constructing the bank, special care was exercised not to cut it away too near the dam because of the possibility of facilitating leakage. It was considered desirable to obtain material for the bank from below top water level to avoid disfiguring the country, and to increase the storage as much as possible.

It will be seen from the 95 ft. O.D. contour of the base of the Cornbrash, that the top of the Cornbrash (35 ft. in thickness) would be 130 ft. O.D. near Abbot's Mill Bridge. This assumes its full thickness at this point, and hard rock in fact was identified 4 ft. below the bed of the stream at 134 ft. O.D. Moreover spring water was observed which caused a small slip at this spot about a year before impounding began. Another spring (known as 'Wilson's Spring') was also observed in a depression in the hillside; this hereafter was to prove a useful indicator of leakage.

The wing curtain, starting at right-angles to the dam at a point near the end of the cut-off trench on the left bank, was taken for 550 ft. upstream until the base of the Cornbrash appeared above top water level. A double row of grout holes was made for 280 ft. of this distance. This wing curtain covered an area of 4,000 yd.^{2} and consumed 300 tons of grout. The success of this wing trench was fundamental to the success of the scheme.

In all, 900 tons of cement were used in 17,300 ft. of bore hole in the main and wing curtains, the cost of which was £2 per yd.^{2} of screen. No measurable leakage ensued.

The material for the bank was mainly Forest Marble (clays and mudstones) for which $\frac{3}{8}$ in. per ft. was allowed for settlement, and the puddle for the core of the embankment was from the Oxford Clay on the right bank, giving a puddle with the following characteristics:

Liquid limit	55 per cent
Plastic limit	17 per cent
Natural moisture content	24 per cent
Shrinkage range	22 per cent

The permeability ranged from $1 \cdot 2 \times 10^{-6}$ to $4 \cdot 1 \times 10^{-6}$ cm per sec.

The junction between the concrete in the cut-off trench and the puddled clay in the core of the embankment was made by a groove formed in the concrete into which the clay pressed (*Figure 32b*).

Pore-Pressures

For the Sutton Bingham dam the main object was to ensure that pore-pressures did not exceed 66 per cent of the height of the embankment during construction until the crest was reached. A special experiment was also made by Mr. A. D. M. Penman to determine pressures effects in the upstream embankment after filling and it may be that continuous records would be interesting to correlate with the material and consolidation of the bank.

Figure 33a shows five porous pots ('tips'), the pressure cells, Nos. 3, 4, 5, 6, 7 buried in the downstream embankment and Nos. 1 and 2 tips in the upstream slope. The water, squeezed out of the clay during the construction

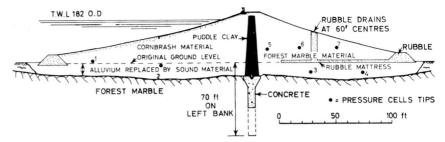

Figure 33 (a). Sutton Bingham Reservoir. Position of pore-pressure cells, 'tips', at Sutton Bingham earth dam are shown thus: Nos. 1 and 2 tips were buried in the permeable Cornbrash limestone material of the upstream embankment. When the reservoir was filled, the pressure registered by No. 1 tip (about 10 ft. under the bank), was nearly equal to the water pressure at 182 O.D. The pressure registered by No. 2, halfway between the toe and the core, (and 28 ft. under the bank), was 12 ft. less than the water pressure at 182 O.D. Pressure cells Nos. 3, 4, 5, 6 and 7, were buried in the Forest Marble material of the downstream embankment. The pore-pressures amounted to 55 to 60 per cent of the height of the embankment when it was completed

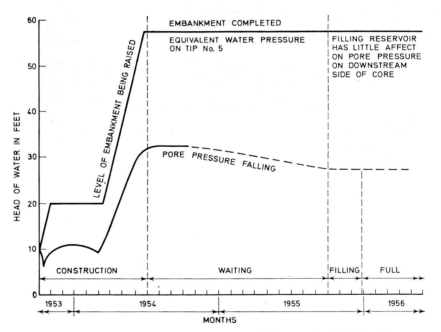

Figure 33 (b). Sutton Bingham Reservoir. Pressure cell No. 5 is typical of the five downstream embankment cells for showing how the pore-pressure rose during the nine months, construction of the dam. When the crest was reached, the pore-pressure slowly fell from 32 to 28 ft. during the next fifteen months which reduced the percentage of pore-pressure to bank pressure from 55 to 48 per cent, well under the stipulated safety of 66 per cent. Thereafter pore pressures remained constant during filling and for a few months afterwards

of the embankment would enter the pots and the pressure would be conveyed through 5 mm polythene tubes to a central gauge-house. (The polythene tubes must be cleared of air and embanking material must be placed with care for heavy bulldozers or compacting machinery can affect the readings in the gauge house. This apparatus was an improved version of that used in 1951 but it has again been improved at the Selset dam in 1962, and at the Scammonden dam in 1968. *Figure 34* shows how the several polythene tubes were taken care of during construction (at Bough Beech, 1967).

Figure 33b shows that the pressures rose to about 50 per cent of the overburden pressure on the downstream embankment until the crest was

Figure 34. Bough Beech Reservoir, Kent. Polythene tubes for nine pore-pressure determinations during construction. The tubes are ultimately collected in a gauge-house when the dam is finished

reached, and then they began to fall for a year until the reservoir was filled, when the pressure fluctuated slightly but never rose materially. The upstream experimental tips' readings rose as the water-level in the reservoir rose and some impermeability was indicated in the Cornbrash which was a matter of interest as it was a new experiment on 'behaviour'.

Argal Dam (near Falmouth, Cornwall)[1]*

Argal was the first (1941) of a series of modern concrete dams built on the granites of the West Country (*Figure 35a, b and c*).

Concrete gravity, curved, radius	500 ft.
Top water level	260 ft. O.D.
Maximum depth of water	28 ft.
Maximum height of dam	44 ft.

* Acknowledgement is due to the Falmouth Corporation through Mr. B. J. Sweeney for the inclusion of these notes.

(a)

(b)

(c)

Figure 35. Argal dam. (a) Gravity dam on Cornish Granite completed in May 1942. (b) Upstream view, before filling (December 1941). (c) The cut-off trench and broad foundation is in disintegrated and hard Cornish Granite. Some grout pipes on the left were drilled to 100 ft. In all there were 4,000 ft. of hole, but the cement consumed was only 65 tons. The concrete aggregate was partly gabbro and epidiorite (imported) and partly hard granite, dug on the site, with Gwithian Blown Sand

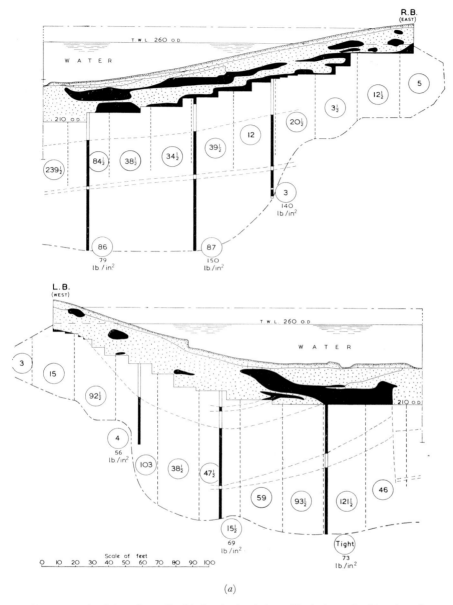

(a)

Figure 36. Argal dam, Cornwall. (a) *Granite foundations.* Black *denotes hard sound granite.* White *denotes disintegrated granite and subordinate china clay.* Stippling *denotes concrete in trench. At the surface there is a 5 ft. layer of top soil and broken granite. For the 6 bore holes, the hard granite is shown in black, disintegrated granite in white. For the grout curtain there are 100 bore holes aggregating 4,000 ft. in length, some of them inclined; the extent of the grout screen is indicated and the amount of cement used per 25 ft. width is given in cwt. There were 4 or 5 injection holes per width. Sodium silicate was used for lubricating the cement in many holes. In all only 65 tons were used, equivalent to $\frac{1}{3}$ cwt. per ft. of bore hole, or only 5 or 6 lb. per ft.^{2} of curtain. Pressures in lb. per in.^{2} are given. (cont.)*

Maximum height of dam . . . . 44 ft.
Maximum depth of trench . . . . 30 ft.
Capacity . . . . 143 million gal.
Length of crest . . . . 330 ft.
Chord-height ratio 8

In the trench there were found lenticular masses of hard sound granite and a good deal of disintegrated granite with subordinate china clay, and the possibility of encountering pockets of china clay in and below the trench was the chief danger. However, from the information of the exploratory and grout bore holes, the granite under the cut-off trench appeared to be fairly honest (*Figure 36a, b and c*).

Below the concrete-filled trench about 100 bore holes were drilled, aggregating some 4,000 ft. in length; some of those at the ends of the dam and a

(*b*) (*c*)

Figure 36 cont.—(*b*) *and* (*c*) *Left and right bank broad foundation and cut-off trench in Cornish granite of variable texture* (*October 1939*). *It is inferred that the granite of the so-called 'batholith' of Lands End is affected by pneumatolysis, that is, stresses due to cooling of the granite formed cracks and fissures through which boron, fluorine, carbon dioxide gases and steam passed and softened the felspar, locally converting the rock to China Clay. Pipes of kaolization may extend over small areas to considerable depths. In this district the China Clay is subordinate* (*see Figure 36a*) *the granite being more disintegrated rather than kaolinized*

few in the middle were inclined. There were 6 deep bore holes to 100 ft. Owing to the compactness of the granite comparatively little cement could be injected even with the use of the lubricant sodium silicate. In all, 65 tons of cement were used in 4,000 ft. of hole, or $\frac{1}{3}$ cwt. per lineal foot, or 5–6 lb. per ft.[2] of the area of the screen. Since the war the dam has been raised by about 10 ft. by pre-stressed concrete.

Derwent Valley Dams (for the water supplies of Sheffield, Nottingham, Derby and Leicester Districts)[5,6]

The Howden and Derwent dams in the Derwent valley are both about 1,100 ft. in length and 114 ft. and 117 ft. high respectively. Constructed with concrete and faced with masonry, they are founded on Yoredale rocks of alternating sandstone and shale which are much wrinkled and folded. These wrinkles extend to 70 ft. below the surface; open joints in the rock at Howden were 9 in. in width and 6 ft. in height.

The Ladybower dam (also in the Derwent valley), constructed in 1940–42, is 140 ft. in height (with a trench extending to about 300 ft. below top water level). The crest is 1,250 ft. in length. It is an earthen embankment built on very thin alternating seams of grit and shale, the Sabden shales (olim Yoredales), which were found to have a deep-seated crumple in the centre which did not flatten out until a depth of 175 ft. below the valley bottom was reached. Two million gal. per day were pumped from the trench during construction. Eight miles of bore holes were drilled which consumed

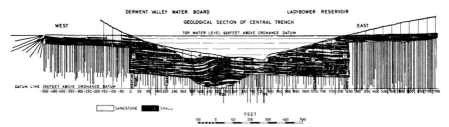

Figure 37. Ladybower reservoir. The trench and cut-off curtain were noteworthy as they extended to great depth below the surface in the Lower Carboniferous grits and shales. Some 8 miles of bore holes were drilled into which 1,600 tons of cement were pumped. In the trench a large crumple was exposed typical of sedimentary strata of alternating grit and shale

1,600 tons of cement. Some of the cement was proved to travel underground for a distance of 200 ft.

Figure 37 shows the crumple in the trench and the area of the grout screen which is about 40,000 yd.^{2}

Trentabank Dam (near Macclesfield, Cheshire)*

The Trentabank earth dam (constructed 1923–1929) is of great interest owing to the presence of large pockets of wet glacial sand revealed in the trench during construction. These were for the most part missed in the exploratory pits and bore holes, and they caused considerable anxiety when discovered during the digging of the trench which was to be excavated down to firm grits and shales (*Figure 38a and b*).

Earthen embankment
Maximum depth of water	65 ft.
Capacity	130 million gal.
Top water level	847 ft. O.D.
Gathering ground	646 acres
Area of reservoir	23 acres
Average annual rainfall	41 in.

The wet sand was dewatered foot by foot in the trench with great difficulty down to the Yoredale Grits and Shales below the Millstone Grits.

* Acknowledgement is due to the Macclesfield Corporation through Mr. J. H. Dossett for permission to include these notes.

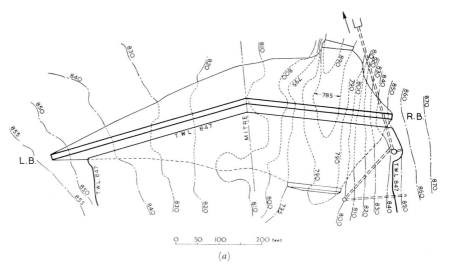

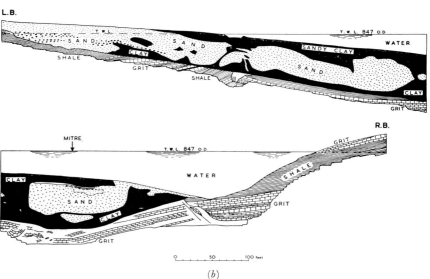

Figure 38. Trentabank dam. (a) The contours of the valley lent themselves to a mitre pointing downstream to gain storage without substantially increasing the length of the dam. (b) This was a particularly difficult trench to excavate as it had large volumes of Sand (stippled) as well as soft Boulder Clay (black). The foundations of the trench are in much disturbed alternating shales and gritstones. There is no grouting, the time of construction being in the early twenties. There appears to have been a buried preglacial valley on the left of the present valley (see bottom half of section)

These, as will be seen from the section (*Figure 38b*) proved to be considerably disturbed, as is so often the case in the rather soft sedimentary strata of the British Pennines.

The shale on the right and left sides of the valley indicates a general dip across the valley. Under the centre the sliding or slumping has been resisted by the underlying gritstone which has resulted in faults and contortions.

The section suggests also a buried pre-glacial valley on the left of the present valley.

The trench attained a depth of 100 ft. in places, no grout was used below and no leakage resulted. As the embankment rested largely on Boulder Clay and Sand it may be considered as an example of a dam built on sand.

The configuration of the valley was such that it was possible to adopt a mitre pointing downstream in order to gain storage without substantially increasing the length of the dam (*Figure 38a*). As the contoured plan shows, the slopes of the embankments of local clays and sands were made very flat.

Selset Dam, Yorkshire (the Tees Valley and Cleveland Water Board)*

The Selset Dam is an earthen embankment with puddled clay core, concrete cut-off trench, length, 3,043 ft., height, 130 ft.

Foundation : 7–10 ft. alluvial gravel, 15–20 ft. soft Boulder Clay, on overlying sandstone and shale.

To gain a satisfactory foundation rather than remove some 200,000 yd.^{3} of Boulder Clay, it was decided to place the embankment upon it by inserting vertical sand drains 18 in. in diameter. These went through the gravel and soft Boulder Clay (10 ft. apart, arranged on a 10 ft. square grid), and were filled with $\frac{1}{8}$ in. diameter crushed whinstone which enabled the water pressure in the soft Boulder Clay to be reduced thereby consolidating the Clay and increasing its shear strength so that it was strong enough to support the embankment as it was built. The water is collected in a 12-in. layer of limestone rubble on the alluvial gravel under both the upstream and downstream embankments.

Instruments were installed in the foundation between a number of sand drains, and they showed that the water pressure was satisfactorily relieved.

As the gravel and Boulder Clay were full of boulders, some very large (for example, 2 yd.^{3}) the vertical holes were sunk by hollow pile equipment; these tubes were able to push the boulders, because of the low shear strength of clay in which they were embedded. If they had been piles they would have been filled with concrete, but in this instance, as stated above, they were filled with whinstone. The total number of the sand drains was 4,071 of an average depth of 20 ft.

Haweswater Dam (Westmorland)†

The depth of the Haweswater Lake was increased by 96 ft. between 1934 and 1942 to provide a water supply for Manchester—distance 80 miles.

*Very many thanks are due to Mr. J. Kennard for verifying the extract and adding additional information to the section dealing with the Selset dam.[7]

†I am indebted to Mr. G. E. Taylor for reading through the notes which refer to the Haweswater dam a short time before his death.[8]

The original dimensions of the Lake were as follows.

Maximum depth	102	ft.
Top water level	694	O.D.
Area of water surface	350	acres

After the dam was built, the following data applied.

New top water level increased by dam	790	ft. O.D.
New area of water surface increased by dam	1,050	acres
Area of natural gathering ground	7,970	acres
Average annual rainfall	85·1	in.
Capacity of reservoir (above draw-off)	18,660	million gal.
Capacity of reservoir with storage at catchments	23,500	million gal.
Gross yield (with catchments not included in above area)	35·7	million gal. per day
Gross yield (with catchments)	80	million gal. per day

The configuration of the valley and the soundness of the rock enabled the capacity of the reservoir to be enlarged to balance nearly the whole of the average annual run off (35·7 million gal. per day) from 7,970 acres, without excessive expenditure.

This was, however, a temporary measure because large catchment areas have since been added with some additional storage in view, so that a more normal proportion of the average annual run off will be balanced.

Type of dam (*Figure 39*)	Concrete buttress
Maximum height	120 ft.
Length of crest	1,540 ft.
Length of overflow (designed for a standard flood of 2 ft. depth)	227 ft.
Average pressure on base	18,400 lb. per ft.^{2}
Maximum pressure on base (toe)	31,300 lb. per ft.^{2}
Minimum pressure on base (heel)	7,400 lb. per ft.^{2}

Although the pressures of this buttress dam on the foundation rock were approximately twice those which would have been due to a gravity dam of the same height, the rock was well able to withstand them many times over.

Geology of dam site

The whole of the Haweswater reservoir and the dam is situated on the Borrowdale Rocks which are of volcanic origin; they have thrust their way into the Ordovician, destroying a large part of that formation between the Skiddaw Slates (early Ordovician) and the Coniston Limestone (late Ordovician just below the Silurian formation).

The Borrowdales vary between hard basic andesites and acid rhyolites and in places there are hard green slates which have been cleaved after eruption. Associated with the rocks there is a great number of minerals and the general ensemble results in a rock of excellent bearing strength for supporting dams (*Figure 40a and b*).

Like all large lakes in Cumberland (*see* table) Haweswater is known as a 'rock-basin', that is, a lake which has been carved out of the solid rock by the pressure of a glacier coming from the mountains and moving down the

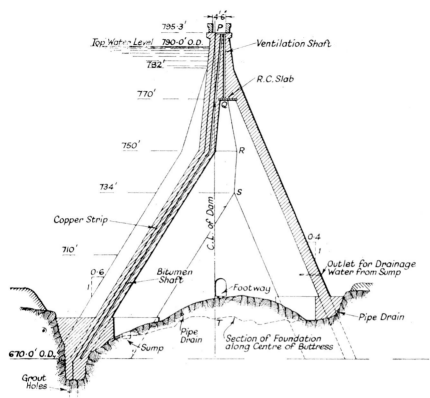

Figure 39. Haweswater buttress dam. The dam is founded on andesite covered by only a small quantity of glacial material

Characteristic data of the Lakes of the English Lake District with geological formations

Name and water authority	Surface level (ft. O.D.)	Depth (ft.) Max.	Depth (ft.) Mean	Surface area (acres)	Volume (million gals.)	Area of gathering ground (acres)	Average annual rainfall (in.)	Chief geological formation
Bassenthwaite	223	70	18	1,300	6,380	84,000	—	Skiddaw Slates (Ordovician)
Buttermere	329	94	54·5	210	3,450	4,500	110	
Crummock Water (Workington and Cockermouth)	322·75	144	87	620	14,500 586*	13,800	80	Skiddaw Slates (Ordovician) and Borrowdales
Derwentwater	244	72	18	1,300	6,300	20,200	—	
Ennerdale Water (Whitehaven)	370	148	62	750	12,350	10,880	85	Skiddaw Slates and Igneous
Loweswater	397	—	—	162	—	2,200	54	
Thirlmere (Manchester)	587·5	158	82	812	8,956*	10,120	91·5	Borrowdales
Ullswater	476	205	83	2,200	49,100	36,000	80	Millstone Grit Borrowdales, Skiddaw Slates
Wast-Water (Sellafield)	200	258	134	720	25,750	11,000	100	Borrowdales and Igneous
Windermere	130	219	78·5	3,650	76,400	56,600	—	Silurian Shales
Coniston Water	143	184	79	1,210	25,000	14,900	—	
Haweswater (Manchester)	790	198**	85	974	18,660*	7,970	85	Borrowdales

N.B. The volumes of the lakes are based on H. M. Mill in the *Geographical Journal* (1895) and Mr. Alan Atkinson in the *Journal of the Institution of Water Engineers*, Vol. 5 (1951).

* Volume above draw-off.
** Depth after raising.

Figure 40. Haweswater dam, Westmorland. (a) Left bank excavation in Borrowdale where the moraine covering over the Borrowdale andesite rock was at the maximum depth of 30 ft. Nevertheless, this was approaching 100 ft. higher than the bottom of the lake upstream. (b) Right side excavation showing small covering of drift on andesite rock

(a)

(b)

valley from S.W. to N.E. The full force of this glacier has excavated a hole in the rock of at least 200 ft. in depth, measured from a top water level of 694 ft. O.D. which is the top water level governed by the rock or moraine material, or both, upon it at the outlet end of the lake and ignoring any sediment which has settled in the bottom since glacial times. The maximum depth is about half way along the lake in the valley, the lake being shallowest at the ends. At the outlet end, on the east, the deposits of morainic sand and clay mask the rock, although it was proved to be quite near the surface, that is, only 30 ft. at the north end of the left bank when foundations of the

Figure 41. Haweswater dam. Mardale Church and village had to be submerged when the level of Haweswater lake was raised

dam were excavated (*Figure 40a*). *Figure 40b* shows a general view of the right end of the trench, and *Figure 41* the demolition of Mardale Church which had to be submerged.

The mechanics of the excavating power of Cumberland glaciers is obscure but the power must be derived from the great height from which they travel, namely, about 2,000 ft. O.D., ascertained, for example, from the unique boulders of granophyre, the rock of Red Pike, found on the Skiddaw Slates of Starling Dodd summit with an intervening valley a mile wide between.

No form of transport other than a glacier could have been available to carry these boulders between the peaks.[9,10]

The cut-off trench in the andesite rock was taken down about 5 ft. below the main excavation through the moraine material, except that in one place of crushed rock and haematite, it was taken down to 45 ft.

Two grout holes, 20 ft. in depth per 35 ft. block of trench, were all that was necessary, some did not take any cement at all even with a pressure equivalent to top water level in the dam; other holes took only 5 cwt.

NOTES ON WELSH DAMS

(See *Figure 23*, page 76 for location)

There are several dams in North and Mid-Wales for the water supplies of Liverpool, Birkenhead, Rhyl and Birmingham in palaeozoic rock valleys, at one time interfered with by glaciers.

In South Wales dams include several earthen embankments for the water supplies of Newport, Cardiff, Merthyr (Taf Fechan), the most recent being those for Newport (Talybont) and Swansea (Usk.) These embankments are all in the more recent Old Red Sandstone and Carboniferous formations.

During the construction of the Usk dam some interesting problems arose with pore pressures; another problem which arises, in South Wales especially, is that of coal mining under dams and reservoirs.

Vyrnwy Dam (North Wales)[11]

The Vyrnwy dam (*Figure 42*), built between 1881 and 1892 for the water supply of Liverpool, was the first of a series of masonry and concrete dams. It is a gravity structure 1,172 ft. in length and 161 ft. in height built on

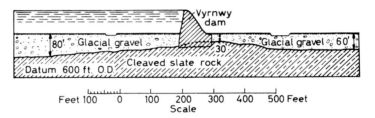

Figure 42. *Vyrnwy dam, North Wales. Longitudinal section of valley above and below the dam*

Silurian slates on a buried rock-bar at the end of a lake carved by a glacier and filled with morainic material. This rock-bar was diagnosed by Deacon and the site was chosen in the valley where the moraine covering was least in depth (30 ft. instead of 80 ft.), the rock being 50 ft. nearer the surface than it was at the original selected site upstream. This resulted in a saving of £400,000.

Alwen Dam (North Wales)

The Alwen gravity concrete dam, built between 1911 and 1916, for the water supply of Birkenhead, is 100 ft. in height, 450 ft. in length, and has a radius of 500 ft. It is founded on Denbighshire Grit of the Silurian, into which the trench is sunk for a depth of 52 ft. On the right bank there is a great depth of Boulder Clay so that the dam is continued into the hillside as an earthen embankment with a puddle core keyed into the Boulder Clay.

Figure 43. Pen-y-Gareg dam. Top water level 945 ft. O.D., height 123 ft., length of crest 523 ft., capacity of reservoir 1,330 million gal. Ordovician

Figure 44. Craig Côch dam. Top water level 720 ft. O.D., height 120 ft., length of crest 513 ft., capacity of reservoir 2,000 million gal. Ordovician. The dam has a radius of curvature of 740 ft.

Caban Côch Dam, Pen-y-Gareg Dam (*Figure 43*), and Craig Côch Dam (Mid-Wales) (*Figure 44*)[12]

These three masonry dams, about 120 ft. in height and between 500 and 600 ft. in length, for the water supply of Birmingham, are all built on Ordovician slates and conglomerates; the Craig Côch dam is slightly curved. The maximum pressure to be sustained by the strata is between 8 and 9 tons per ft.² when the reservoirs are empty.

Claerwen Dam[13]

Claerwen dam is a curved concrete gravity structure, built between 1948 and 1950, also for the water supply of Birmingham, 220 ft. in height and

Figure 45(a). See legend opposite

Figure 45(b). See legend opposite

(c)

Figure 45. Claerwen dam. (a) General view of the so-called mudstones of the Ordovician looking from the left to the right bank (August, 1948). (b) In the hard Ordovician mudstones there was a fault containing a width of 7 or 8 ft. of clay unbared in the formations. Some investigation was necessary to find out the extent of the fault but the junction with the clay in the fault with the mudstones was proved to be watertight and no leakage was anticipated. (c) This 220 ft. high gravity concrete dam was built on Ordovician mudstones of high bearing strength

1,066 ft. in length. The dam is on Ordovician mudstones. These strata are similar to slates but less cleaved and have more bedding planes and joints. A fault of 7–8 ft. of soft clay was unbared but proved to have a solid junction with the mudstone. The crushing strength of the mudstone was 400 tons per ft.^{2} across cleavage planes and 200 tons per ft.^{2} in line with cleavage planes. Only 60 tons of cement grout were consumed in 18,000 ft. of bore holes (*Figure 45a, b and c*). The concrete aggregate is gritstone.

The Clywedog Dam[14] *

The main function of the Clywedog buttress dam, completed in 1968, is to ensure low flows in droughts, and hence it is important to the several water authorities which draw water from the Severn. It also acts for the alleviation of floods and for the generation of electric power, *Figure 46*.

The site is on the River Clywedog which joins the River Severn near Llanidloes, Montgomeryshire, 15 miles in a straight line north of Rhayader (*see Figure 23*, page 76).

The site is in a steep and narrow valley with side slopes approaching 45° and a chord/height ratio of 600/237, under 3, known as the Bryntail Gorge. (Volume of reservoir 40,500 acre-ft. Area of gathering ground 12,098 acres and the estimated catastrophic flow 18,000 cusec.)

Ordovician which in the gorge at the site of the dam consists of Grauwacke siltstones and shales. A major fault dipping at 70° crosses the gorge a few hundred feet south of the dam and is disturbed by old mine workings in the crush zone for barytes and galena.

* Acknowledgement is due to Mr. N. J. Cochrane for reading these notes.

Figure 46 (a). Clywedog Dam. This is a buttress dam 237 ft. in height in a gorge on the Ordovician. The buttress toes are arranged to sit on hard Grauwacke gritstone, hence the curvature downstream and this fits in well with the surface contours. The left hand end shows the damage done to the cableway by fanatics during construction, and the place was under police control when this photograph was taken (5.6.67). The right bank spillway is also shown

Figure 46 (b). Clywedog Dam. Stilling pool and two tunnels. The space between the buttresses is covered but is hollow inside

The dam buttresses sit on beds of different elasticity, but the toes sit on massive Grauwacke grits; the dam is curved, however, downstream to enable them to do so. (The subsidiary retaining earth dam upstream, Bwlch-y-gle, is also curved downstream, but because of a road approach.)

Other advantages were that the buttresses as they are at right angles to the curvature become more parallel with the valley sides and, therefore, excavations for each tended to be less. Finally the curve is generally considered by the designers to be aesthetically better than the straight line.

Usk Dam (South Wales)[15]

The Usk dam for the water supply of Swansea is built on the Old Red Sandstone formation which consists, at the site of the dam, of a reddish brown and green sandstone with occasional clay bands dipping 50 degrees to the south-east.

The maximum height is 109 ft. and the crest is 1,575 ft. in length. Boulder Clay, consisting of gravel, Devonian boulders and clay was used for the banks, clay for the puddle core, concrete in the trench. The trench was taken down to a maximum depth of 77 ft. and there was a grout curtain 30 ft. in depth and 1,600 ft. in length in which 3,562 ft. of bore holes consumed only 134 tons of cement, that is, 0·75 cwt. per ft., or 5–6 lb. per ft.^{2} of area of curtain.

As excavation proceeded for the broad foundation of the bank, a deposit of about 10 ft. of Boulder Clay was revealed overlying silt over an area of about 20,000 yd.^{2} As this would cost about £25,000 to remove it was found possible to relieve the water pressure in the permeable silt underlying the Boulder Clay by 30 vertical 8-in. holes filled with sand penetrating the Boulder Clay and going down into the underlying permeable silt. The vertical drains were connected with horizontal drains 30 × 12 in. filled with sand at the surface. As the silt had a low permeability value (6 × 10^{-5} cm per sec) pore-pressures were recorded in it during the construction of the dam and some of them went up to 80 per cent of the overburden pressure. These dropped at the following rates.

October 31, 1952 .	. 70–80 per cent of overburden	⎫ The weight of the fill
January 31, 1953 .	. 60–66 per cent of overburden	⎬ was taken as 2–3
April, 1953 . .	. 58–62 per cent of overburden	⎭ times that of water

As undue delay was threatened two horizontal drainage blankets were placed in the embankment as construction proceeded; these consisted of a layer of gravel and broken stone each 3 ft. in thickness.

NOTES ON SCOTTISH DAMS

(See *Figure 47* for location)

Although the geology of Scotland lends itself more to concrete dams than to those of earth, there are some important earthen embankments in the Lowlands and Southern Uplands for impounding the water supplies for

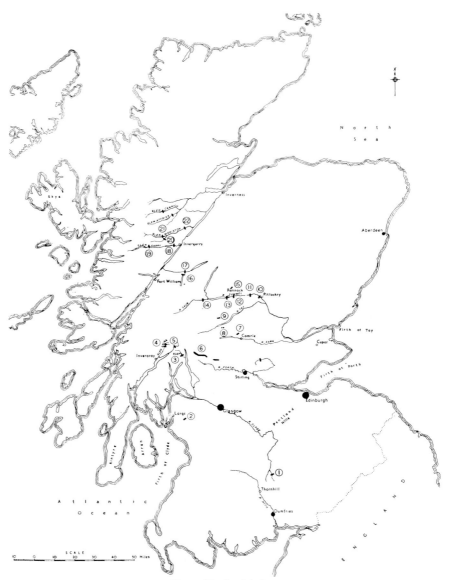

Figure 47. Scottish dams

(1) Daer.
(2) Knockenden.
(3) Sloy.
(4) Upper and lower Sron.
(5) Allt-na-Lairige.
(6) Loch Katrine.
(7) Lednoch.
(8) Breachlaich.
(9) Lochan-na-Lairige.
(10) Pitlochry.
(11) Clunie.
(12) Dunalastair.
(13) Rannoch.
(14) Gaur.
(15) Errochty.
(16) Treig.
(17) Laggan.
(18) Garry.
(19) Quoich.
(20) Loyne.
(21) Cluanie.
(22) Moriston.

Edinburgh (Pentland Hills), Glasgow (Loch Drunkie), there is a large dam on Boulder Clay at Daer for supplying water in the Queensberry country, and an interesting earth dam, also for supplying water for Dundee.

In the south-west of Scotland, the igneous and metamorphic formations lent themselves to the construction of the Galloway Power Company's curved concrete dams, and the Kinlochleven Blackwater dam and Lochaber Laggan and Treig dams.

To the north, the metamorphic rocks are ideal for many developments by concrete dams along the River Tummel for hydro-electric power, although at Pitlochry there was considerable disturbance of the bed-rocks by glaciation.

Further to the north again the metamorphic complex proved to be well suited for many buttress dams of different types, and a granite intrusion proved sufficiently reliable for such a modern structure as a pre-stressed concrete dam.

There are no multiple arch dams and due partly to the wideness of the valleys there is only one true thin arch cupola or double curved dam, 'Monar' near Inverness.

EARTHEN EMBANKMENTS

Knockenden Dam (near Largs)[16]

Shortly before the construction of the Knockenden dam a similar earth dam was built of Boulder Clay on Boulder Clay at Muirfoot near Largs. This dam, when it had reached a height of 70 ft., started sliding (by horizontal movements) and could not be brought up to the full height of 90 ft. as intended. (*Figure 48 (a)*).

As the Knockenden dam, which had reached a height of 30 ft., was almost identical with the dam which had moved, exceptional trouble was taken to ensure that the full height of 90 ft. could be safely achieved; for example, granular material was substituted for Boulder Clay in part of the downstream slope, a berm of 100 ft. was introduced halfway up the upstream slope and several 6-in. bore holes were made in the embankment to obtain samples to test the degree of consoldation. (*Figure 48 (b)* and *(c)*).

It was in two of these that a foreman reported that water was overflowing 2 ft. above the level of the embankment. This strange occurrence, attributable to compression of water in the pores of clay, is believed not to have been encountered before in Great Britain.

Hence, four special bore holes, 3 in. in diameter perforated with $\frac{1}{4}$ in. holes, were inserted in the bank for measuring the pore pressures, and the water levels were measured regularly during the completion of the bank to the full height of 90 ft.

Water overflowed on one occasion from one of these holes but generally it remained some little way down the bore holes; the highest reading was when the pressure was 53 per cent of the 90 ft. bank load (bank 130 lb. per ft.^{3}, water 62·3 lb. per ft.^{3}). As 66 per cent pressure or under is considered to be a safe uplift pressure, the Knockenden bank was deemed to be satisfactory.

Figure 48 (a). Muirfoot Dam. When the dam was within 20 ft. of the intended top water level, the embankment began slipping. it was finished with a flat 'berm', seen best to the left of, and below, the valve tower and the height was limited to 70 ft. after remedial steps were taken

Figure 48 (b). Knockenden Dam. This dam was identical with Muirfoot but was finished at 90 ft. in height. Owing to the slide of the Muirfoot dam, ten six-inch boreholes were drilled in the upstream embankment when Knockenden was 30 ft. high, and in two of these, after a few hours, water rose above ground level from the projecting boreholes indicating that water pressure in the pores of the Boulder clay in the bank was greater than the static head. This is the first British experience of pore-pressure danger and the measurement of pore-pressure, even for quite low dams, particularly during construction with modern rapid earth moving plant has now become standard practice

Figure 48 (c). Knockenden Dam. Upstream slope. The dam was successfully built with imported granular material to the full height of 90 ft.

The measurement of pore-pressures during the construction of an earth dam has now become standard practice in Great Britain as it is most desirable with modern rapidly moving plant that a check should be kept on the consolidation, and that the pore-pressures should not exceed 66 per cent of the height of the bank during construction.

It has also been found that the dissipation of pore-pressure may be very slow and these experiences confirm the old time opinion that earth embankments should be constructed as slowly as possible.

Daer Dam (near Thornhill)[15, 17]

Daer dam is of Boulder Clay, 135 ft. in height and 2,600 ft. in length on Silurian. There was, however, little clay in the material and no pore pressures were recorded. The trench was taken at least 3 ft. into sound rock 8–22 ft.

Figure 49. Daer Dam, Lanarkshire. This earthen embankment was made from material overlying the Silurian in the area of the reservoir. There are large rockfill toes both upstream and downstream. The volume of the dam is about $2\frac{1}{3} \times 10^6 yd^3$. Downstream slope $2\frac{1}{2}$, $2\frac{1}{4}$ to 1 and the upstream slope 3, $2\frac{3}{4}$ to 1

below the surface (one zone, 35 ft.) and cracks in Silurian were grouted in the trench. A grout curtain of 14,000 ft. of bore holes consumed only 72 tons of cement.

The Backwater Dam (near Dundee)*

The reservoir is some 20 miles (32 km) north west of Dundee and impounds 5,500 m.g. ($25 \cdot 00 \times 16^6$ m^3) to give initially a gross yield of 9 m.g.d. (40 800 m^3/day) for the area of supply of the East of Scotland Water Board.

* I am very grateful for Mr. W. G. N. Geddes's kindness and trouble in making suggestions in the above notes.

The foundation work for the dam was carried out between 1964 and 1966. The dam was built in two seasons 1966–1968 and was opened in October 1969. The dam is of boulder clay with horizontal granular drainage blankets at 20 ft. (6m) intervals vertically. Maximum height, 140 ft. (43 m); crest, 1,800 ft. (549 m) in length and volume 1.6×10^6 yd.^{3} (1.22×10^6 m^3) the material for which was borrowed immediately upstream of the dam (*Figure 50a*).

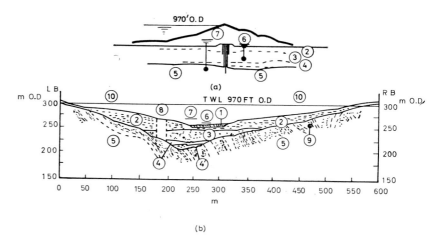

Figure 50. The Backwater dam. (a) *Outline of dam under construction. The foundation consists of glacial material; impermeable clay 2, very permeable sand and gravel 3, clay deposits 4. These strata extend to 165 ft. below the surface resting on 5. semi-permeable metamorphic grits and phyllites. Artesian water level in the sands and gravels gave a record of 835 ft. O.D. whereas that in the metamorphic was recorded at 885 ft. O.D., or about 50 ft. higher.*

(b) *Geological formation*
  1. *Denotes recent alluvium.*
  2. *Glacial till*
  3. *Band and gravel complex* ⎫
  4. *Lake deposits* ⎬ *Glacial deposits*
  5. *Schistose grit and phyllites metamorphic rocks.*
  6. *Artesian water level 835 ft. O.D. in Glacial deposits.*
  7. *Artesian water level 885 ft. O.D. in Metamorphic 'bed-rock'.*
  8. *Grout test area.*
  9. *River diversion tunnel.*
  10. *TWL. 970 ft. O.D.*

The site is two miles (3 km) north of the major Grampian Highlands fault running SW–NE separating the Old Red Sandstone on the south from the metamorphic rocks on the north. The surface geology at the dam site is shown in *Figure 50 (b)* and the table on page 111. The number in brackets after the type of strata is a reference to *Figure 50b*.

Having ascertained the geological conditions in the valley by borings and surface examinations a comprehensive grout curtain was formed down to the Metamorphic rocks preceded by a full-sized experimental grouting test. This curtain was required to meet similar conditions to those encountered at Sylvenstein earth dam in Germany and Serre Ponçon earth dam in France and it is the first curtain of its kind in Great Britain.

Formation	Strata	At centre of dam		At test section	
		ft	m	ft	m
Superficial Glacial-deposits	Recent Alluvium around river-bed (1).	10	3	0	0
	Glacial till, boulder clay, from a glacier, impermeable (2).	26	8	55	17
	Sand and gravel from melting glaciers. (3). Contains water, under pressure.	55	17	80	24
	Glacial lake deposits of fine silts (4).	20	6	30	9
	Till, clays (2).	16	5	0	0
	Total depth	127	39	165	50
Metamorphic rocks	Schistose grits and phyllites interbedded. The phyllites are weathered and weak. The grits are harder. The dip is variable with many clay-filled joints. The formation contains water under pressure (5).				

Figure 50 (c) shows some of the metamorphic rocks overlain by the Superficial glacial deposits in the excavations for the overflow channel

Figure 50 (c). Backwater Dam. Overflow channel under construction near top water level on right bank. On the right is 'bed-rock' i.e. metamorphic schistose grits and phyllites showing 20 ft. Opposite and beyond the men are glacial deposits which are 165 ft. thick overlying the bed-rock in the bottom of the valley which is more than 100 ft. below this view

Large Scale Grouting Test

The experimental test was made using 33 No. grout tubes, 10 ft. (3 m) apart on three concentric rows. A central 50 in. (1270 mm) diameter steel lined shaft equipped with inspection port-holes was formed after grouting to enable examination of the ground to be made and to take specimens to show the adequacy of grouting.

Three types of grouts were used: coarse clay cement; medium deflocculated bentonite, and fine silicate base. Details of these are given in the following table.

Item	Clay cement (Coarse)	Deflocculated bentonite (medium)	Silicate based (fine)
Composition	Montmorillonite (Liquid limit 330) (Plastic limit 67) and Ordinary Portland Cement	Montmorillonite & Portland cement plus monosodium-ortho-phosphate for defloc-culating and sodium silicate for stabilising the suspension	Silicate aluminates
Viscosity			
(second marsh)	40–45	40–45	30–35
(centipoise)	12–14	12–14	6·0–7·0
Shear (lb./ft.^{2})	370–500	58–88	89–109
(g/cm^2)	156–211	24–37	38–46
Setting time	at 28 days	at 30–60 min.	at 20–30 min.
Grout injected ft.^{3}/ft. holes in outer rows	15	9	nil
ditto m^3/m	1·39	0·83	nil
Grout injected ft.^{3}/ft. holes in middle rows	3·5	12·5	8·0
ditto m^3/m	0·32	1·16	0·74
Grout injected ft.^{3}/ft. holes in centre row	nil	nil	22·0
ditto m^3/m	nil	nil	2·04
Total volume of grout injected ft^3	32,000	30,000	14,500
m^3	906	850	410

Manchette Grouting Method. In the experimental test block and indeed in the major grout curtain afterwards, it was possible to pump all three grouts in proper sequence at any depth, in any borehole at will, by the use of Manchette tubes.

These steel tubes $1\frac{1}{2}$ in. to $2\frac{3}{4}$ in. (38–70 mm) in diameter were lowered into the several boreholes (which were filled with weak grout to keep them from falling in) and the grout under pressure in the tubes forced through small holes covered with a rubber sleeve (the manchette) and penetrated through the weak grout into the strata. When pressures were reduced the rubber sleeves returned to their closed position. Pressures of around 400 lb./sq. in. (2,800 kN/m^2) fractured the weak surrounding grout and the grouting pressures thereafter were usually controlled to 100–150 lb./sq.in. (690–1,030 kN/m^2).

Results of Grouting. Glacial till (2) on *Figure 50b.* From the various permeability tests the permeability is not more than 1×10^{-5} cm/sec. It was seen through the portholes to be very impermeable, compact and almost unfissured. Fine lenses of more permeable material, however, were grouted under pressure with silicate aluminate which heaved the strata at the surface by $\frac{1}{2}$ to 1 in. (12–25 mm) the uplift being much less than 1 per cent of its depth below the surface. This is considered to be an indication of the grouting efficiency. In the central 270 m of the valley down to about 25 m below ground surface level, five rows of grout tubes were spaced at 10 ft. (3m) centres through the Till.

In the remaining right and left ends of the Manchette curtain, the number of rows of tubes reduced to three and finally to a single row. All tubes and rows were 10 ft. (3 m) apart. The curtain terminates at both abutments with a solid 6 ft. (1·8 m) wide concrete cut-off sunk to a maximum depth of 15 m.

Sand and gravel complex (3), *Figure 49b.* Below the till core five rows in the central 270 m of the valley were drilled 10 ft (3 m) apart through the sand and gravel to 25 m. All three types of grout were injected into the sand and gravel complex. The thicker clay cement grout was injected initially into the outer rows and this was followed by deflocculated bentonite grout and silicate aluminate grouts. The centre row had small initial injections of clay cement and deflocculated bentonite grouts followed by the penetrating silicate aluminate grout in larger quantities. As far as observations went an average coefficient of $1-10^{-5}$ cm per sec was attained.

Schistose grits and phyllite, (5) *Figure 50b.* The junction between the superficial metamorphic rocks is very permeable, weathered and decomposed. It was therefore injected at first by clay cement grout 5 ft. (1·5 m) above and 10 ft. below the interface. In the centre of the valley three rows were injected and under the glacial till one row at the abutments. Subsequently silicate aluminate was found to be necessary for the whole length of the junction. In addition grouting in 10 ft. stages was carried out to 50 ft. (15 m) below the junction by conventional methods and tested by pumping measurable quantities of clean water under pressure into the boreholes.

Water

One of the interesting features of this site is that there are at least two major underground water levels (6) and (7) on *Figure 50a,* one in the glacial sand and gravels and the other in the underlying schistose and phyllite bedrocks. In the upper sands and gravels (3) in *Figure 50a,* a pumping test for seven consecutive days was made in an 18 in. borehole which recorded 30,000 gals./day without lowering the water level 'near the borehole'. For excavation purposes dewatering the site was thus impracticable.

It was of interest to note that when inspection of the portholes in the shaft of the grouted test block was made a total of 5 litre/min. leakage was collected. Of this total 70 per cent was from the sand and gravel complex and 30 per cent from the limited depth of schistose grits and phyllites penetrated. This showed the necessity of grouting the sand-gravel complex and the glacial-metamorphic interface.

CONCRETE DAMS

Laggan Dam

Laggan is a concrete gravity dam with a slight curvature. It is 175 ft. in height and has a crest of 700 ft., the chord-height ratio therefore is about 4.

As it is in a narrow valley with a very large gathering ground, flood water is discharged through six siphons in addition to flowing over the full length of the crest (*Figure 51a and b*).

Figure 51(a). Laggan dam

Pitlochry Dam*

It is intriguing to read an account by Mr. A. A. Fulton describing the Pitlochry hydro-electric development in Scotland.[18]

> 'While the dam proper is not a large structure, a substantial concrete cut-off wall was needed to render watertight a raised beach of sand and gravel on the east bank of the river. The length of this cut-off was greater than that of the dam itself, and the excavation had to go down deeper to reach rock because the gravel bank concealed an old river channel at a lower level than the present one.'

The dam is 54 ft. in height with a crest of 475 ft., and the cut-off wall on the east bank extends 115 ft. in depth and is 900 ft. in length.

The shape of the valley at the dam site comprises a steep hill on the right bank, a wide valley in which the river runs, a not so steep hill on the left

---

*Mr. J. Guthrie Brown has most kindly sent complete details of the trench section (*Figure 54*) and has confirmed the accuracy of these notes on the Pitlochry dam.

bank, further left a low plateau for 100 yd., with a 'slight depression' and thence a gradual rise on the extreme left of the valley.

The slight depression is of considerable interest and *Figure 53* shows how picturesque and innocent it seems to be until it is compared with the geological section (*Figure 54*). It shows that a valley of quartzite with two

Figure 51(b). Laggan dam

depressions is buried by varying depths of permeable gravel and sand. Whereas the quartzite rock under the dam is quite near the surface, and on the right bank only at the most 30 ft. below top water level, on the left bank it attains a depth of 115 ft. below top water level, or some 150 ft. below the original ground level.

Figure 52a and d shows the dam, with the quartzite almost at the surface with the gravel deposit under the trees on the right looking upstream, while *Figure 52b* shows the left (east) sand and gravel bank where the rock is getting deeper at the extreme right of the photograph.

Figure 52c shows, in a vivid way, the long walkway over the cut-off wall in a deep cutting necessitated for excavating the slight depression (beyond the lamps in the photograph). Here an excavation of over 100 ft. in depth was necessary down to the rock through the gravel which was replaced with concrete up to top water level (301 O.D.).

A grout curtain under the dam and right bank consumed some 1,800 cwt. of cement, the area of the grout screen being about 40,000 ft.²

Figure 52(a).

Figure 52(b). [*see legends opposite*

(c)

(d)

Figure 52. Pitlochry dam. (a) Showing the well known fish pass on the left and on the right the trees on the glacial drift covering rock, the surface of which is below the river level. (b) On the left bank the quartzite rises 10 ft. or so above the river bed and is covered with gravel and sand (below the trees). (c) The walkway in the cutting over the 900 ft. long cut-off wall, on the left bank. The cutting was necessitated by the gravels and sands attaining a depth of 115 ft. below top water level. These had to be excavated and replaced by concrete to prevent water escaping, from left to right, out of the reservoir. (d) The dam has two movable floating drum gates each 90 ft. in length; the powerhouse is shown on the left and the famous fish pass on the left of the powerhouse. The quartzite is within 5 or 10 ft. of most of the river bed here

Figure 53. Pitlochry dam. The dam is on the right behind the trees; the deep buried valley is in the middle; the high ground between is sand and gravel with solid quartzite only a few feet (under 50 ft.) below water level. Under the indentation in the centre of the picture, the rock is as much as 115 ft. below the water level, indeed, a pretty but dangerous valley

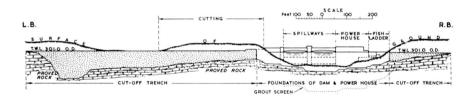

Figure 54. Pitlochry dam. The section of the valley, looking downstream as in Figure 53. The dotted part represents concrete-filled trenches necessitated by a thick deposit of permeable material on the rock extending to ground surface. If this material had not been removed and replaced by concrete, considerable leakage would have ensued. The approximate depth of the grout screen under the dam is shown

The cutting is shown in Figure 52 (c); the dam and power house in Figures 52 (a) and (b)

(Reproduced by courtesy of Mr. J. Guthrie Brown of Sir Alexander Gibb and Partners)

Monar Arch Dam*

Loch Monar is about 30 miles west of Inverness and the dam is in a valley which provided a theoretical chord/height ratio of 3·3. The dam is the first double curvature dam in Great Britain because the valley was found to be geologically sound, and so was the engineering, for the tendered

Figure 55 (a). *The Monar dam. This dam, with a chord/height ratio of about 3·0, is a cupola double curvature arch-dam, the first in Britain, and founded on fine-grained granulite of pre-Cambrian age. When some 80 ft. of water were impounded during 1963–4 the horizontal movement of the dam relative to the granulite was 15 mm downstream.*

Completed 1963
Length of Crest   528 ft.   Max. thickness 19 ft.
Maximum height 128 ft.   Min. thickness 12·5 ft.
Angle   114°   Volume (concrete) 36,000 yd^3

(*Reproduced by courtesy of the North of Scotland Hydro-Electric Board*)

cost was 9 per cent lower than a comparable gravity dam. *Figure 55a* shows the dam under construction.

The site is symmetrical within the Moine psammitic, fine grained, granulite of pre-Cambrian age, with only a thin covering of drift. The

---

* Mr. C. M. Roberts and Mr. L. H. Dickerson have kindly read these notes.

granulite being a metamorphic rock of quartz and felspar, although strong, is jointed and contains thin peletic, or derived clay layers, pegmatites and quartz veins. Grout was therefore applied upstream, downstream and under the cut-off trench to increase the shear strength of the joints and crushing strength of the granulite; essential for a double curvature dam such as the Monar. The cement consumption per foot of bore hole was about 3 lb. Uplift records were provided by 42 relief holes with V-notches and drains and, in addition, 8 pressure relief holes.

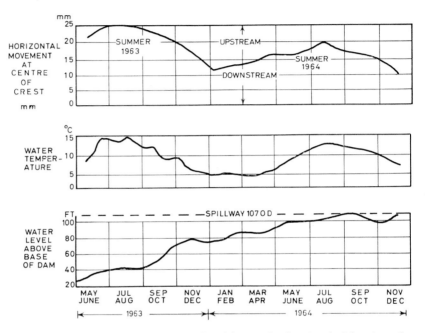

Figure 55 (b). *The Monar dam. Inverted pendulums anchored at 120 ft. below the surface have shown a deviation of 15 mm in the winter of 1964 when the reservoir was nearly full. When the dam was filled, the movement appeared to vary with the temperature of the water*

Dyke

During construction a lamproschist dyke was discovered outcropping 75 ft. upstream from the dam and dipping downstream at 38·5 degrees. By drilling, it was established that the dyke passed beneath the dam at a depth of 60 ft. below the foundations under the river. The dyke is 4 ft. to 8 ft. in thickness, dark green in colour, and is essentially a chlorite-calcite schist with thin irregular bands of quartz and calcite and xenoliths (strange stones —erratics), particles of the country rock embedded in the schist.

It is deduced that such dykes, which run in a SW–NE direction, were intruded in late Caledonian times and were subjected to faulting towards the end of that period. The rocks now exposed at the site were then at a considerable depth, the over-lying rock having been eroded during successive

Ice Ages with Loch Monar and its gorge being carved out in the last Ice Age 10,000 years ago.

By experiment and inference it was decided that the schist was substantially impervious and that it alone would not affect deviation of the dam by more than 3 mm. An example of the total movement of the dam relative to the stable rock beneath the lamproschist dyke was measured by four inverted pendulums (*see Figure 55b*) installed in the foundations and anchored below the dyke at a depth of 120 ft. from the surface. As the dyke appeared near the upstream of the face and was filled with gravel and morain material, the outcrop was specially investigated and a hole was discovered but since it was not of the karstic variety all was well.

Nevertheless the engineers felt the necessity to say:

'The experience gained from this site has shown that an arch dam should be chosen only where the nature of the foundation has been explored to a degree not usual for other types of dam. It seems desirable to excavate the foundations before designing the concrete and deciding on the extent of rock fortification required.'

THE RIVER BEULY BASIN

The undermentioned concrete gravity dams with the Monar arch dam form part of the North of Scotland hydro-electric development in the River Beuly basin.

Dam	Height	Crest	Volume	Geological Formation
Aigas	27·4	91	0·05	Left bank. Old Red Sandstone conglomerates. Right bank. Moine schists overlain by conglomerates at higher levels; river bed boundary of the Moine thrust plane. Grout 100 ft. below excavation through thrust plane fault.
Beannachan	23·1	93	0·02	Faulted mica-schist.
Kilmorak	29·3	116	0·04	Conglomerates with lenses of sandstone grouting to 100 ft. below foundations of dam.
Loichel	19·2	170	0·02	Schist.

The Chliostair Dam, Harris

While Monar is the only large double curvature arch dam and Allt-na-Lairige the only prestressed concrete dam in Britain the Chliostair dam in the N.W. corner of the Isle of Harris is the first thin arch dam, as it was built in 1961 (*Figure 56*).

Situated on granite gneiss in part of the Lewisian Complex of pre-Cambrian age, the dam is about 50 ft. high and the thin arch 140 ft. in length is between two massive concrete abutments.

There is an igneous dyke running diagonally across the foundation which necessitated additional excavation.

On the left bank there is an outcrop of hard rock in soft rock which seemed to be detached and according to the late Prof. Richey it had been 'plucked' away from the parent rock in the Ice Age. It was cracked and too large to remove entirely and was incorporated in the foundation by grouting.

On the right bank there was a series of horizontal slabs of rock, resulting from ice pressure and in order to join these together pressure contact grouting and rock-bolting were tried but without success. However washing out the clay parting seams at 20 lb./in.^{2} enabled contact grouting to be possible and 70 tons of cement and sand were thereby injected.

Figure 56. Chliostair Dam, Isle of Harris, Scotland. Constructed in 1961 the dam is the earliest thin arch in Britain. Granite-gneiss, much disturbed by glaciers in the Ice age formed hard unstable slabs of rock with parting seams filled with clay which had to be washed out and replaced with grout and made to form part of the foundation. There was also a 10 ft. wide igneous dyke to be negotiated. (Reproduced by courtesy of The North of Scotland Hydro-electric Board)

Some excavation work had to be done later in this grouted area which the contractor found difficult as the grout had penetrated well in between the slabs so the grouting operations were considered to be successful.

The Orrin Dams*

The main Orrin dam is a concrete gravity structure about 15 miles west of Inverness (*Figure 47*, page 106). Height 167 ft., crest about 1,025 ft. in length. It is founded on psammitic granulite having the texture of fine sand.

The rock is generally sound but at 200 ft. from the dam, upstream, on the left bank there is a zone of shattered rock and a mineralised seam containing calcite and pyrites. A leakage path here would be suspected

_____
* Mr. L. H. Dickerson has kindly read the notes on the Chliostair and Orrin Dams).

particularly as 100 ft. further from this zone there is brick-red staining in a major and in minor subsidiary joints. The granulite had been altered, silica being replaced by calcite, for reasons unknown. After filling there was little leakage and the cement grout consumption, to a depth of 130 ft. in the granulite, with holes 10 ft. apart, and maximum pressures 50 lb./in.^{2}, was only 50 tons.

The subsidiary Orrin earthfill embankment is 80 ft. in height and 1,025 ft. in length with a 3 ft. to 6 ft. wide concrete core wall keyed 5 ft. into granulite. The core is constructed in lengths of 45 ft. with copper-bitumen water stops and the embankment is formed out of moraine material with the upstream face covered with pre-cast concrete slabs.

Allt-na-Lairige Dam*

Allt-na-Lairige dam was built on granite in 1954–56, 50 miles N.N.W. of Glasgow, and has the following characteristics.

Capacity of reservoir	810 million gal.
Top water level	993 O.D.
Area of gathering ground	5·3 sq. miles (with catchwaters)
Average annual rainfall	125 in.
Maximum height of dam to rock	73 ft.
Length of overflow weir	252 ft.
Length of crest	1,360 ft.
Length of pre-stressed section	966 ft.

The geology of the reservoir site is described by Mr. J. A. Banks[19,20] as follows.

'The Allt-na-Lairige reservoir basin is a high-level glacial valley, broad in the base, and fairly regular in cross-section. All the works lie within the outcrop of a large intrusion of granite. The granite is of variable grain size with porphyritic feldspar in places; it is a sound, resistant, and impervious type of rock. There is some close jointing and the rock could hardly be described as massive but, on the whole, the joints are widely spaced.

At the site of the dam the rock is exposed in the bed of the river and outcrops at intervals on both sides of the valley. There are shallow depressions between these outcrops which are filled with morainic material and moorland peat, but over the whole range of the dam the average depth from the ground to rock surface is only about 5 ft. The line selected for the dam is on a low rock ridge running roughly at right-angles to the lines of the valley and the rock surface appears to fall in both the upstream and downstream directions.'

The ability of the granite to take the pre-stressed load was ascertained experimentally by burying jacks in concrete in a pit 18 ft. deep and pumping the jacks up to 4,400 tons which displaced the granite only by $\frac{1}{10}$ in. As this experiment was carried out on better granite 1½ miles from the site, the cables at the dam were embedded for a depth of 27 ft.

The Scotch granite was excellent for this unique undertaking; in Cornish granite, for example, the possible presence of china clay might have made the project somewhat hazardous.

*Mr. J. A. Banks has very kindly read these notes on the Allt-na-Lairige dam.

Figure 57. (a) *Allt-na-Lairige dam. This is the only large pre-stressed concrete dam (73 ft. in height) which has been constructed in Great Britain. It is on very hard granite which sustains a total pull of about 47,600 tons over the 966 ft. length of the pre-stressed dam. The central tower is the control valve house and is of gravity section.* (b) *The overflow weir and channel of granite which is hard enough to be left for flood water to pass over it*

(Figure 57(a) reproduced by courtesy of the North of Scotland Hydro-Electric Board)

Figure 57a shows a general view of the pre-stressed dam; the gravity section at the central intake control block gives a good idea of the saving in concrete (15 per cent in this case) of the pre-stressed section.

Figure 57b shows the overflow weir (252 ft. in length) with the overflow channel, the granite of which is sufficiently hard to form a natural floor.

THE GLEN SHIRA DAMS*

Lower Sron Mor Dams[21]

In Glen Shira, near Inveraray, there are three interesting dams situated on altered igneous (epidiorite) and altered sedimentary rocks (phyllites and schists) known as the Dalradian Metamorphic Complex.

The phyllites (Ardrishaig Phyllite) have been changed from calcareous shales and hence are soft. The schists, however, resulting from material comprising schistose grits, quartz schists and mica schists, are harder and fairly massive.

The special geological interest here is that two dams, one concrete and one earth, seem to be in the same valley side by side within 50 yd. of each other, but separated by a high hard mass of rock. Actually the concrete dam is in an epigenic (newly made) valley and the earth dam is in an older pre-glacial valley filled with 'rubbish' left by a glacier.

The concrete dam impounds a depth of about 50 ft. and the earth dam 40 ft. of water, the capacity of the resulting reservoir being 350 million gal.

Figure 58 gives a plan and diagrammatic elevation of the two dams, and *Figure 59a and b* gives typical sections of the foundations of both dams. *Figure 60* is a view of the verrou between the dams.

Figure 61 shows the earth rockfill dam on soft morainic material, and *Figure 62a and b* the concrete dam and the hard rock on which it is built.

As the capacity of the reservoir impounded by these dams is relatively small (350 million gal.) although it commands a large gathering ground (22 sq. miles) with high rainfall (105 in. per annum), the water which it cannot hold is pumped up only some 140 ft. into the large Upper Shira reservoir which has the following characteristics.

Upper Sron Mor Dam

Concrete buttress (round-head)

Capacity	4,700 million gal. ⎱ giving a valuable electric
Top water level	1,108 O.D. ⎰ head of 1,100 ft.
Maximum height	148 ft.
Length of crest	2,250 ft.

The configuration of the mica schists of the floor of the valley of this fine dam is well seen in *Figure 63a and b*. *Figure 64* shows a verrou between two valleys in embryo. The grout curtain consumed 0·67 lb. per ft.^{2} of cement.

*Mr. J. A. Banks and Mr. J. Paton have very kindly read these notes on the Glen Shira dams.

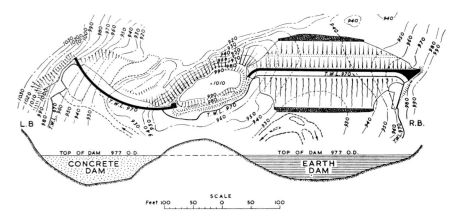

Figure 58. Lower Sron Mor dams. Showing two dams in the same valley, or in two adjacent valleys, with a 'verrou' between them. The left-hand dam is a concrete structure founded on hard schistose grits, whereas the right-hand dam was founded in soft phyllites which would seem to be the older of the two valleys

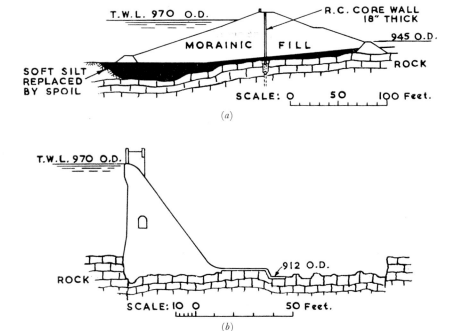

Figure 59. Lower Sron Mor dams. (a) Showing the section of the earth bank on soft silt which was replaced by rock spoil some 20 ft. deep. The embankment was constructed with material from the glacial moraine in the valley. (b) Showing the section of the concrete curved dam on hard quartz and mica schists, and is very different from that 50 yards away, which necessitated an earth embankment

Figure 60. Lower Sron Mor dams. The left-hand dam is a curved gravity concrete structure in a rocky gorge of schistose grits, whereas the right-hand dam is of earth and rockfill in an ancient valley which has been filled up with glacial material. In the middle, between the dams, is a hard high cliff of maiden rock between the valleys

Figure 61. Lower Sron Mor dams. The earthen dam (left) in the soft morainic material and the concrete dam in the rocky gorge below (right). The Upper Sron Mor dam is in the distance

(Reproduced by courtesy of the photographer W. Ralston Ltd., and the North of Scotland Hydro-Electric Board)

(a)

Figure 62. Lower Sron Mor dams. (a) The curved concrete dam, the natural rocky verrou (see Figure 46) and the earthen embankment shown in the left-hand upper corner. (b) Showing the overflow and stilling pool in foreground

(b)

(a)

(b)

Figure 63. Upper Sron Mor dam. (a) The right abutment end of the upper dam on schists. (b) The left abutment of the upper dam on schists with verrou on left of photograph (see also Figure 64)

Figure 64. Upper Sron Mor dam. The upper dam with reservoir impounded by the two lower dams in foreground. The rocky verrou is well seen and has been taken advantage of in siting the dam; on its left is Figure 63a and on its right Figure 63b

SOME TYPICAL SCOTTISH BUTTRESS DAMS

For the construction of these buttress dams, concrete coarse aggregate was generally near at hand, but sand was brought from a distance.[22]

Name	Length of crest (ft.)	Maximum height (ft.)	Length–height ratio	General geology
Sron Mor. Round head buttress. Main upper dam (*Figure 65*).	2,250	148	15	Soft phyllite with bands of quartzite, quartz schist, and limestone.
Errochty. Diamond head buttress (*Figure 66a, b and c*).	1,310	127	10	Quartzite and some granulite.
Lednoch. Diamond head buttress (*Figure 67*).	950	130	7	Epidiorite.
Loch Sloy. Massive gravity buttress (*Figure 68a and b*).	1,170	165	7	Fissured mica schist.
Lawers. Lochan-na-Lairige. Massive gravity buttress (*Figure 69a and b*).	1,100	130	9	Ben Lawers schists or phyllites with beds of quartzite and quartz schists present.
Lubreoch. Massive gravity buttress (*Figure 70*).	1,740	100	17	Quartz mica schist of the Ben Lawers group.
Giorra. Massive gravity buttress (*Figure 71*).	1,500	100	15	Siliceous granulites with a few thin partings of mica schist.

(Additional information has been given through the kindness of Mr. P. L. Aitken of the North of Scotland Hydro-Electricity Board)

Figure 65. Sron Mor dam. Downstream view of buttress on phyllite with bands of quartz, quartz schist, and limestone in the Dalradian Metamorphic Complex of altered shales (phyllites). Altered grits (schists, schistose grits, quartz schists, mica schists). The probable deformation under the buttresses is ¼ inch

(a)

(b)   (c)

Figure 66. Errochty dam. (a) *General view of buttress dam on quartzite and granulite.* (b) *Downstream view of diamond head buttress on quartzite. As excavation proceeded bands of clay were encountered which had a lower bearing pressure than the quartzite.* (c) *Upstream view of dam under construction on quartzite. For the grout curtain some 25,700 ft. of bore holes were drilled which consumed 9,133 cwt. or 7·6 lb. per ft.^{2} of grout screen*

Figure 67. Lednoch dam.[23] *The broad foundation rests on an intrusion of epidiorite (below overburden and fissured rock); there is a cut-off trench 10 ft. deep under buttresses and 5 ft. deep between buttresses and a grout curtain to a depth of 80 ft. The water-cement ratio of the grout was 8 to 1 by weight, and 95 tons or about 0·28 cwt. per yd.^{2} were consumed in the screen. The foundation grouting consumed 152 tons. The maximum height of the dam is 125 ft. and the crest length is 950 ft*

(Reproduced by courtesy of the North of Scotland Hydro-Electric Board)

(a)   (b)

Figure 68a and b. Loch Sloy dam.[24] *This dam, 1,170 ft. in length and 165 ft. in height is a gravity buttress type, built on fissured mica schist which became more solid at depth. Although the foundation was built on mica schist which proved generally sound at 20–30 ft. below the surface, there were pockets of sand and pebbles which had to be excavated down to a depth of 85 ft. below the surface before sound rock was reached*

Figure 69a and b—Lawers. Lochan-na-Lairige

Figure 70. *Lubreoch dam. Elevation looking north-west. This long dam, up to 100 ft. in height, is on quartz mica schist*

(Reproduced by courtesy of the North of Scotland Hydro-Electric Board)

Figure 71. Giorra dam. View from the north end
(Reproduced by courtesy of the North of Scotland Hydro-Electric Board)

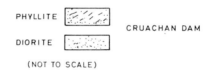

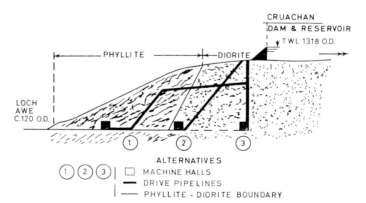

Figure 72. Cruachan dam. Alternative sitings for a 9 ft. diameter supply pipe to the turbine hall from the dam. The phyllite is weak compared with the diorite and from experience in phyllite, alternative No. 1 was rejected although the machine hall was nearest to the mouth of the tunnel. No. 3 entailed the longest tunnel. No. 2 was adopted

Cruachan Dam* (Pumped Storage Scheme)

The Cruachan Dam is located (*see Figure 47*, page 106) 7 miles N.W. of (4) Upper Sron reservoir and 7 miles W. of Dalmally. It is a buttress dam, 153 ft. in height, crest 1,037 ft. in length and is similar to the Sloy dam. The Cruachan dam is founded on diorite on which there was little overburden.

Downstream of the dam the diorite ends, and although no major problems arose with the dam, much thought was given to the siting of the supply pipe to the turbines in the machine hall as the diorite was replaced by phyllite which in the contact zone had been baked by the diorite (*Figure 72*). The dimensions of the supply pipe, as adopted, are of the order of 9 ft. diameter for the lower quarter of the length and 16·5 ft. for the upper three-quarters. The advantages and disadvantages for three alternatives (*Figure 72*) which were considered are:

Advantages	*Disadvantages*
1. Short distance to the machine hall both during construction and operation.	Phyllites too weak for supply pipe pressure system with the order of dimensions suggested above. (Based on experience of phyllite in the Inverawe tunnel nearby.) Long high pressure system.
2. Shorter high pressure system than No. 1. Shorter distance to machine hall than No. 3.	Longer high pressure than No. 3. Longer access to machine hall than No. 1.
3. Shortest high pressure system.	Longest access tunnel to machine hall both in construction and operation.

The first alternative might seem to be the best on engineering grounds but it is the worst on geological grounds, and the high pressure supply system contemplated in largely unreliable phyllite, was rejected. Either of the other alternatives, subject to contractors' prices, would be acceptable, as the supply systems of both are wholly in diorite. In tendering the contractor chose the second alternative.

The diorite under the dam is jointed but not weathered and very little cement grout was consumed in the grout curtain. Deformation measured by inverted pendulums 84 ft. below dam foundation suggested a maximum of $\frac{1}{2}$ mm, probably due to concrete shrinkage effect which seems to bear

* Mr. L. H. Dickerson has kindly read this note on the Cruachan Dam

out the reliability of diorite in the Cruachan District. Even so construction proved difficult and dangerous enough in the sloping high pressure tunnel system and machine hall. (Short, W. D., in discussion, *I.C.E.* 37 671 (1967).)

SOME SCOTTISH ROCKFILL DAMS

Three interesting rockfill dams are also to be found in Scotland.

Loch Treig (part of the Lochaber Power Scheme, 1921-30)

Earth and rockfill with reinforced concrete slabs on the downstream face with a vertical concrete core-wall in an excavation extending to 100 ft. below the surface (*Figure 73*). The water level in the loch was raised 39 ft.

Figure 73. Loch Treig dam. This is an early rockfill dam of the Lochaber Power Scheme (1921-30) which raises the water level in Loch Treig by 39 ft. Upstream face is seen at low water (1959)

Loch Quoich (part of the Garry and Moriston Hydro-Electric Schemes, 1945-55)

This rockfill dam raises Loch Quoich by over 100 ft. and is on metamorphic mica schists with little drift covering (*Figure 74a and b*).

Breachlaich Dam

This dam, under construction in 1959, is to raise Lochan Breachlaich (*Figure 75a–d*).

(a)

(b)

Figure 74. Loch Quoich.[25] (a) *The rockfill (mica schists) was obtained from a quarry ½ mile upstream, 286,000 yd.³ plus 100,000 yd.³ obtained from tunnel and foundations. Rock with a high percentage of hornblende, mica, and material below ⅜ in. in size was excluded. The layers, 2 ft. in thickness, were rolled with 20 passes of a 10-ton roller and sluiced with twice the volume of the layer with water and rolled with 10 passes of a vibrating roller.* (b) *Upstream face of concrete having a slope of 1 in 1·3 was placed by the vacuum concrete process*

(Reproduced by courtesy of the North of Scotland Hydro-Electric Board)

(a)

(b)

Figure 75. Breachlaich dam (see opposite for legend)

(c)

(d)

Figure 75. Breachlaich dam. (a) This interesting rockfill dam is near Loch Tay and is 80 ft. in height and 1,300 ft. in length. Downstream embankment is seen under construction. (b) Upstream embankment under construction. (c) Part of reservoir area showing square tunnel mouth which will lead the water to St. Fillans generating station which is also supplied from Lednoch reservoir. (d) Quarry for rockfill

REFERENCES

[1] LAPWORTH, H. Geology of dam trenches. *Trans. I.W.E.*, **16,** 25 (1911).
[2] HOLLINGWORTH, S. E., TAYLOR, J. H., and KELLAWAY, G. A. Large scale superficial structures in the Northampton Iron Field. *Quart J. geol. Soc. Lond.*, (1943) 100.
[3] WALTERS, R. C. S., WALTON, R. J. C. and PENMAN, A. D. M. (in discussion). Water supply for the Yeovil District (Sutton Bingham Scheme). *J.I.C.E.*, **8,** 71 (1957).
[4] LAPWORTH, C. F. A concrete dam. *Wat. & Wat. Engrg.*, 45 1(1942).
[5] SANDEMAN, E. The Derwent Valley Waterworks. *Min. Proc. I.C.E.*, **206,** 152 (1920).
[6] HILL, H. P. The Ladybower reservoir. *I.W.E.*, **3,** 414 (1949).
[7] CONTRACT J., Soil mechanics as an aid to tendering (Selset) 967 Nov. (1956), and KENNARD, J. & M. Selset Reservoir: Design and Construction. I.*C.E.* **21,** 277 (1952).
[8] TAYLOR, G. E. The Haweswater dam. *I.W.E.*, **5,** 355 (1951).
[9] HOLLINGWORTH, S. E. The influence of the glaciation on the topography of the Lake District. *I.W.E.*, **5,** 405 (1951).
[10] MARR, J. E. The geology of the Lake District. Cambridge University Press. 1916.
[11] DEACON, G. F. The Vyrnwy works for the water supply of Liverpool. *J.I.C.E.*, **126,** 24 (1896).
[12] MANSERGH, E. L., and MANSERGH, W. L. The works for the supply of water to the City of Birmingham for Mid-Wales. *Min. Proc. I.C.E.*, **190,** 3 (1911).
[13] MORGAN, H. D., SCOTT, P. A., WALTON, R. J. C., and FALKINER, R. H. The Claerwen dam. *J.I.C.E.*, **1,** 249 (1953).
[14] FORDHAM, A. E., COCHRANE, N. J., KRETSCHENER, J. M., BAXTER, R. S. The Clywedog reservoir project. *I.W.E.*, **24,** 17 (1970).
[15] SHEPPARD, G. A. R., and AYLEN, L. B. The Usk Scheme for the water supply of Swansea. *J.I.C.E.*, **7,** 246 (1957), and
LITTLE, A. L. Compaction and pore-water pressure measurements on some recent earth dams: Usk, Daer, etc. *I.C.O.L.D.*, III 205 (1958).
[16] BANKS, J. A. Problems in the design and construction of Knockenden dam. *J.I.C.E.*, **1,** 423 (1952).
[17] KERR, H., and LOCKETT, E. B. Daer water supply scheme. *J.I.C.E.*, **7,** 46 (1957).
[18] FULTON, A. A. Civil Engineering Aspects of Hydro-electric development in Scotland. *J.I.C.E.*, **1,** 272 (1952).
[19] BANKS, J. A. The employment of prestressed technique on the Allt-na-lairige dam. *Fifth Congress of Large Dams* II (1955) 341.
[20] BANKS, J. A. Allt-na-lairige prestressed concrete dam. *J.I.C.E.*, **6,** 409 (1957).
[21] PATON, J. The Glen Shira hydro-electric project. *J.I.C.E.*, **5,** 593 (1956).
[22] WILSON, E. B. Aggregates for British dams. *I.C.O.L.D.* R.33 **I,** 545 (1961).
[23] ALLEN, A. C. Features of Lednoch Dam including the use of fly ash. *J.I.C.E.*, **13,** 279 (1958–59).
[24] STEVENSON, J. The construction of Loch Sloy Dam. *J.I.C.E.*, **1,** 100 (1952).
[25] ROBERTS, C. M., WILSON, E. B., THORNTON, J. H. and HEADLAND, H. The Garry and Moriston hydro-electric schemes. *J.I.C.E.*, **11,** 41 (1958).

BIBLIOGRAPHY

Backwater dam
GEDDES W. G. N. and PRADOURA, H. H. M. Backwater Dam in the County of Angus, Scotland Grouted Cut-Off *I.C.O.L.D.*, **I,** 253 (1967).
ISCHY, E. and GLOSSOP, R. An introduction to alluvial grouting. *I.C.E.*, **21,** 449 (1962) and **23,** 722 (1962).
SCRIMGEOUR, J. and ROCKE, G. Backwater Reservoir for Dundee Corporation Waterworks. *I.W.E.*, **20,** 325 (1966).

Monar Dam
FULTON, A. A., DICKERSON, L. H., I.C.E. discussion on their paper, 'Design and construction features of hydro-electric dams built in Scotland since 1945', *I.C.E.*, **29,** 713 (1964) and **33,** 474 (1965).
HENKEL, D. J., KNILL, J. L., LLOYD, D. G., SKEMPTON, PROF. A. W. Stability of the foundations of Monar dam. *I.C.O.L.D.*, **I,** 425 (1964).
ROBERTS, C. M., WILSON, E. B., WILTSHIRE, J. G. Design aspects of the Strathfarrar and Kilmorack hydro-electric scheme. *I.C.E.*, **30,** 449 (1965).

Chliostair Dam
JARVIS, R. M. Some features of the construction of a small arch dam in Harris. *Edinburgh & East of Scotland Assoc. I.C.E.*, (Feb. 1962).

Orrin Dam
GOWERS, A. G. Some points of interest in the design and construction of Orrin dam, Ross-shire. *I.C.E.*, **24,** 449 (1963).

Cruachan Dam
YOUNG, W. and FALKINER, R. H. Some design and construction features of the Cruachan pumped storage project. *I.C.E.*, **36,** 407 (1966) and SHORT, W. D. (in discussion) *I.C.E.*, **37,** 671 (1967).

16

AUSTRIAN DAMS

Figure 76 shows the sites of some of the large dams which have recently been constructed in Austria and adjacent countries of Italy, Germany and Jugoslavia.

It is probable that more than half the water power resources of Austria have already been developed, but the larger power schemes are concentrated in only a few areas. In addition there are several smaller reservoirs for power and water-supply.

The following dams are listed in 'Large dams in Austria', published in 1964, in English, and presented to the participants at the Eighth Congress of I.C.O.L.D. by the Austrian National Committee of I.C.O.L.D.

DAMS IN WESTERN AUSTRIA

Years of Construction	Name of dam	Type	Max. height m	Crest length m	Top water level O.D.W. m	Main formation at dam site	River	Locality (see Figure 76)
1921–5	Spullersee (South dam)	G	36	280	1825	Lias Marl	Ill	near (1)
1923–5	Spullersee (North dam)	G	26	186	1825	Lias Marl	Ill	near (1)
1928–31	Vermunt	G	50	488	1743	Gneiss	Ill	near (1)
1939–48	Silvretta (Lateral dam)	G	80	432	2030	Mica-Schist	Ill	near (1)
	Silvretta (Lateral dam)	G	31	140	2030	Amphibolite	Ill	near (1)
	Silvretta (Biele dam)	R	25	733	2030	Moraine	Ill	near (1)
1943–5	Gmuend (Gerlos)	A	39	69	1190	Quartz-Schist	Inn	(6)
1950–7	Bächental	Dome	34	70	952	U Trias Dolomite	Inn	near (3)
1955–8	Lünersee	G	28	380	1970	U Trias Dolomite	Ill	near (1)
1961–5	Kops	A	120	420	1809	Gneiss	Ill	near (1)
		G	43	195	1809	Gneiss	Ill	near (1)
1962–5	Gepatch	R	150	600	1767	Augen-gneiss Talus	Inn	34 km s. of (2)
1962–7	Durlassboden	R	77	475	1405	Schist	Inn	near (6)
1924–7	The Achensee	natural lake 133m deep, 10km in length			929	Lower and upper Trias limestones	Inn	(3)

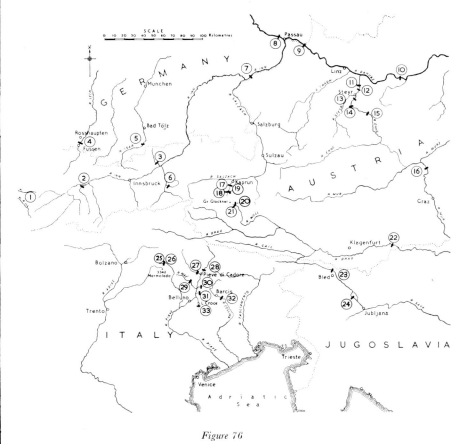

Figure 76

(1) *River Ill, three dams (Rodund, Lünersee, Vermunt)*
(2) *Imst*
(3) *Achensee*
(4) *Rosshaupten*
(5) *Sylvenstein*
(6) *Gerlos*
(7) *Braunau*
(8) *Passau (Kachlet)*
(9) *Jochenstein*
(10) *Ybbs-Persenbeug*
(11) *Mühlrading*
(12) *Staning*
(13) *Rosenau*
(14) *Ternberg*
(15) *Grossraming*
(16) *Pernegg*
(17) *Limberg*
(18) *Mooser*
(19) *Drossen*
(20) *Möll*
(21) *Margaritze*
(22) *Schwabeck*
(23) *Moste*
(24) *Medvode*
(25 and 26) *Fedaia dams*
(27) *Valle di Cadore*
(28) *Piave di Cadore*
(29) *Pontesei*
(30) *Vaiont*
(31) *Gallina*
(32) *Barcis*
(33) *Croce*

143

GEOLOGICAL NOTES ON DAMS IN WEST AUSTRIA

Spullersee Dam

These are a pair of attractive gravity dams, over 45 years old in a high mountainous district. The south dam is on vertical Lias Marls with a strike parallel to the dam axis with a fault of 250 m in the vicinity.

The smaller, north dam, is also on Lias Marl with a steep upstream dip and a strike oblique to the dam axis (*Figure 77*).

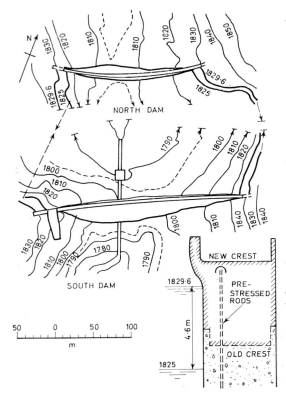

Figure 77. The Spullersee Dams. These are situated in the Ill Valley in Western Austria and were built in 1921–25. They are slightly arched concrete gravity dams to retain a top water level of 1,829·6 O.D., valuable in hydro-electric development. The geological interests of this site are as follows: The South dam is on vertically bedded Lias Marls with a strike parallel to the dam axis. The North dam is on nearly vertically bedded Lias Marl with a strike oblique to the dam axis. The two dams are on a ridge of impermeable Lias Marl as shown by the contours to the north and south. The raising of these 45-year old dams, by 4·6 m, in 1964, by pre-stressed cables embedded in vertical Lias Marl is shown diagrammatically in the lower diagram

In 1964, the dams were raised by prestressing to increase the storage by 4·6 m. in level, from 1,825 O.D. to 1,829·6 O.D.

The details of prestressing are:

	South Dam	North Dam
Present height of dam (max)	36 m	26 m
Raised height of dam (max)	40·6 m	30·6 m
Crest length	280 m	186 m
Number of anchors (4 m apart)	75	42
Total length of anchors	2,320 m	1,091 m
Average length of anchors	31 m	26 m
Prestressing, tons	132	64
Number of 7 mm wires	35	17
Diameter of holes	10·1 cm	7·6 cm

The holes are filled with cement and fly ash with an intrusion aid. The crests of the dams are raised by concrete, 3·7 m in width and 4·6 m in height with parapet walls on the upstream faces and a line of anchors about a metre distant.

Vermunt Dam

A high concrete dam of 50 m built in 1931 on biotite gneiss of the Silvretta nappe near the Swiss frontier. The condition of this dam was examined after more than 30 years operation and was found to be sound, middle-aged in appearance.

However, leakage had increased from $\frac{1}{2}$ litre per second in 1939 to one litre per second in 1963 and there is a tendency for uplift pressures to lessen.

Silveretta (lateral and side) Dams and Biel Dam (Rodund Scheme)

These three interesting dams (*Figure 78*), were constructed in 1939–1948 to maintain a water level of 2,030 m O.D. in a mountainous region near the Swiss frontier. The *Biel dam* is formed on a moraine and follows the contours of this Würmian moraine which is a satisfactory foundation for an earthen embankment. The maximum height of the dam is 25 m and the curve of the crest, concave, facing upstream, is said to blend well with the mountain scenery. The embankment slopes are 1 to 2 H, the right end of the crest on the mountain side, runs for 800 m until it reaches ground on the left bank at a level of about 2,030 m O.D.

This low ground for a distance of about 800 m reaches the right bank of the *Silvretta dam*, which is on folded mica-schist, changing to augen-gneiss in the centre and to amphibole on the left bank, overlain with 21 m of moraine and gravel.

At the left bank the concrete abuts against a 'verrou', the Kleiner Ochsenkopf, of fine fibrous amphibolite which appears a little above 2,030 m O.D. Beyond the 'verrou', there follows the short lateral concrete dam for 140 m up to some 31 m in height also on amphibolite.

Compare the Lower Sron Mor dams (page 126) where, by force of circumstances, there is a concrete dam on hard schistose grits and a moraine-filled dam on soft phylites with a 'verrou' between them.

Lünersee Dam

This dam, near the Liechtenstein frontier is built on a slender ridge, like the Spullersee dams on Lias Marl but on Upper Trias (Hauptdolomit), polished by glacial erosion with bedding planes dipping down-stream. It is a concrete gravity dam of variable height (max. 28 m), which runs for 380 m along a ridge of Dolomite of variable level, with a 'verrou' to be seen only in section.

The crest is wavy in plan to follow the natural contours of the Dolomite causing the convexity of the dam to be generally downstream. Scenically it is a most attractive dam.

Kops Dam (*Figure 79*)

An arch dam (1961–5), 120 m in height, on the crystalline gneiss of the Silvretta nappe. The left bank is sound Aplite gneiss with bands of amphibolite, the latter prevailing on the right bank, quartzite schists are subordinate. Some talus up to 20 m exists in the valley floor. The right end of the dam has a concrete block to resist some 43,000 tons horizontal thrust.

Adjoining the arch-dam, right end is a gravity dam with a maximum height of 43 m in which an overflow weir is provided. It is considered one of the most remarkable dams in Austria.

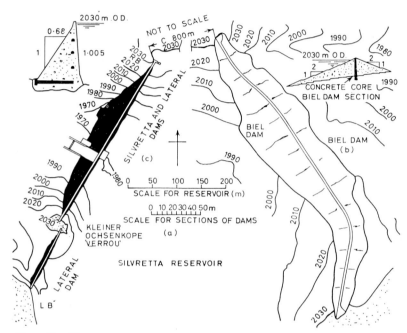

Figure 78. Silvretta reservoir. (a) *The Silvretta reservoir is enclosed by two dams, Silvretta and Biel, in the mountainous district of the Ill valley near the Swiss border T.W.L. 2,030 m O.D.* (b) *The Biel dam of gravel fill and R.C. core, 13 m maximum height, follows the contours of a compact moraine of Würmian age on mica-schist and diverts the water to the N.W. which would normally flow to the N.E.* (c) *The Silvretta dam is a concrete gravity dam with a maximum height of 80 m (governed by the height of the Biel dam), the right bank end being in highly folded mica-schist of thin seams becoming augan-gneiss in the centre of the valley, these rocks covered up to a depth of 21 m of moraine and gravel deposits. The left bank touches an island, Kleiner Ochsenkopf, of garnetiferous amphibolite. Thereafter the concrete lateral dam passes over amphibolite with a deep gorge filled with moraine material*

Gmuend (Gerlos) Dam

Built in 1943–5 this dam is the first high solid arch dam without gallery constructed in Austria. It is in the Opferstock Gorge of quartz schists, quartz sandstone schists with 2 or 3 m of gravel in the bottom of the valley. The grout curtain averages about 5 m in depth. As the left

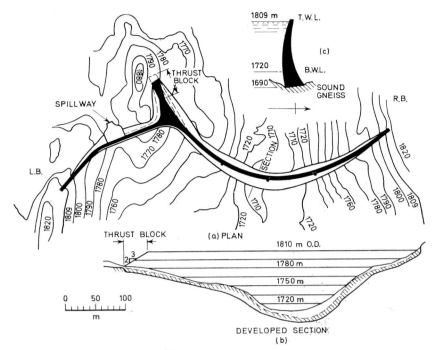

Figure 79. Kops Dam, Austria. The plan (a) shows the surface countours for the Kops dam and the site of the dam where it was ultimately fixed. To retain the required storage and water level, 1,809 O.D. the right bank, as will be seen from the contours, is ideal. The left bank level of the gneiss, however, is 40 to 50 m too low to construct any end to anything, let alone a dam to take a horizontal thrust of 43,000 tons. To retain the advantages of the suitability of the sound gneiss in the rest of the valley and the economies of a thin arch dam, a concrete block of 70,000 m³ capacity is necessary. In (b) this gives the odd appearance of an arch dam, 415,000 m³, in volume and impounding over a 100 m of water without an end. At right angles to this abutment is another dam, volume 50,000 m³, is necessary, up to 43 m in height, of gravity section to retain the water in another valley. In this case the 'verrou' (top left) which is found so often in glaciated regions has been turned into a thrust-block or an integral part of two dams; some interesting thinking must have been done before this bold structure was decided upon. (c) Shows the section of the thin arch which is on sound Aplite-gneiss

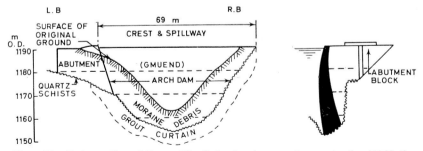

Figure 80. Gerlos or Gmuend Dam. The Gerlos dam is a war-time construction (1943–5). A thin arch dam in a narrow quartz valley; deceptive in shape, for on the right bank quartz-rock is at the top water level required (1,910 m O.D.), but on the left bank it is 20 m below. None of the superficial moraine was suitable to withstand the horizontal thrust of a thin arch dam 39 m in height, with a crest of 69 m in length, and a radius of crest of 32 m, so a concrete block, 2,050 m³ was built as an artificial end to the dam, on the left bank

bank slope compared with the right bank is very flat, a substantial block of concrete, 2,850 m^3 attaining nearly 20 m in height was built into the left bank to take the thrust of a symetrical arch. (*Figure 80*)

The gathering ground is 158 km^2, from which the annual run-off is 173 × 10^6 m^3. The catastrophic flood is assumed at 216 m^3/sec. which would entail 1·23 m over the whole length of crest of 69 m (which can be reduced to 0·7 m when the gravel-scour and supply pipes are in operation). The gravel scour concrete/rubble pipe, 2·1/2·0 m in diameter, was laid for 1,115 m in the bed of the reservoir, with dredging. The dam is now known as the *Gmuend dam*.

Between 1963 and 1969 slides in the quartz schists occurred downstream of the dam necessitating extensive strutting (*see* page 70).

Bächental Dam

This is a cupola or dome-shaped dam in a valley where the chord/height ratio is less than 70/34, say 2. It is on well-bedded, little jointed hard Upper Triassic dolomite (Hauptdolomit). A sand-trap is provided at this dam for the Dürrach diversion to the Achensee station.

Gepatch Dam

The first large rockfill dam in Austria (1962–65), 150 m in height (*Figure 81a*) is constructed on a ridge of Augen-Gneiss crystalline rocks in the Ötz valley, the rim of a glacial basin, overlain by talus, moraine and gravel to 60 m in thickness.

The clay core (founded on augen-gneiss) was obtained 600 m downstream and consists of 80 mm maximum grain size with 1 per cent bentonite added on the upstream side: gradation up to 200 mm and more for the filter zones (*Figure 81b*).

The embankment is gneiss quarried 1·3 km downstream and was compacted in 2 m lifts by a 17 ton vibrating roller. There is a morning-glory bell-mouth 13·3 m diameter.

Durlassboden Dam

A rockfill dam near the Gmuend (Gerlos) reservoir. The left bank is steeper than the right bank: the formation on the left bank is solid green carbonate Epidotechlorite Schist with over-burden of moraine; on the right, a slide of graphite, phylites and carbonate quartz.

In the bottom of the valley there is 135 m of sand, gravel, silt, and at 50 m below the floor, impermeable silt below which is the grout curtain of clay/cement, bentonite and algonite gels, estimated to have reduced permeability to $1·1^{-4}$ cm/sec.

The Achensee Lake

This is a natural lake, 10 km by 1 km and 133 m (maximum) in depth, area 6·8 km^2, storage utilised 72 × 10^6 m^3 from a level of 929 m O.D. to

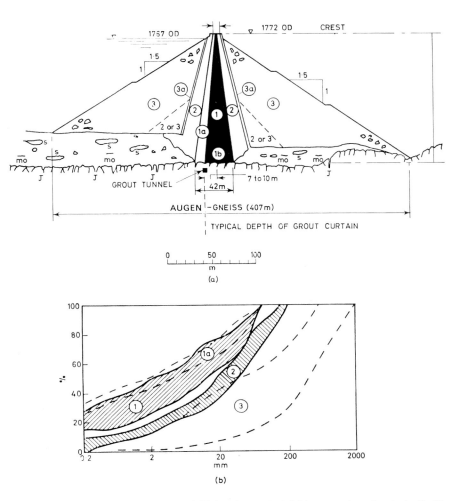

Figure 81. Gepatch Dam. This is a rockfill dam (150 m in height) on moraine redeposited. Boulder clay, talus and gravel up to 60 m in depth, often called glacier rubbish, on augen-gneiss into which the core and filters penetrate. The details of the section (a) are interesting. In the fluvio-glacial block material overlying the augen-gneiss there are sands and lenses of gravel '(s)', as well as moraine, material ('mo)' which was proved to be sound enough to bear the 150 m dam (except in certain parts where there are layers of peat). A grout tunnel 7 to 10 m upstream of the centre of the core penetrated sheets of augen-gneiss with accasional joints. The extent of the width of clay core (42 m) at the base (or 80 m including filters). (b) Details of core and transition filters:

1. Core, maximum grain size 80 mm screened, found 600 m downstream ditto but with 1 per cent of Bentonite added.
1b. A layer of 100 cm of muddy clay on augen-gneiss.
2. First filter, on both sides of core, from river deposits with mixed grain up to 200 mm.
3. Quarry-run Gneiss in embankment from 1·3 km downstream, compacted with 17 ton crane vibrator, two-metre lifts, 4 passes.
3a. Second filter, Quarry fines

918·85 O.D. or about 11 m. The mean gain in head by using this lake is 356 m. The Achensee Lake is on the Lower and Upper Trias limestones limestones overlain by mud, moraine and gravel from the Inn river.

The lake has recently been augmented from the newly built dam at Bächental.

KAPRUN DAMS IN CENTRAL AUSTRIA

One of the most important developments of recent years is south of Kaprun, where there are six dams in the Hohe-Tauern Alps in a gathering ground of 127 km^2 just below Gross Glockner (3,797 m O.D.), the highest mountain of Austria (*Figure 76*). As the River Salzach rises at a high level in these mountains, which are chiefly composed of metamorphic crystalline rocks, it is a valuable source of power and provides excellent dam sites. The average run-off is 250 × 10^6 m^3.

For the power station at Kaprun there is a small auxilliary dam known as the *Bürg dam* in an ideal site in a gorge of calcareous mica-schist, dipping northwards, downstream.

DAMS OF THE KAPRUN SCHEME (RIVER SALZACH)

Years of Construction	Name of dam	Type	Max height m	Crest length m	Top Water O.D. (m)	Main formation at dam site	Locality see Figure 76
1946–7	Bürg	G	19	73	847	Calc-mica-schist	Near (17)
1940–51	Limberg	A	120	350	1,672	Calc-mica schist	(17)
1951–5	Mooser	R & G	104	462	2,036	Mica-schist	(18)
1952–5	Drossen	A	112	357	2,036	Mica-schist	(19)
1950–2	Möll	A	93	164	2,000	Calcareous mica-schist	(20)
1951–2	Margaritze	R & G	40	150	2,000	Calcareous mica-schist	(21)
1962–4	Diessbach	R	36	204	1,415	U. Trias Dolom	40 km N. of Kaprun

The basic formation of most of this area is thus granite core overlain by chlorite green schists, dolomites, calc-mica schists and phylites of Carboniferous Triassic age. Upon these strata, glaciers ploughed their way and left their rubbishy deposits leaving behind three excellent reservoir sites.

The Wasserfall Reservoir (Limberg dam)

This reservoir is formed by the *Limberg dam* which impounds the water of the Wasserfall reservoir (*Figures 82a and b*). The Limberg arch dam is on a natural rock-barrier in a gorge of an 'extinct' lake, the valley floor is covered with detritus but most of the sides are sound rock exposed and polished by the passage of the glacier which has truncated the sloping sides of the valley, as often seen for example in Cumberland. Calc-mica-schist is the predominant rock and an arch dam was adopted; the chord/height

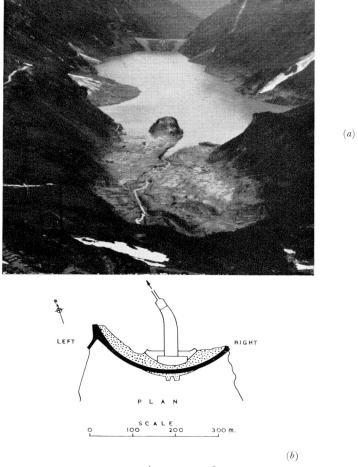

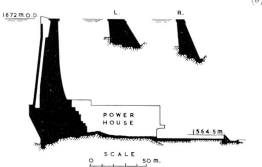

Figure 82. Kaprun: Limberg dam. (a) The Limberg arch dam in metamorphic crystalline rocks has a chord-height ratio of 2·7. The Wasserfall reservoir, capacity 86×10^6/m³ was nearly empty in June 1957. (b) Sections through arch dam on the metamorphic crystalline rocks of the Tauern Alps. The dam was completed in 1951

ratio at the site being 2·7. The left bank was found to have some graphitic schists in the upper part, and filled with moraine material which has been replaced by concrete.

There are several minor faults on the left bank. On the right bank, joints were uncovered parallel with the valley resulting in rock falls and thereby suggesting potential paths for leakage. A group of joints across the valley, which dipped steeply upstream, increased resistance to sliding and to permeability; apparently another instance where the bottom of a valley is more favourable than the valley sides.

Pendulum measurements indicated a deformation of the order of 20 mm downstream with a rise in water level of 100 m.

The Mooserboden Reservoir (The Mooser Dam and The Drossen Dam)

It is of great interest that in this glaciated region there are two reservoirs for which two dams for each reservoir are required to retain the water (*Figures 83 and 84*). In the case of the Mooserboden reservoir, capacity 87×10^6 m³, there are two gorges which necessitate two arch dams (Mooser and Drossen) of different types which are both over 100 m in height with a verrou of rock between them.

The *Mooser dam* is a gravity arch or curved dam for the chord/height ratio of about 4·3 is built on basically calc-schist with mica-schist, green schist, quartz-schist, dolomite and other rocks. Faults weaken the structure as a crush zone on the right bank was uplifted. The slender ridge of rock, which slopes upstream and downstream subjected the calc-schist to weathering and karstification which necessitated deep grouting down to 120 m.

The *Drossen dam* is an arch dam covered by 13 m of overburden and talus and the schistose rocks are more complicated and disturbed than those at the Mooser site and faults caused further complexity. However, any type of dam could have sufficed as the chord/height ratio is about 2·2, but the arch dam was the most economical. Analysis by the Trial Load method was confirmed by both model experiments at the Tauernkeftewerke Laboratory and by Professor Oberti at Bergamo. The soft green schists, however, under the Drossen dam had to be excavated and replaced by a plug of concrete.

The Möll-Margaritze Reservoir (The Möll Dam and Margaritze Dam)

The main foundation for the *Möll dam* (93 m in height) is calc-mica schist, highly polished, with some faulting and crushed schists which had to be removed and which necessitated consolidation grouting. With a chord/height ratio of 1·5, an unsymetrical arch was adopted and, as in the case of the other dams, the concrete was vibrated both at Möll and at Margaritze.

Figure 83. Kaprun: Mooserboden reservoir. (a) The Mooser gravity arch dam upstream face in foreground and the Drossen arch dam beyond, the crystalline rock verrou between. (b) The Mooser gravity arch dam downstream face in foreground and the Drossen arch dam upstream curved face beyond, the tunnel mouth in the verrou can be discerned. (c) Mouth of the tunnel in the metamorphic crystalline rocks at the left end of the Drossen dam leading through the verrou to the Mooser dam

(Photographs by courtesy of P. B. Mitchell)

The *Margaritze dam* is a gravity/arch or curved dam (40 m in height) with a chord/height ratio of 4 (with a 400 m 'verrou' between it and the Möll dam) which consists of a flat polished furrow in the solid calc-mica-schist on the right bank. No leakage was observed, presumably owing to the muddy bottom of the reservoir (*Figures 85a and b*).

Some of the water in the gathering ground of this reservoir is led in tunnel for 11·6 km through the calc-schist and pumped into the Mooserboden reservoir and main generating plant.

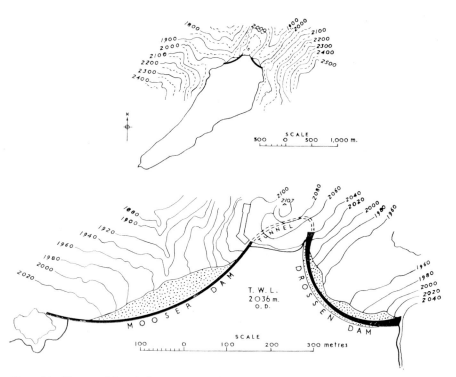

Figure 84. Kaprun: Mooserboden reservoir. For this reservoir the water is impounded by two dams on the metamorphic crystalline rocks considerably interfered with by glacial action; the top water level is 2,036 m above the sea. There are two valleys to be impounded; on the left, the Mooser dam is a curved gravity type in a wide valley having a chord/height ratio of about 4, while on the right the Drossen dam is an arch in a narrow valley with a chord/height ratio of about 2. A tunnel has been cut between the dams for access purposes under the verrou, 2,107m O.D., a common feature in glaciated regions

Diessbach Dam

This dam is situated on a sill of Upper Trias limestone with intercalations of Lias limestone, permeable to a depth of 25 m, below this impermeable.

The embankment is of local Dacksteinkalk fill; grain size 5 cm to 60 cm compacted with an 8-ton vibrating roller, downstream slope 1 to 1·54 and upstream slope 1 to 1·7 H with asphalte concrete layers 14 cm in all. Grout curtain at upstream toe.

*Figure 85. Kaprun: Möll reservoir.
(a) The Möll dam (height 93 m) and Möll reservoir (2,000 m O.D.) looking across the Posterzen glacier on top right to Gross Glockner mountain (3,797 m O.D.). The water melting from this glacier not only increases the yield of the system but also the yield in summer. (b) Like that of Mooserboden, this reservoir in the same area but not in the same valley has to be formed by two dams side by side with a verrou between them. These are a gravity dam, the Margaritze in a wide valley with a chord/height ratio of about 4·0, and the Möll dam, a thin arch, in a narrow valley with a chord/height ratio of under 1·5). In (a) the Margaritze dam is to the left of Möll. Its position is shown in (b)*
(Photograph by courtesy of G. C. S. Oliver)

(a)

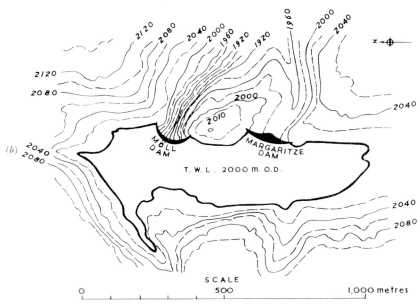

(b)

DAMS IN EASTERN AUSTRIA

Years of construction	Name of dam	Type	Max. height (m)	Crest length (m)	Top water level OD (m)	Main formation at dam site	River	Locality on map (see Figure 76)
SOUTH-EAST (KLAGENFURT AREA)								
1956–7	Rotgüldensee	R	18	112·0	1,710	Gneiss		⎫
1954–7	Grosser Mühldorfersee	G	46·5	432·7	2,319	Gneiss		⎬ Developments 80 km N.W. of Klagenfurt
1954–7	Kleiner Mühldorfersee	G	41	158·5	2,379	Augen-Gneiss		
1954–7	Hochalmsee {	G	24·5	236·6	2,379	Gneiss		
		R	8·8	120·5				⎭
1954–7	Radlsee	R	16·7	211·7	2,399	Gneiss		
1954–7	Freibach	E	41	150·0	729·2	Trias-Jurassic Limestone		25 km S.E. of Klagenfurt
NORTH-EAST (ENNS-DANUBE AREA)								
1941–50	Ternberg	G	28	⎫		Trias Sandstone and Marl	Enns	14
1942–50	Grossraming	G	37	⎬ These dams form part of the Salza power supply		,, ,,	Enns	15
1941–51	Staning	G	20			Miocene	Enns	12
1941–52	Muhlrading	G	14			,,	Enns	11
1950–54	Rosenau	G	21	⎭		Cretaceous	Enns	13
1947–49	Salza	A	52	121	771	U. Trias Dolomite	Enns	80 km S.E. of Salzburg
1948–50	Ranna	A	45	126	493	Gneiss Quartzite ⎫	North	5 km E. of (9)
1950–52	Dobra	G	52	220	437	Gneiss ⎬	of	55 km N.N.E. of (10)
1953–57	Ottenstein	A	65	240	495	Granite ⎭	Danube	5 km N. of (10)

Rotgüldensee Dam (Figure 86)

This is built on a compact gneiss ridge with few joints and little seepage and consists of a small earthen embankment, rather than a concrete wall, because of the abundance of rock-slide debris and moraine material and the availability of an 8-ton vibrating roller.

As clay is scarce and the dam will have to be raised at a later date from 1,713 to 1,740 m O.D. some 27 m, a sloping core 1·5 m in width with a

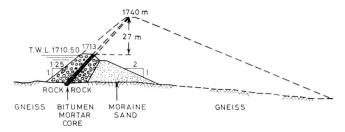

Figure 86. Rotgüldensee Dam. Owing to the abundance of banking material and to the absence of clay, a rockfill type (18 m in height) was adopted with a view to possible raising by 27 m in the future. This is shown diagrammatically only to indicate the ambitious nature of the project. The foundation rock is sound gneiss

slope of 1 to 1 of bituminous (B80) mortar stone-dust, sand and crushed gravel up to 30 mm, placed hot was adopted (with a suitable join to the spillway with asphalte and copper sheeting).

The embankment slopes are upstream 1 to 1·25 H with quarry-run rock and 1 to 2 H downstream with coarse waste and moraine sand.

Grosser and Kleiner Mühldorfersee Dams, Hochalmsee and Radlsee Dams

These dams are all near each other at the high elevation of about 2,000 O.D. Although the gneiss for all of them is sound enough for the support of the dams, in some of the foundations incisions, faults, disintegrations and permeability troubles were encountered and each of these demerits was suitably dealt with. Circumstances also dictated the types of dams adopted. For the *Müldorfersee dams*, 10 per cent of concrete was saved by making hollow gravity inspection galleries, and vertical faces to reduce ice pressure. The *Hochalmsee dam* is partly a concrete gravity dam and partly rockfill. The *Radlsee dam* is a rockfill dam with a concrete core, designed to suit foundation conditions. All the dams were constructed between 1954 and 1958.

Freibach Dam

This earthen embankment on Triassic-Liassic Limestones and shales has a wide central rolled core and the upstream slope is flatter than the downstream slope and the slopes in both are variable (*Figure 87a*). The thick core of clay material passes a 30 mm sieve and is mixed with 2 per cent

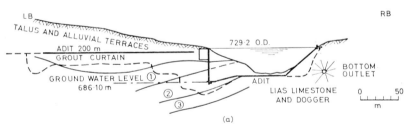

Figure 87 (a). Freibach Dam. The shaley Lias Limestone formation and the soft overburden material determined the choice of an earthen embankment, 41 m in height. There are several interesting features of this earth dam. The access gallery down the right slope of the valley (shown in black) can be also utilized for grouting the Lias Limestones and Dogger area. The gallery through the core under the centre of the dam (shown in black) from which grouting or other treatment of the underlying soft strata can be controlled. A vertical shaft (in black), concrete apron and bentonite grouting was necessary between the spillway and submerged left bank. An adit 200 m in length (in black) for grouting the talus and gravel under the left slope of the valley was found to be necessary, after the reservoir had been filled, with provision for extension. The leakage water (at 686·1 m O.D.) is pumped back into the reservoir. The galleries, shaft and adits are shown by thick straight lines

1. *Varved silt.*
2. *'Rot Shutt'.*
3. *Liassic sediments*

Yugoslav bentonite. The upstream slope varies from one to 1·75, 2·25, 2·75, 3·0 H and the downstream slope from one to 1·6 and 2·0 H, at the toe like some of the old English dams resembling a slope of $\log_e x^n$. Next to the core is the modern equivalent of old English 'selected material' i.e. clay filter and various zones of sand and fine gravel upstream and downstream; the highest grain size being 100 mm.

On the left bank (*Figure 87b*) rockfill and gravel overburden were grouted and a long tunnel of 200 m is constructed for grouting the Jurassic limestones to a depth of 30 m; this was found to be necessary after the reservoir had

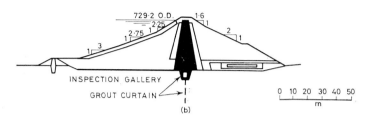

Figure 87 (b). Freibach Dam. The impervious core and horizontal blanket under the upstream embankment are shown in black. The embanking or weighting material consists of unscreened talus. Various grades of gravel and talus act as filters under the upstream slope and horizontal base downstream. The upstream slope is stone-pitched and the downstream slope is clay with grass. The inspection gallery under the clay core in Figure 87 (a) is shown in the section and a grout curtain is provided for. Some compensation for this difficult site is the proximity of most of the material for the embankment and core.

been filled. The steeper right side of the valley was grouted from a tunnel in the Dogger and black Lias Limestone. Seepage water is pumped back into the reservoir. A feature of the dam is the tunnel in the base of the core in an earth embankment.

Salza Dam

On Upper Trias Reef Limestone is slightly dolomitic near the Salza falls. The narrower part of the gorge was originally suggested but it was so disrupted that the final site was moved upstream where the left bank is very strong and the right bank somewhat weaker.

The grout curtain is taken to 20–25 m below base of dam at $1\frac{1}{2}$ to twice the hydrostatic pressure.

Dobra Dam

This is built on ortho-gneisses with amphibolite and mica-schist with thin talus covering. The crushed gneiss on the right bank although sound at depth was even worse than expected from a trial boring; highly crushed and broken gneiss, rich in schist and loam, filled fissures which could not be made impermeable nor consolidated. Excavation was therefore carried down to solid rock and the right embankment end thickened to withstand the arch thrust. Prestressing was used which pressed the abutment 4 mm into the bank.

Ottenstein Dam

An arch dam with a chord/height ratio of about 3, built on porphyric 'Rastenburg' granite; an aplite vein on the left bank and an old fault on the right bank were proved to be inocuous. The left bank however was found to be undercut by a former bend of the River Kamp, this necessitated excavation and the filling of the hole with 3,000 m³ of concrete.

(The foregoing accounts of dam sites show that Austria has one thing in common with other countries, namely that 'each site must be treated on its own merits'—Herbert Lapworth).

Danube Development (Austria–Germany)

There are, so far, three barrages on the Austrian Danube, they are: (1) Ybbs-Persenbeug between Linz and Vienna wholly in Austria; (2) Braunau below the junction of the rivers Salzach and Inn between Salzburg and Passau, jointly developed between Austria and Western Germany; and (3) Jochenstein between Passau and Linz, also jointly developed between Austria and Western Germany.

(a)

(b)

Figure 88. Danube: Jochenstein barrage. (a) Downstream view of German side on left bank of Danube of the new barrage founded on 'mixed' granite. Owing to the rapid flow there was no sediment on the river bed. The control weir is in the foreground, the power-house in the middle and there are large locks for shipping on the left bank. (b) Downstream view of Austrian side on right bank of Danube, taken from locks, showing power-house and six control weirs.

Some details of the Jochenstein barrage may be of interest as it is built on exceptionally good 'mixed' granite which is quite near the bed of the river, there being very little overlying alluvial material.

(c)

(d)

Figure 88. Danube: Jochenstein barrage (c) Long training embankment protecting the German village of Obernzell as the Danube was raised about 20 m. (d) Upstream view of barrage with operational crane and barges in foreground

Figure 89. Danube: Barrage at Kachlet above Passau. This barrage is wholly in Western Germany and is older (compare with Figure 88b)

The Jochenstein barrage, begun in 1952, was completed in 1959; it is situated a mile upstream of the Austrian village of Englehartszell opposite the (West) German village of Jochenstein.

The gathering ground here is 77,025 km^2 and it gives an annual average flow of 45,000 × 10^6 m^3 or 1,430 m^3 per second; thus, between 1901 and 1949 the flows were: mean flow, 1,430 m^3 per second; minimum flow, 370 m^3 per second; maximum flow, 4,200 m^3 per second.

The barrage, the maximum height of which is 32 m, consists of gates on the right (Austrian) side of the river, turbines working under an average head of 10 m in the centre of the river, and on the left (German) side locks in duplicate 24 m in width and 230 m in length (*Figure 88a, b and d*).

As the river level has been raised by about 20 m, the German village of Obernzell has been protected by a long embankment (*Figure 88c*).

Upstream, a little to the north-west of Passau, it is interesting to compare the older barrage at Kachlet (*Figure 89*), built some years ago on the same formation of mixed granite and with locks of approximately the same dimensions as those at Jochenstein, together with a power-house in the centre and gates on the right bank—a much more imposing structure than the larger duty barrage downstream.

BIBLIOGRAPHY

General
Large Dams in Austria. Austrian National Committee of the I.C.O.L.D. Österreichische Wasserwirtshaftsverband, Wien. Published in English and presented to the 8th Congress *I.C.O.L.D.*, Edinburgh (1964).
CLAR, MUELLER, PACHER, PETZNY and STEINBOECK. On the practice of foundation rock investigation in Austria. *I.C.O.L.D.*, R.50, **I**, 927 (1964).
GRENGG, HERMANN. Observations faites sur les barrages l'Autriche. *I.C.O.L.D.*, R.44, **II**, 811 (1964).
Durlassboden Dam
KROPATSCHEK, H. and RIENÖSSL, K. Travaux l'étanchment du sous-sol du Barrage de Durlassboden, *I.C.O.L.D.*, R.42, **I**, 695 (1967).
Gepatch Dam
LAUFFER, H. and SCHUBER, W. The Gepatch rockfill dam in the Kauner Valley. *I.C.O.L.D.*, R.4, **III**, 635 (1964).
Kaprun Dams (Limberg, Möll, Margaritze, Mooser, Drossen)
ASCHER, H. H., HORNINGER, G., and STINI, J. *Festschrift: Die Hauptstufe Glockner-Kaprun Tauernkraftwerke A.G.* (1951).
——— ——— ——— *Festschrift: Die Oberstufe Glockner-Kaprun Tauernkraftwerke A.G.* (1955).
GÖTZ, J. *Das Tauernkraft. Glockner-Kaprun* (1956).
STEYR, A. G. *Ennskraftwerke* (1956).
TREMMEL, E. Analysis of pendulum measurements (Limberg). *I.C.O.L.D.* II 233 (1958).
WINKLER, E. M. in *Hydro-electric power system of Tauern Kaprun. Salzburg, Austria, Engineering Geology Case Histories*. Editor P. D. Trask, No. 2 (1958) Geological Society of America.
Spullersee Dams
RUTTNER, A. The heightening of dams by the use of prestressed anchors and the problems arising from this method of construction. *I.C.O.L.D.*, C.22, **V**, 639 (1967).
Vermunt Dam
STEFKO, E. and INNERHOFER, G. Condition of Vermunt dam after more than 30 years operation. *I.C.O.L.D.*, R.36 **III**, 645 (1967).
River Enns development
STEYR, A. G. *Ennskraftwerke* (1956).

17

BELGIAN DAMS*

There are four large dams in Belgium on the rivers Vesdre and Warche, within 30 miles east and south-east of Liége, namely, La Gileppe dam near Jalhay, Vesdre dam near Eupen, and the Warche dams at Butgenbach and Robertsville.

They are all in the Cambrian-Devonian part of the Ardennes and the two first named, built in 1875 and 1950 respectively, are interesting to compare because of their similarity and because both dams are today as good as new.

La Gileppe Dam

La Gileppe dam is 47 m in height with a crest-height ratio of about 5, masonry, of gravity section, near the Cambrian-Devonian junction which is almost vertical when viewed at the ends of the dam (*Figure 90a*). It is adorned by a magnificent stone lion weighing 300 tons, 44 ft. high on a pedestal 26 ft. high (*Figure 90b*), and there are two overflow channels at each end of the dam which serve for the purpose of official car parks (*Figure 90c*).

The siting (in 1870) appears to have been admirable, and M. J. de Clercq says that the geological formation is as follows:

Eifelian (Middle Devonian) to 550 m in thickness	Quartz schists composed of conglomerate of grit and psammites, schists, red earthy and Arkose.
Gedinnian (Lower Devonian) to 2,200 m in thickness	Conglomerate with grit and psammites in quartz phyllades and schists.
Revinian and Salmian (Cambrian) to 8,600 m in thickness	Phyllades quartzites, quartz-phyllades.

The Devonian are downstream, the Cambrian are upstream, and the dam is sited on the grit and psammites with schists. As the axis of the dam is parallel to the bedding of the strata, the same rocks are continued across the valley which guarantees an adequate foundation, the schists providing the necessary watertightness.

*Through Monsieur A. Achten I am grateful for most of the foregoing information on Belgian dams to Monsieur J. de Clercq, l'ingénieur en Chef Directeur des ponts et Chaussées Ministère des Travaux publics et de la Reconstruction.

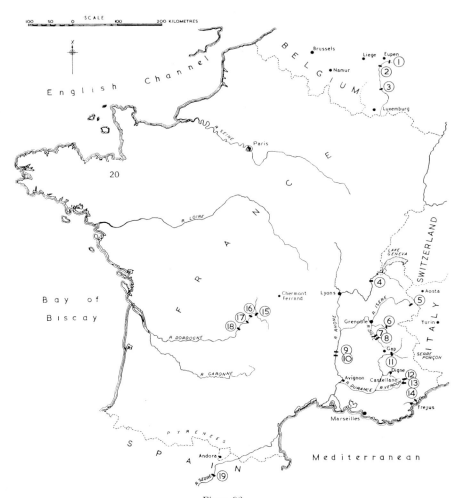

Figure 92

(1) *Vesdre.*
(2) *La Gileppe.*
(3) *Esch-sur-Sure.*
(4) *Génissiat.*
(5) *Tignes.*
(6) *Chambon.*
(7) *St. Pierre-Cognet.*
(8) *Sautet.*
(9) *Montélimar.*
(10) *Donzère-Mondragon.*
(11) *Serre Ponçon.*
(12) *Castillon.*
(13) *Chaudanne.*
(14) *Malpasset (Fréjus).*
(15) *Bort.*
(16) *Marèges.*
(17) *Aigle.*
(18) *Chastang.*
(19) *Oliana.*
(20) *Rance Estuary.*

Four exploratory bore holes only, parallel with the thalweg upstream to downstream, were drilled and proved the top of the gneiss to be 926 m, 935 m, 937 m, and 929 m O.D. respectively (*Figure 93*).

No bore holes were drilled across the valley to ascertain the level for founding the heel of the dam, as it was assumed that the top of the gneiss would not be below 926 m O.D., covered by 25–30 m of alluvium and drift.

When large portions of the dam at each end were constructed, it was found that the gneiss rock was much deeper in the middle of the valley

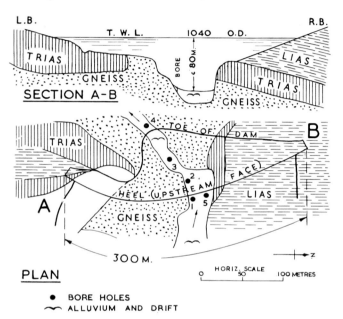

Figure 93. Chambon dam. Only four bore holes (*Nos. 1, 2, 3 and 4*) were drilled to ascertain the level of the top of the gneiss covered by drift. At point No. 5, 20 m from No. 1, the top of the gneiss was found to be 11 m deeper and generally the level of the top of the gneiss varied considerably (see Figure 94)

than anticipated, in fact it was almost vertical where the upstream face or heel of the dam was to have been (*Figure 94*). Thus, at 20 m distant from where the top of the gneiss was 926 m O.D., it was found to be 915 m O.D., that is, 11 m lower.

As the glacial drift and alluvium were water-bearing, water inundated this deep excavation and the situation became *angoissante*! The 5 per cent batter on the face had to be altered and a reinforced concrete plinth had to be made on the gneiss to support the newly designed upstream face. In addition, the position of the downstream face or toe had to be altered.

The parts of the right and left sides of the dam which had already been constructed had to be joined up, by changes of slope, to the new upstream batter of the central part. This patching is visible to the trained eye when the reservoir is empty (*Figure 95*).

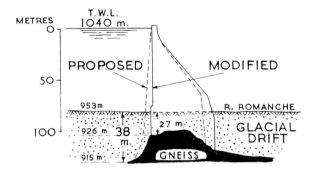

Figure 94. Chambon dam. When the two ends of the dam had been constructed it was found that part of the upstream toe would have to be some 38 m below the surface under water-bearing glacial drift. The dam had to be altered during construction; the sloping upstream face was built with a vertical face supported on a reinforced concrete plinth to reach the rock 11 m nearer the surface

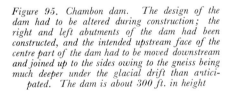

Figure 95. Chambon dam. The design of the dam had to be altered during construction; the right and left abutments of the dam had been constructed, and the intended upstream face of the centre part of the dam had to be moved downstream and joined up to the sides owing to the gneiss being much deeper under the glacial drift than anticipated. The dam is about 300 ft. in height

A delay of one year ensued and the cost was greatly increased, all for want of a few extra bore holes before construction had begun.

Grouting

The gneiss was considered to be an excellent foundation and some experience was gained in constructing the overflow tunnel through it on the right bank; it was reported to be very sound and stood up with little support.

Grouting pressures under the central part of the dam were between 40 and 50 kg per m² in the gneiss (*Figure 96*).

The overlying Triassic massive limestone with large fissures, in which the top half of the left bank end of the dam is founded, absorbed 315 kg per lineal metre, the penetration of the grout not being increased by pressure.

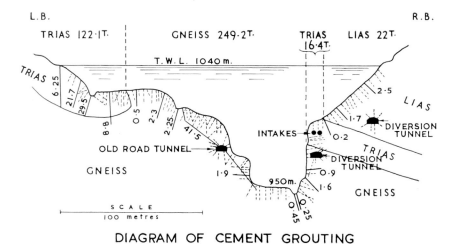

DIAGRAM OF CEMENT GROUTING

Figure 96. Chambon dam. The total amount of grout used was 410 tons. The total length of bore holes was 2,430 m. The Trias limestones absorbed more cement than the Lias schists, namely, about 315 kg per m in the Trias and 69 kg per m in the Lias. Typical quantities of cement in tons in individual bore holes are shown

The overlying crushed clay calc schists of the Lias, on the right bank, absorbed about 69 kg. per lineal metre.

Total length of bore holes	. 2,430 m
Total amount of cement used	. 410 tons
Amount of cement used per metre, lineal (A.M.L.) . . .	. 169 kg per m

Tignes Dam (Isère)

The Tignes dam impounds the headwaters of the Isère, 100 miles east of Lyons. It is a few miles west of the Italian border near the Petit St. Bernard Pass.

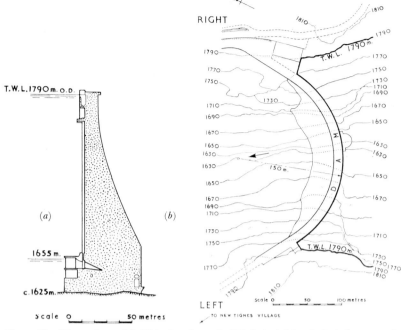

Figure 97. Tignes dam. (a) This dam is nearly 550 ft. in height. It is built on massive Triassic quartzite in a gorge of which the width at top water level is twice the depth. The width at the base of the curved dam is only about 140 ft., whereas, the equivalent gravity section would be at least 370 ft. (b) The dam is in a gorge for which the chord/height ratio is 2. The radius of the thin arch type, suitable for this ratio, is 150 m. Above top water level the contours are flatter on the left side of the valley than on the right which is much steeper (see Figures 98 and 99)

Figure 98. Tignes dam. At the right end of the dam there is a large paved open space for cars and the quartzite clearly exposed, dipping at 45 degrees downstream. The pipe down the mountain on the top right-hand corner is a catchwater

Figure 99. Tignes dam. (a) The right end of the dam abutment is on hard impervious Lower Triassic quartzite, dipping at 45 degrees downstream. (b) The left abutment of dam (height about 500 ft.) with exposed boss of hard quartzite above dam, behind which is a partly buried valley. (c) The left abutment above which is a rock verrou and beyond, the new village of Tignes on lower ground (the people and the car on the dam give an idea of the scale)

The reservoir was constructed between 1947 and 1952 and gained notoriety in the English press because the 350 inhabitants of the village of Tignes were most reluctant to leave and live in the new village built for them on the left bank above top water level downstream.

The dam, a curved thin arch type, is not only the highest one in France, but it is also situated at the very high altitude of 1,790 m (5,875 ft.) above sea level.

The chief dimensions of the Tignes dam (*Figure 97a and b*) are as follows:

Height	165 m
Radius of arch	150 m
Base width of arch	43 m
Crest width (carrying a roadway 26 ft.)	10 m
Crest length	400 m
Volume of dam	630,000 m³
Chord-height ratio	2

The reservoir commands rather a small gathering ground of 66 sq. miles or, with diversions, 100 sq. miles from two streams on the right bank (comprising 4½ miles of tunnel).

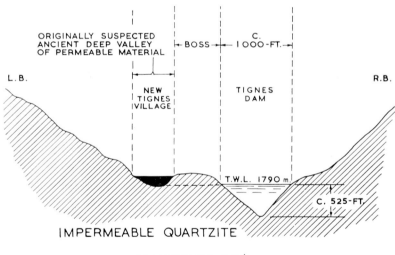

Figure 100. Tignes dam. An ancient valley of unknown depth filled with permeable material had to be investigated before the Tignes dam could be built. Fortunately, the valley proved not to be very deep, but its base determined the top water level of the reservoir (1,790 m)

The capacity of the reservoir, which is useful for turbine generation, is 230×10^6 m³. The area of the water impounded is 270 hectares.

The foundation for the dam is in Lower Triassic quartzite which is very hard and compact; the general dip being 30–45 degrees downstream (*Figure 98*).

The Middle and Upper Triassic limestones, which supplied aggregate for the concrete of the dam, outcrop about a mile upstream.

The chief geological interest, however, is the gorge where the dam is situated. The right-hand side of the gorge is in the massive quartzites near the high and steep north-eastern side of the valley (*Figure 99a*). Whereas the left-hand side of the gorge to the south-west consists of: (*a*) an exposed boss of hard quartzite some 100 yd. in extent which forms a very suitable abutment (verrou) (*Figure 99b and c*); and (*b*) further to the south towards the left-hand cliff of the valley, a large deposit of Drift, the surface of which forms a depression (encoche) some hundreds of yards across.

Thus, the disquieting question presented itself. Was there another gorge, buried under this Drift, filled with material which would give rise to leakage and so render the dam and reservoir useless?

Fortunately, the bottom of this depression of Drift proved that the quartzite rock was not very much below the surface, although its level ultimately governed the top water level in the reservoir which was finally adopted (*Figure 100*).

On filling the dam in 1953, it was reported that at a level of 1,783 m O.D., 7 m below top water level some 58 litre per sec. (1 million gal. per day) leakage occurred on the right bank through the quartzite.

Sautet Dam (Isère)

The dam is constructed in a narrow gorge (*Figure 101a*) in Lower Lias limestone. It is situated on the River Drac, which joins the River Isère near Grenoble, and has the following characteristics:

Type	Gravity arch
Top water level	765 m O.D.
Crest	766 m O.D.
Maximum height	126 m
Width of gorge at crest (766 m O.D.)	65 m
Width of gorge at 641 m O.D.	8 m

The gorge has been cut by a glacier forming a new course for the river, known as an epigenic valley.

The strata, of evenly bedded Lower Lias, consist of alternating beds of hard limestone, sometimes of crystalline and soft argillaceous clays. The bedding is generally horizontal, but there are minor folds, numerous joints and small faults (measurable in terms of inches). Above the Lower and Middle L as appear Upper Lias clays or schists (*Figure 101b*).

During the construction of the dam, troubles occurred with the foundations for the power-house immediately downstream of the dam, through large blocks of material, chiefly on the left-hand side of the valley, flaking off.

The horizontal bedding of water-bearing limestones with the impervious shales coupled with a network of vertical joints and minor faults and rolls, which were cut into by the excavation for the power-house foundations on the left bank, caused large wedges or blocks of strata several dozen metres in extent to become detached.

Hence the site of the power-house, originally intended to be placed on the left bank, had to be altered and put partly under the right bank with very heavy protective works to keep the strata of the gorge in place (*Figure 102a*).

These protective works are shown in *Figure 102b*, taken from the bridge looking vertically downwards showing the struts and, looking upwards, from the bottom of the valley in *Figure 102c*.

(a)

(b)

Figure 101. Sautet dam. (a) View from bridge showing the dam in a gorge of Lower Lias limestone, bedded horizontally. The intake building is on the top right-hand corner. The triangular wedge on the crest of the dam divides the overflow. (b) Gorge below Sautet dam showing unstable Lias limestone. Middle and Upper Lias limestone is above the 'Υ'. Height of section is about 300 ft. above stream

This shows, as Lugeon pointed out, that it is imprudent to cut into a cliff for the whole distance between natural faults or joints of the virgin rock.

A good deal of grout was consumed in the foundations of the dam, namely, 3,000 tons in 6,000 m of bore hole, or 500 kg per lineal metre. The pressures ranged from 5 to 6 kg per cm^2 to 40 kg per cm^2. The bore holes for injecting the grout were made from four headings as well as from the diversion tunnel.

The system of grout holes about the gorge at the dam site is shown in *Figure 103*. This is of exceptional interest as an example (and an early one too) of grouting Lias limestone (1925–35).

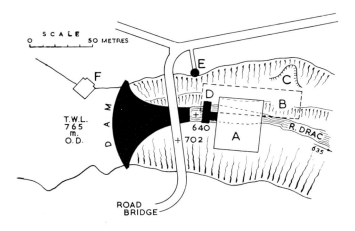

A FINAL POSITION OF POWER HOUSE
B ORIGINAL INTENTION
C SLIPPED PART ABOVE INTENDED POWER HOUSE
D BUTTRESSES TO PROTECT NEW BUILDING
E LIFT WELL ACCESS TO POWER HOUSE
F INTAKE TO TURBINES

(a)

(b)

Figure 102. Sautet dam. (a) The dam (and plug) is shown in black and the slipped limestone (during construction) made it necessary to alter the position of the power-house from the left bank to the right bank (from B to A). Large concrete buttresses were also constructed shoring up the sides of the valley at D. (b) Plan of the power-house taken vertically from the bridge, 300 ft. above, showing (1) protective R.C. beams across valley, (2) roof of power-house below on right bank, (3) the building high up on the right bank is a disused overflow. (Cont.)

Figure 102 cont. (c) To support the weak Lias limestone, 300 ft. below the surface, and to protect the roof of the power-house, exceptionally massive concrete arches and beams were constructed across the gorge. The limestone is seen at the top right-hand corner supported by a concrete strutt (B). Above, is a concrete arch (A) the underside of which is shown black. Below is a massive triangular strutt (C) going behind (B). The letter (D) on Figure 102a corresponds with the right-hand top corner of this view

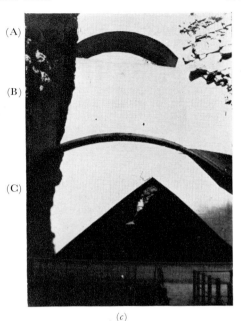

(c)

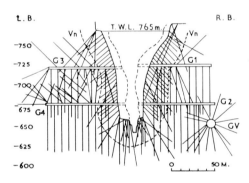

Figure 103. Sautet dam. Grouting system in Lias limestone. Owing to the high cliffs the bore holes were drilled from adits (G1–G3 and G2–G4) and diversion tunnel works (GV). Vn (shown dotted) is the limit of a particularly dense screen of grouting

(Reproduced by courtesy of Masson et Cie)

Sautet Reservoir (Isère)

The Sautet reservoir, 48 km south-east of Grenoble, on the River Drac, is of extraordinary interest inasmuch as the leakage from it was originally estimated at 2,800 litre/sec., subsequently revised to 2,500 litre/sec. This leakage appeared several days after the reservoir was filled in 1935. In 1957 it was 1,900 litre/sec.

The main dimensions of the reservoir are as follows:

Capacity	130×10^6 m^3
Top water level	765 m O.D.
Area of the water surface	350 ha.
Maximum depth	126 m
Area of gathering ground	990 km^2

The reservoir is one of the first, if not the first, to be recorded as suffering from severe leakage due to the water submerging an old river valley filled with permeable debris left by a glacier.

The old buried river course of the Drac is filled with alluvial stones at the base and clayey morainic material at the top; it was traced by P. Lory, as branching northwards away from the present valley between ½–2 km upstream of the dam leading into the present valley of the River Sézia (*Figure 104*).

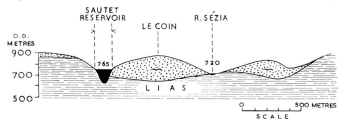

Figure 104. Sautet reservoir. As the top water level is 765 m, and the hill under Le Coin is very permeable, leakage of 50 million gal. per day occurred into the River Sézia at 720 m O.D.

Hence, leakage flows through this very permeable material into the River Sézia about a kilometre to the north of the present Drac valley submerged by the Sautet reservoir.

The water appeared in the Sézia at a level of about 720 m O.D. As the top water level of the reservoir is 765 m O.D. there would be a pressure head up to about 45 m.

The leakage water is clear so there is no anxiety that erosion is excessive or that it will endanger Le Coin Hill under which the leakage flows. As the reservoir is used for electric power no vital harm is done by the loss of water, particularly as the Sézia flows into the Drac lower down and the water is regained for power (of course at some loss of head) (*Figure 105*). Nevertheless, attempts to stop the leakage, either by grouting, injecting cement into the grit and gravel-filling of the ancient valley, or by plastering the side of the reservoir with clay have been considered. The latter, however, would be washed back into the reservoir when the water level was lowered (*Figure 106a and b*).

To inject clay into the filling of the old valley was impracticable owing to its great width and depth, and any form of watertight screen would be very costly and not justifiable.

Some threat of erosion occurred in the Sézia valley where the leakage appeared necessitating some protection works (*Figure 106c*).

Natural silting has evidently taken place because after 22 years (1935–57) the leakage has fallen from 2.500 to 1,900 litre/sec.

St. Pierre-Cognet Reservoir (Isère)

The St. Pierre-Cognet reservoir, completed in 1957, is some 8 km downstream of the Sautet reservoir on the River Drac which joins the Isère near Grenoble some 48 km to the north-west (*Figure 105*).

The chief characteristics of the reservoir are as follows:

Capacity . . . . 29×10^6 m³
Top water level (as built) . . . . 580 m O.D.
Maximum depth . . . . 70 m
Area of gathering ground = 1,118 + 362 (catchwater)
　　　　　　　　　　 = 1,480 km²

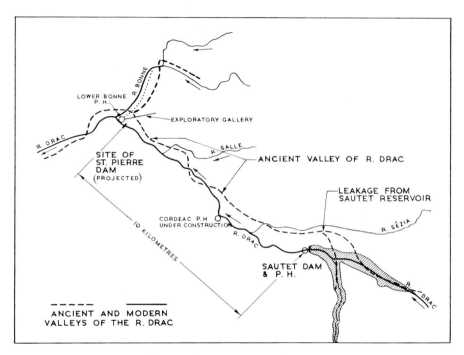

Figure 105. *Sautet reservoir. There was a loss of some 50 million gal. per day into the adjoining valley of the Sézia, owing to the existence of an old pre-glacial valley (shown by broken line) filled with permeable material*

The great interest of the St. Pierre-Cognet reservoir is that, for many years, its construction was considered to be impracticable due to the experience of the Sautet reservoir constructed in 1935, from which the leakage was proved to be over 50 million gal. per day.

The Sautet leakage was through one or more pre-glacial period valleys of the Drac (the existence of which were unsuspected at the time) as they were filled with morainic material hundreds of feet in width and depth (*see* Sautet reservoir on page 177).

Like that of the Sautet dam, the site of the St. Pierre dam is in an epigenic gorge, a later course of the river, cut by a glacier, through the Upper Lias limestone (Bajocian). These beds dip at approximately 45 degrees upstream and are well bedded and little disturbed.

The thalweg or axis, and therefore the deepest part of the *old* valley, meandered a great deal, but the new valley, superimposed upon the old by

(a)

Figure 106. Sautet reservoir. (a) Part of dam (lower left). Control building over intake for turbines (lower right). Above roof of control building, indentation shows pre-glacial valley. (b) Part of glacial valley showing coarse permeable conglomerate which extends below water level. The path in the foreground is a little above top water level. (c) The reservoir in the distance when looking over filled-up valley of the old River Drac, Le Coin Hill to the right and Sézia valley between the fence and the hill in the middle distance. The slopes of the valleys are from left to right

(b)

(c)

180

the glacier of Würmian age, is straight and cuts more than once through the axis between the Sautet reservoir and the site of the new St. Pierre-Cognet dam.

Hence, as water in the St. Pierre-Cognet reservoir was to be over 75 m in depth and would submerge the old moraine-filled valley, similar to that submerged in the Sautet reservoir, the practicability of the St. Pierre-Cognet reservoir was very much in doubt, particularly when it was ascertained that the old valley filled with permeable material led into another valley, the Bonne.

The exact nature of the material filling the old valley and its extent were unknown, but it is known that such deposits left by glaciers are heterogeneous and that their nature is unpredictable. However, as the site would be such a useful one for a dam, it was decided to drive an exploratory heading, horizontally, at a level of 563 m O.D., 17 m below the top water level of the intended St. Pierre-Cognet dam (*Figure 107*).

After penetrating 236 m of solid rock, the heading entered, as predicted, into the ancient bed filled with glacial material ('alluvium'), which was sufficiently cohesive to stand up well enough during tunnelling with a few wooden frames.

Unfortunately, the ancient valley was not narrow and the heading was driven for 416 m through the alluvium before it met the other (rocky) side. Four bore holes determined the shape of the ancient valley, the lowest point of which was 535 O.D., or 44 m below top water level of the projected dam 580 O.D.

The glacial material met with was far from homogenous; it consisted of patches of clay, alternating with very permeable material (unconsolidated pebbles). These different materials were not in any specific order and varied from foot to foot. From this information the dimensions of the old valley suggested that the construction of a cut-off trench in it would be extremely costly for the maximum depth would be 45 m and the length at top water level 506 m.

Following the conclusion that the construction of the proposed dam at St. Pierre would be much too expensive, it was suggested that an attempt should be made to prove the permeability of the old valley by introducing water into the exploratory heading, at a pressure corresponding approximately to that which would be produced when the proposed reservoir was full. The trial would be very searching; the water would exert full pressure on the right-hand end of the projected dam, and should escape into the Bonne by the underground route already approximately forecast. If the water level were made equal to that of the top water level of the projected St. Pierre-Cognet reservoir, an even greater pressure would be created than that which would be produced by the effective water level in the reservoir.

Cutting a very long story short, a full size experiment was carried out by leading water from an intake at a high level of the River Bonne into the adit, which was not much below top water level, and the open bore holes. This was done for 40 days and the rate of leakage was ascertained at some

I am especially indebted to M. Ailleret for all the interesting data contained here on the Sautet and St. Pierre-Cognet dams.

4½ million gal. per day (240 litre/sec.). This was also confirmed by model experiments with adjustment for top water level.

As this leakage would not affect the safety of the dam and would not be material for hydro-electric power, the dam was proceeded with.

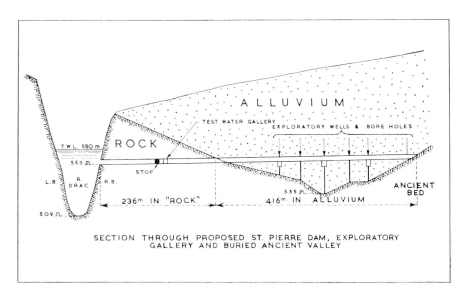

Figure 107. St. Pierre-Cognet dam. It was at first thought to be impossible to build the St. Pierre-Cognet dam because with a water pressure of 45 m considerable leakage might be anticipated through a very large pre-glacial valley filled with permeable material. Before construction, a water test was made by pumping water through an adit and bore holes into the buried valley and as the quantity absorbed by the alluvium was under 4½ million gal. per day (not considered to be excessive for this particular case) the dam was built

The dam is shown in *Figure 108a* with the straight epigenic valley cutting across the old meandering pre-glacial valley of the Drac.

Figure 108b shows, on the left, the indentation which marks the course of the old pre-glacial valley, and *Figure 108c* shows the permeable glacial material filling the old valley; the Lias limestone on the left dipping at 45 degrees upstream and the permeable material on the right (showing whiter in the illustration) with trees.

The reservoir was completed in 1957 and M. Ailleret wrote (23.12.57): 'the loss has not been stabilized as the level in the reservoir has varied, but nevertheless when the reservoir is full the loss seems to be somewhere between 180 and 200 litre/sec.' Thus the leakage of 240 litre/sec., which was forecast, has been very well confirmed.

St. Pierre-Cognet Dam (Isère)

The St. Pierre-Cognet dam, a thin arch concrete type, is about 40 km south-east of Grenoble and was constructed in 1955–57.

Figure 108. St. Pierre-Cognet dam. (a) The dam is 75m in height in an upper Lias limestone gorge in a straight valley cut by a glacier. (b) The bend on the left indicates part of the older pre-glacial valley. (c) On the left, is a section of the side of a pre-glacial valley which has a gentle slope down to the water in the reservoir. The strata of the old valley are dipping upstream at 45 degrees. Above the old valley slope is a cliff of glacial drift (white at top left-hand corner). The permeable drift extends under the water on the right. The surface slope of the old valley in this view appears to be about 22·5 degrees dipping into the water

(a)

(b)

(c)

The main dimensions of the dam are as follows:

Crest	585 O.D.
Top water level	580 O.D.
Maximum height	75 m
Chord	100 m
Chord-height ratio	1·4
Volume of concrete	42,000 m³
Radius of arch	65 m
Thickness of arch (top)	3 m
Thickness of arch (bottom)	12 m

Figure 109. St. Pierre-Cognet dam. The dam under construction, showing Lias limestone cliff on right bank

The dam is placed in an epigenic (newly made) gorge in the Upper Lias limestone (Bajocian age) having a general dip of 45 degrees upstream (*Figure 109*). This limestone was found to be sound at a depth of usually 5 m; the right bank, however, was more fissured than the left, and was treated with cement. The gorge is 200–300 yd. downstream of a 'fossil' valley, that is, the old course of the River Drac, now filled in with permeable morainic material left by the glacier which had cut the gorge (*see* St. Pierre-Cognet reservoir on page 179).

For grouting, for securing watertightness under the dam across the valley, the following details of the 'mask', screen, or curtain are available.

Number of holes	49
Length of holes	2,117 m
Cement used	300 tons

Cement per m	137 kg
Area of screen	10,000 m²
Cement per m²	30 kg

For securing a satisfactory broad foundation junction with the Lias limestone, the amount of cement used is interesting.

'Contact' grouting

Number of holes	168
Cement used	40 tons
Cement per m² of support	24 kg

To convert the sections of the thin arch dam into a monolith, the amount of cement used for keying the sections was as follows:

Cement	30 tons
Area of sections	3,400 m²
Cement per m² of jointing area	9 kg

Other data include:

Area of gathering ground	1,118 km²
Area of gathering ground of River Bonne (catchwater)	362 km²
Total area of gathering ground	1,480 km²
Mean annual flow	46 m³ per second
Estimated flood, 10 years	500 m³ per second
Estimated flood, 100 years	800 m³ per second
Estimated flood, 1,000 years	1,800 m³ per second

The disposal of flood water is provided for by top and bottom gates in the dam: (*a*) two bottom gates which deal with 350 m³ per second, each under a normal top water level of 580 m O.D.—700 m³ per second; (*b*) three spillways controlled by roller gates to pass 700 m³ per second under a head of 580 m O.D.—700 m³ per second. A total of 1,400 m³ per second.

If for any reason the bottom gates (*a*) become damaged or cannot act, the top gates (*b*) are able to take care of 1,700 m³ per second under a head of 5 m, that is, 585 m O.D.

During construction the two diversion tunnels were designed to take: right bank—220 m³ per second; left bank—370 m³ per second. A total of 590 m³ per second, with a coffer dam having a crest of 536 m O.D., being about 26 m high above the bed of the River Drac.

Monteynard Dam

Another thick arch dam on the Drac constructed recently is 150 m in height with a crest 210 m long. It is on Lias Limestone, Toarcian, which dips nearly at 45 degrees parallel to the valley. These strata are much faulted, the five principal faults are named Bérénice, Aglaé, Clotilde on the left bank and Joséphine and Julie-la-Rousse on the right bank. Most of them are vertical and parallel to the gorge. These faults (not necessarily their names) explain why a thin arch dam was not adopted.

Castillon Dam (Basse Alps)*

The Castillon dam (*Figure 110*) is some 50 miles north-west of Cannes and 4 miles north-east of Castellane. It impounds the River Verdon, a tributary of the Durance.

The reservoir commands an area of 675 km², covers an area of 500 ha. and holds 149×10^6 m³.

The dam has the following characteristics:

Type . . . . . . Thin arch, concrete, radius of face	70 m
Top water level	880 m O.D.
Maximum height	100 m
Length of crest	200 m

Figure 110. Castillon dam. General view

The interesting point about this dam is that it is constructed on fissured limestone of Upper Jurassic age.

The right abutment is on very broken limestone due to an anticline, a limb of which inclines towards the valley, although the general dip is upstream.

The right side of this fractured valley which is to be subjected to a pressure of 100 m of water would (*a*) not only leak like a sieve, but (*b*) would not be at all suitable to sustain the great thrust of a thin arch dam. In addition, although the left bank strata dip upstream, and to some extent into the hillside, the joints and fissures in the limestones would probably (*c*) cause considerable leakage.

As we are not concerned in this account with adumbrating the views of English engineers regarding (*a*), (*b*) and (*c*), that is, the impracticability of building dams and reservoirs on fissured limestone, we feel we cannot do better than give a free translation of the full account by M. Erhmann and M. Suter of how the problem was tackled, supplemented by M. Gignuox and R. Barbier.

---

* I am indebted to M. A. Decelle for the site pamphlet of the Castillon and Chaudanne dams (1955), and to the le Chef du Service Travaux, Electricité de France, Marseille.

The dam is in a straight gorge where the topographic conditions are very favourable, but the geological conditions are much less favourable, especially on the right bank.

The gorge in the Verdon valley runs in a north-to-south direction in Upper Jurassic limestones and, on the right, these dip obliquely towards the thalweg (*Figure 111a*).

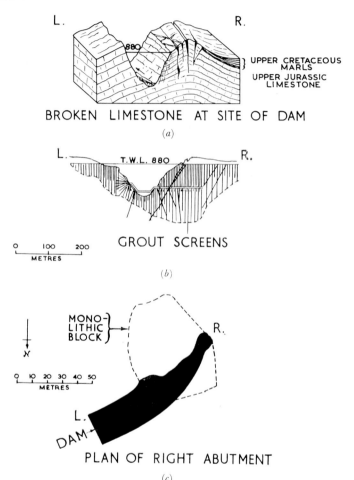

Figure 111. Castillon dam. (a) *Anticline of broken Upper Jurassic limestone to be converted into a monolithic block.* (b) *The extent of the grout screen.* (c) *As the limestone was so cracked very thick grout (and concrete) were used to retain thinner grout within the volume bounded by the dotted line*

At some 40 m from the end of the dam there is an anticline, much fissured, named on site the 'Cassure au Large', or the Cheiron anticline.

The dip towards the axis of the valley and the anticline have caused a dislocation of all the beds of limestone right down to the floor of the valley;

thus, in this zone the fissures are extremely numerous and some of them exceed a metre in width.

To ascertain the extent and significance of the fissured zone and to study the possibility of reconstituting a solid abutment for the dam, numerous exploratory works were made, the chief of which were: (*a*) construction of shafts and galleries; (*b*) investigation of known fissures; and (*c*) construction of bore holes with water and cement injection trials.

As a result of this preliminary work, a fairly economical and technically sound solution of the problem posed by the state of the ground for the right abutment was found.

The investigations showed that there could be no question of excavating into the disturbed bank in order to find sound rock on which to place the right-hand abutment of the dam. If such an attempt were made, highly dangerous slips would have occurred. It was decided, therefore, to consolidate the rock and to reconstitute it by injection, converting it into a monolith for the right abutment.

In order to determine the extent of the zone to be treated, bearing in mind the nature of the rock and the zone of weakness in it, a broad estimate was made and it was calculated that a volume of some 200,000 m^3 should be reconsolidated, that is, recompacted and rendered watertight; such a block would form a truncated pyramid of which one of the faces would form the abutment of the dam. In the interior of this block it was decided to carry out the following.

(*a*) Insert concrete in all fissures where access was possible. This entailed filling some 2,000 m^3 of rock, thus the interstices were about 1 per cent.

(*b*) Inject grout for which a preliminary screen of grout was put round the block to prevent loss of material into the open, down the hillside or into fissures of strata which were outside the specified block.

These operations thus formed a kind of box and the materials used were a mixture of cement, inert materials, and bentonite (*Figure 111c*). Into the interior of this box a mixture of cement and inert materials was successfully injected without loss of material and the finest of fissures were filled.

The work was carried out from three exploratory galleries (strengthened with concrete); the method avoided long headings and separated excavation and concrete work from boring and grouting. Also, these galleries, serve for inspection, and if necessary enable further grouting to be effected.

The amount of grouting necessary for the consolidation of this block was as follows:

Length of bore holes	16,315 m
Liquid grout	2,650 m^3
Cement (dry)	1,300 tons
Surface treated	13,100 m^2

In addition to the block the amount of grout in the screen in the Liassic limestones under the dam was as table on opposite page.

As shown on the diagram (*Figure 111b*), some bore holes extended down to no less than 200 m below top water level where the limestone was proved to be more solid. Eighteen rigs were used in the drilling of these holes.

The resulting leakage was only 100 litre/sec.

	Cement used (tons)	Length of bore holes (m)	Cement, cwt. per m. of bore hole	Area of screen (m)	Cement, cwt. per m.² of screen
Right bank	3,000	6,075	10	33,200	1·8
Centre	995	1,350	14	4,350	4·6
Left bank	1,200	2,800	8·5	8,900	2·7
Totals	5,195	10,225	—	46,450	—

As the right bank, downstream of the abutment, was steep, having numerous vertical joints and faults with bedding planes dipping towards the valley, it was decided to strengthen these rocks by three concrete counterforts between the levels of 780 and 820 m O.D.

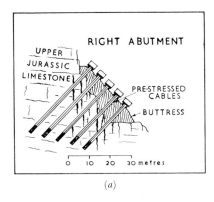

(a)

(b)

Figure 112. Castillon dam. (a) Owing to the weakness of the Jurassic limestone in the vicinity of the right side abutment, not only had heavy buttresses to be built against the limestone, but they had to be kept in place by pre-stressed cables carried down into sound limestone. (b) Natural Jurassic limestone strengthened by concrete buttresses, some of which carried pre-stressed cables embedded in the natural rock to hold a block of grouted rock together

These counterforts acted in two different ways: (a) they supported superficial blocks of limestone; and (b) they consolidated the rock block, by facilitating the insertion of five cables, pre-stressed to 1,000 tons, into inclined bore holes. These pre-stressed cables were bedded in cement for 40 m in the natural rock, and the block was therefore put into compression (*Figure 112a and b*).

Chaudanne Dam (Basse Alps)

The Chaudanne dam is less than 50 miles north-west of Cannes and $2\frac{1}{2}$ km east of Castellane. It is on the River Verdon a tributary of the Durance just below the Castillon dam, commanding, with Castillon, an area of 885 km². The reservoir covers an area of 70 ha. and holds 16×10^6 m³.

The dam (site pamphlet handed to the author through the kindness of M. Decelle, 1955) has the following characteristics:

Type	Thin arch concrete, radius of face	56 m
Top water level		790 m O.D.
Maximum height		70 m
Length of crest		95 m

Like the Castillon dam, the Chaudanne dam is constructed on Upper Lias limestone.

A slip in November, 1950, of some 2,000 m³ occurred on the right bank, necessitating heavily reinforced concrete buttresses and bore holes for

Figure 113. Chaudanne dam. Some of the work in progress (November, 1950) for strengthening the right abutment with concrete buttresses. The area of the slipped material is behind and above the white line; the volume of the rock involved was about 2,000 m³

grouting (*Figure 113*). It occurred during construction and was caused by the fissured and joined limestones rather like those on the right bank of Castillon.

A detailed account of the repairs is given by M. Haffen in I.C.O.L.D. IV (1955) 1106.

Chastang, Aigle, and Marèges Dams (Corréze)

The Chastang (264 m O.D.), Aigle (342 m O.D.) and Marèges (417 m O.D.) dams are all on the River Dordogne, and with the upper dam at Bort (542·5 m O.D.) form a comprehensive hydro-electric system. Part of this development also includes (above Aigle) a magnificent 'route de tourisme' for many miles by the side of the reservoirs.

The Chastang dam is founded on solid granulite, the Aigle dam on solid gneiss, and the Marèges dam on solid granite.

Gignoux and Barbier reported that under the alluvium these rocks, for the most part, are so sound that it was unnecessary to scrape them, even superficially. Below the surface and the alluvium, they required only moderate grouting because they are not affected by alpine folding as they are less diaclastic (broken) than those in the Alps.

Figure 114. Aigle dam. (a) Ski-jump overflow on the roof of the power-house. The left-hand abutment is on hard gneiss. With these rocks a stilling pool is unnecessary. (b) Right-hand abutment on solid gneiss

Chastang

Chastang is the lowest dam down the valley (commissioned 1951), and although the right bank had some granulite at the surface, there were some 13 m of alluvium and scree to be removed from the left bank before sound rock was found.

The granulite rock was found to be sound and in large blocks, although fractures consumed a good deal of grout. Many fissures on the left bank, however, were filled with clay.

The Chastang dam is 85 m in height and is a gravity arch type. Being a long way down this wide valley the length of the dam at the crest is 350 m, the chord-height ratio being over 4.

Aigle

The Aigle dam (1940–45) is 90 m in height and is a thick arch type with a vertical face.

The sound rocks in the valley influenced the decision to use the ski-jump overflow, the end of which can be seen above the power-house in *Figure 114a*; it is designed to carry 880,000 gal. per second. The view also shows the left-hand abutment of the arch; *Figure 114b* shows the right-hand abutment.

Marèges Dam*

The Marèges dam, constructed about 1935, is 90 m in height and is a thin arch cupola type (*Figuresa 115a and b, and 116*).

The absence of foundation difficulties, owing to the nature of the granite, encouraged by M. Coyne to adopt a cupola dam upon which he made important experiments.

(a) (b)

Figure 115. *Marèges dam. (a) Left end abutment of cupola dam and foot, on sound granite. (b) Right end abutment on granite with ski-jump*

Before 1935 the gravity dam was the method habitually adopted in France, and it is said that its solid appearance gave the public the idea that it was of invulnerable strength.

Hitherto there were only two arch dams in France, namely the Zola dam, Aix-en-Provence, and Considère's small reinforced concrete multiple arch dam 'Roche qui boit', Selune.

As the Marèges dam was to be 90 m in height (chord 240) with a chord-height ratio of under 3 and an overhang of 7 m, the pressure was estimated at 60 kg per cm^2 which the granite was well able to withstand.

*I am grateful to M. Coyne for agreeing the notes on the Marèges dam.

As it was thought, from stress calculations, that there would be movement at the base of the dam, Coyne incorporated large reinforced concrete props with the arch to rest on the base to take the place of the thickening of the dam dictated by theory; these are shown in black in *Figure 116*. They were expected to rise on filling or in cold weather, or both, and a series of five electric acoustic stress detectors per prop were inserted between the props and the base. When the reservoir was filled the props remained stationary and there was no lifting and hence any thickening at the base of the dam was unnecessary. (Many stress detectors were also inserted in the dam itself which generally confirmed the trend of the theoretical values of stress although some of them were 30–40 kg per cm² lower.)

The sound granite in the bottom of the valley, like that of other dams, prompted the ski-jump overflow (*Figure 115b*) near the right bank, which is

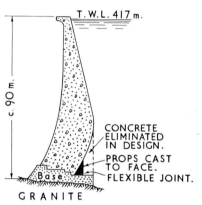

Figure 116. Marèges dam. Founded on good granite, this was the earliest cupola dam (1935) to be constructed in France. The props at the foot of the dam were experimental and stress detectors were attached to ascertain if any thickening at the base of the dam was necessary

designed to carry 700 m³ per second. Instead of having a discharge channel down the dam to perhaps an expensive stilling pool at the bottom of the valley, the jump stops half-way down the dam in mid-air. Its end is turned up and the velocity of the water is such that it clears the dam for a considerable distance without risk of eroding the base of the dam.

It was in connection with this dam that there was an early example of a pre-stressed concrete pipe. The pressure pipe 4·4 m in diameter, for the turbines have to withstand 100 m water pressure when water hammering occurs, necessitating reinforced concrete pipes 1 m in thickness, as, for economy, steel had to be avoided. Tensioned steel hooping was adopted which reduced the thickness considerably.

Bort Dam (Corrèze)*

The Bort dam impounds the head-waters of the Dordogne, about 1 km above the town of Bort-les-Orques and was constructed between 1945–51.

Type	Gravity arch, radius	624 ft.
Top water level		542·5 m O.D.
Maximum height		120·0 m (394 ft.)
Bottom water level		422·5 m O.D.

* Acknowledgement is due to M. A. Apostol, the Minister plenipotentiary, French Embassy.

The site of the dam is located in a defile some 300 yd. in length (*Figure 117*) where the rocks have a general dip downstream. These rocks are metamorphic and igneous, and are composed of the following.

(1) Chrystalline schists of variable quality consisting of mica schists with phyllite at the upstream end of the defile. They are very soft and of low bearing strength.

(2) Quartzite mica schists, hard and compact towards the middle of the defile.

(3) Mica schists, partly altered to felspathic rock, turning into very hard compact gneiss.

Figure 117. Mica schist forms the lower part of the valley and gneiss the upper part. Between these formations about half way up the dam there is a crush zone, a wedge of clay which caused great difficulties in construction. Upstream at about a mile on the left of the valley there is a buried pre-glacial valley filled with material through which the leakage possibilities had to be investigated

(4) The hard gneiss having slipped against the mica schists has caused buckling and crushing, and has produced a so-called 'fault' filled with clay.

(5) Gneiss. *Figure 118* shows the relative position of the gneiss, clay and schist with the dam.

A major slip in the left bank upstream of the dam, now under water, occurred during construction, but *Figure 119* shows generally the ragged nature of the defile.

Major difficulties occurred on the left bank in bridging over the weak clay between the mica schist and gneiss for the dam to sit on. *Figure 120* shows the clay under the dam outcropping between 510 and 480 m O.D.

Minor difficulties occurred on the right bank to ensure a satisfactory base for the broad foundation. *Figure 121* shows the disturbed appearance of the gneiss clay mica schist area on the right bank and *Figure 120* the clay under the dam outcropping between the levels 460 and 480 m O.D.

Figure 118. Bort dam. Diagrammatic section at centre of dam relative to strata

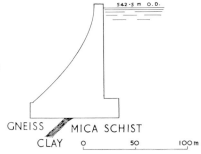

Figure 119. Bort dam. Roof of power-house and ski-jump in foreground. Beyond, on the left bank, the ragged gneiss on clay is seen, which, upstream of the dam, caused a large slip during construction

The major slip on the left bank

The dam had to be sited in the defile as much as possible on the very hard gneiss and on fairly hard mica schist, but for the best site in the defile there was, unfortunately, on the left bank, a crush zone of very weak clay between the gneiss and mica schist (*Figure 120*).

On the left bank the gneiss extended from top water level (542·5 m O.D.) down the side of the hill to the level of 500 m O.D.

The bad zone of clay extended from 500 m O.D. down to about 475 m O.D. when fairly sound mica schist was found below, down to the base of the dam at about 425 m O.D. (*Figure 120* (A)).

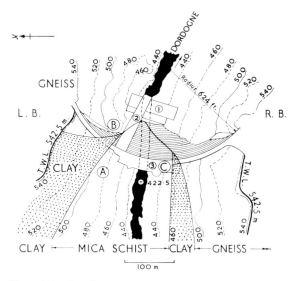

Figure 120. Bort dam. Great difficulties were caused by the clay encountered between the gneiss and mica schist. That at the left bank (B) had to be bridged over, and that on the right bank (C) had to be taken out and replaced by concrete. During construction there was a large slip at (A). (1) Denotes power-house; (2) denotes overflow chute ski-jump; (3) denotes sluices and intakes

Figure 121. Bort dam. Right end against gneiss. weak clay and mica schist below

During construction in July–August, 1947, a slip of several thousand m³ occurred in a downstream direction. To combat this, a concrete arch was built right across the excavation to arrest the slip. This arch was damaged the following year and had to be strengthened by buttresses 3 m in thickness (*Figure 122a*).

Bridging the fault on the left bank

The junction of the overlying hard gneiss with the weak crushed clay was distinct, whereas the lower junction of the crushed zone with sound mica schist was impersistent.

The regular surface of the gneiss clay junction crossed the valley at 45 degrees and headed downstream at a slope of 1 to 1, from 542·5 m O.D.

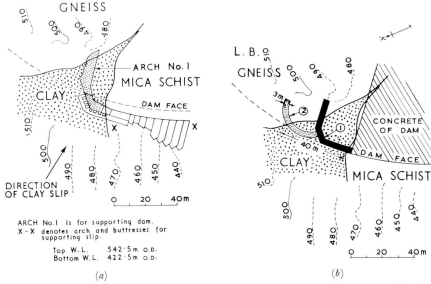

Figure 122. Bort dam. (a) On the left bank between 475 and 425 m O.D. there was a slip which had to be counteracted by a concrete arch strengthened by buttresses between XX. (b) A little way up the left bank there was an expanse of some 40 m of crushed strata equivalent to clay wedged between gneiss and mica schist. Two heavy concrete arches were constructed: (1) abutting the concrete face on the dam and gneiss; and (2) abutting the first concrete arch and the gneiss. The figure shows the plan of these two arches which were coincident with the face of the dam. The intervening clay behind the arches was bridged over with heavy reinforced concrete upon which the rest of the dam was built. Top water level is 542·5 m O.D.; bottom water level 422·5 m O.D.

to 422 m O.D., below which it was believed to incline more towards the vertical (*Figure 120* (B)).

The mica schist under the centre of the dam is upstream and the gneiss downstream, as in *Figure 118*.

As the wedge of bad material on the left bank extended for a width of up to 40 m under the dam, and reached a depth of 40 m, it was decided to 'bridge' this large wedge-shaped crush zone by two concrete arches (*Figure 122b*). These were constructed in the neighbourhood of the slip on the left bank, and under the protection of the concrete buttresses described above and shown in *Figure 122a*.

One of these arches (known as Voute No. 1) springs from the concrete of the dam halfway up the left-hand side of the valley. It was founded on good mica schist below the bad material, at no great depth, and curved round so that the other end of the arch could be founded on sound gneiss at a reasonable depth.

To support the rest of the upstream face of the dam, another arch (known as Voute No. 2) was constructed at a depth of 40 m to reach sound gneiss at one end, and the concrete of Voute No. 1 at the other.

These arches are 3–4 m in thickness and carry some 1,500 tons of rail (32 kg per lin. m) embedded in concrete which, as M. Jeanpierre puts it, had the effect of 'sewing' up the fault.

Right bank difficulties

As the faulted zone of crushed material was narrower on the right bank than on the left, it was excavated out to a depth of 15 m and replaced by reinforced concrete (*Figure 120*(C)).

Tests on rocks

According to M. Jeanpierre, specimens 10×10 cm showed the crushing strength of the sound mica schist (including some of the crushed zone) to be between 23–130 kg per cm^2; gneiss, 90–200 kg per cm^2.

Large-scale tests of the crushing strength of the gneiss and mica schist were made in a 2×2 m exploratory tunnel beginning in the mica schist, penetrating the crush zone, and ending in the gneiss.

Four screw jacks registering up to 300 tons designed to apply a pressure of 100 kg per cm^2 were placed opposite each other on prepared surfaces plastered with cement. They were sited at 7 m on each side of the fault zone on the vertical walls of the adit.

The curve of deformation in the gneiss is shown from Jeanpierre's paper in *Figure 123a*. When pressures up to 75 kg per cm^2 were applied for 20 hours, the deformation returned to nearly zero, and even after a pressure of 94 kg per cm^2 applied for 10 days, the deformation for all practical purposes returned to zero; at any rate the gneiss was proved to be sensibly elastic. Not so, however, the mica schist, which appeared to respond very differently from the gneiss, the deformation was 30 times as great showing that it was certainly not elastic as substantial deformation or permanent set was recorded (*Figure 123b*.)

The disparity in the strength of the mica schist in the full-scale experiment with the 'sample' experiments was baffling; the small sample of good mica schist would seem to be at least $\frac{1}{2}$–$\frac{1}{3}$ as good as the gneiss, whereas in the large-scale experiment, at 30 kg per cm^2 the deformation of the mica schist was $6\frac{1}{2}$ times that of the gneiss.

It was conjectured that the test on the mica schist at 7 m from the bad zone (which was known to be irregular) was too near and did not represent the average strength of the mica schist.

Alternatively, some damage might have been done to the mica schist in driving the tunnel (notwithstanding that the point of application of the pressure had been prepared carefully by hand). However, it was decided to limit the pressure on the foundation to 15 kg per cm² (approximately 15 tons per ft.²).

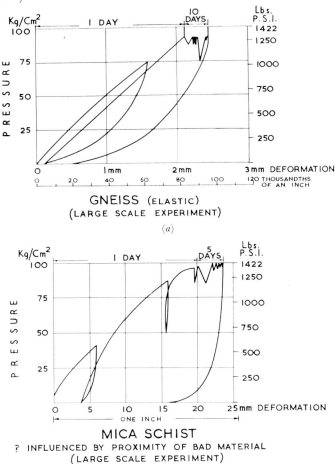

Figure 123. Bort dam. (a) Large scale loading test (4 m²) carried out on gneiss. The deformation even after 100 kg per cm² (approximately 100 tons per ft.²) was under ½ mm. (b) Large scale loading test carried out on mica schist (possibly influenced by proximity of bad material). The deformation after 100 kg per cm² was 16 mm (compare with (a))

Settlement measurements

Although the high central part of the dam was on a sound foundation of mica schist and the lower parts, the abutments, were on gneiss, the medium part on the valley sides, particularly the left, was on bad material. As the dam here was 80–100 m in height, exceptional care was taken in measuring settlement (by water levels) and tilting (by stainless steel pendulums carrying weights of 25 kg) during construction, after filling and after emptying.

Typical records over the difficult section on the left bank revealed the following settlements (height about 100 m to base of foundation).

	Toe gneiss	Centre over fault	Heel mica schist
Full (mm)	2	6	4
Empty (mm)	0	6	10

The settlement of the high central part of the dam per unit of height was comparable to these amounts.

Considering the difficulties in dealing with so large a zone of crushed mica schist clay, these small amounts of settlement caused great satisfaction in view of the disturbing results of the initial trials, the plasticity of the mica schists, and the weak fault zone.

Bort Reservoir (Corrèze)

The Bort reservoir intercepts the River Dordogne which flows from north to south in the north-western part of the Central Massif.

Below the dam and Bort-les-Orgues is a tributary of the Dordogne, namely the Rhue. The Rhue tributary which drains 650 km² is tapped at a high level (above 542·5 m) and led by a diversion tunnel (7¾ miles in length and 14 ft. 9½ in. in diameter) into the reservoir, the total area of the gathering ground thus amounts to 1,717 km².

Capacity of the reservoir	407 × 10⁶ m³
Top water level	542·5 m O.D.
Maximum depth	120 m

Before the reservoir was constructed an investigation on the glacial deposits was made by Professor Jacob about a mile upstream of the dam on the left of the valley. These deposits filled an ancient pre-glacial valley to 214 ft. below top water level and consisted of more or less impermeable sandy clay. However, the scheme was proceeded with as it was ascertained that any leakage would have great frictional resistance in following a path of 3,000 ft. under a head of 120 m or 394 ft. at the most (*Figure 117*).

Génissiat Dam (Ain)*

The scheme for converting the Rhône into a navigable river from the Lake of Geneva to the Mediterranean, and at the same time utilizing the fall in the river to generate electricity and to provide irrigation is nearing completion (1970). (*Figure 132*, page 212.) Some of the most difficult parts of the scheme have been successfully completed, particularly those in the north between the French–Swiss frontier and the Génissiat dam.

*I am especially indebted to M. Delattre for reading these notes on the Génissiat dam.

This interesting and difficult section owes its origin to Emile Harlé who, in 1906, suggested that if a dam were built at Génissiat to impound about 80 m of water, a single reservoir with a top water level of 330·7 O.D. extending to the French frontier, there need be only one set of locks for navigation and one generating station for power at Génissiat instead of a series of small schemes, as formerly proposed, at different places along the Rhône (*see Figure 131*, page 210).

The Génissiat dam is thrown across a Cretaceous limestone gorge 5 miles south of Bellegarde, and about 31 miles south-west of Geneva. *Figure 124a and b* shows the upstream and downstream faces of the dam. *Figure 124c*

Figure 124. Génissiat dam. (a) Upstream view of Génissiat dam across the Rhône. The dam is founded on Cretaceous limestones above which are Tertiary marls and clays and glacial clays. Some of the tops of the six vertical intake towers are seen on the face of the dam. (b) Downstream view of dam and roof of power-house. The dam is a gravity one founded on Urgonian Cretaceous limestone seen on the right

shows the strata of the left bank in photograph (a) and *Figure 124d* is a view upstream at the dam. *Figure 125* shows a section of the dam which impounds some 13,700 million gal. with a maximum height of 104 m on limestone, the reservoir extending 14 miles up the river.

Structure of the region

Figure 126 shows the geology in the vicinity of Génissiat, and *Figure 127* shows a typical geological section from west to east through the dam site.

It is seen that the Rhône here lies in a syncline, about 10 miles wide, between the Jurassic mountains, Crêt du Nu and Vuache (over 4,000 ft. above sea level). Within this trough of Jurassic limestone lie the limestones and marls of the Lower Cretaceous, namely, the Urgonian limestone on which the dam is founded.

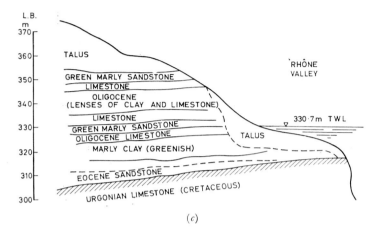

Figure 124. Génissiat Dam. (c) The geological sequence at the left bank of the dam relative to water level, O.D. 330·7 m is Eocene, Oligocene, Miocene on Cretaceous Urgonian limestone. The water submerges the Eocene and Oligocene. The Urgonian Limestone is exposed above top water level on the right bank and is seen in the gorge below the dam in Figures 129 (a) and (b). The strata on Figure 130 (b) are typical of those above 330·7 m O.D. up to Bellegarde and the Valserine. (d) Looking upstream from the dam which impounds 80 m of water, there is a fair width of water although the dam itself is in a gorge of Urgonian Limestone. In the distance it is seen how the two sides of the valley come together and form another gorge. The Urgonian Limestone has risen to top water level and hence the flooded width of the valley is not materially increased

The Urgonian is part of the Lower Cretaceous and equivalent in age to our Atherfield Clay, which is overlain by the Lower Greensand and Albian (Gault and Upper Greensand), and underlain by the Hauterivien which is contemporaneous with the Weald Clay.

The Urgonian limestone (the name Urgonian is derived from Orgon near Arles, lower down the Rhône valley), and the underlying Hauterivien (derived from Hauterive on Lake Neuchâtel), are overlain by several hundred feet of Tertiary clays, which are, in turn, overlain by glacial drift. It is the karsticity of the Urgonian limestone which threatened the feasibility of the scheme. Five miles to the east of the Génissiat dam site there is an old pre-glacial valley, the ancient course of the Rhône, filled with Drift. To the

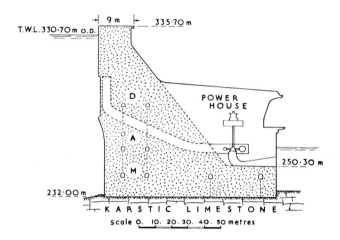

Figure 125. Génissiat dam. The dam is a straight gravity type attaining a maximum height of 104 m. The power-house is incorporated with the dam. The limestone is cavernous, but it is covered with soft Tertiary and Glacial clays, which filled the caverns

west of the site, at a distance of less than one mile, there is another glacier-filled valley (the old Valserine) although not so deep, and this had to be thoroughly investigated before work began.

Structure of the dam site

Although the siting of the dam was mainly influenced by hydro-electric requirements, geological conditions ruled out certain alternatives.

Conflicting opinions arose, particularly between M. Martel, the French authority on caves in limestone, and M. Lugeon, the eminent Swiss geologist and specialist on large dams.

On the one hand, these limestones were said to be riddled with large caves and fissures resulting in the limestone being completely permeable (Martel); on the other hand, being covered by a thick mantle of clay the caves and fissures would be blocked, resulting in the limestone being completely impermeable (Lugeon).

In the course of excavations and preliminary explorations, large fissures and ancient caves in the limestone were, in fact, encountered, but these were filled with clay and sand and some of the fissures actually sustained water under pressure when tested for watertightness.

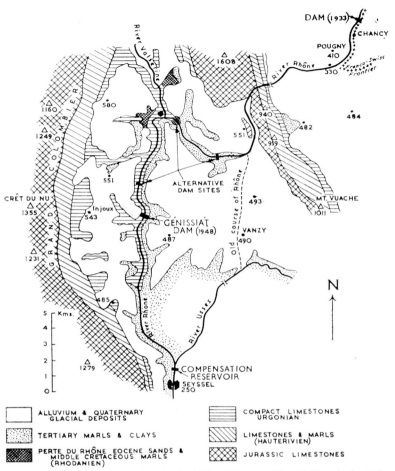

Figure 126. Génissiat dam. Geology in the vicinity of the dam. The old Rhône is about 6 km to the east of the dam. The Valserine in the north disappears near the Perte du Rhône; the old valley of the Valserine is about 1 km west of the dam

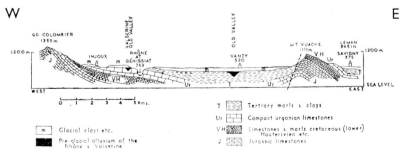

Figure 127. Génissiat dam. Geological section through the site of the dam. There is alluvium in the bottom of the Rhône at the dam and the underlying Hauterivien which is normally impervious was found to be very much disturbed and in consequence permeable

Among other exploratory works, a well was sunk to about 130 ft. below the bed of the Rhône and headings were driven across the valley and upstream of the proposed site, which proved the rock at this horizon to be substantially impermeable and unconnected with the Rhône.

Proving the dam site

During the construction of the works on the left bank, near the level of 318 m above sea level, up to the crest of the dam (335 m), some sand of an old Oligocene lake was laid bare, lying unconformably on the hard limestone of the Urgonian. As a water pressure of some 40 ft. would be put on the junction of the sand and limestone, a high-level trench, some 200 ft. in length, had to be driven into the hillside and filled with concrete.

The diversion works for 2,000 m^3 per second (70,500 cusec), necessitated exceptionally large excavations in the Urgonian limestone, but confidence was gained as these limestones were found to be impervious at a distance of about 60 m (200 ft.) from the side of the valley and showed no infiltration from the river. These calcareous rocks were ideal for the construction of large tunnels without form-work.

During the course of the excavation for the diversion works, the karstic structure of the Urgonian limestone was revealed on a magnificent scale. One such cavern of exceptional dimensions was penetrated on the left bank diversion and proved to be some 15 m in length. This enormous hole was filled with sand and clay of Oligocene, Miocene, and Quaternary age and it is considered that these 'sediments' were not brought into the cavern by the action of running water, but by the natural pressure of the overlying formations, extending hundreds of feet above the cavern, in much the same way as cement and clay are injected artificially under pressure into fissures below dam trench foundations. This fact gave further ground for confidence in the watertightness of the limestone and confirmed the opinion of those geologists who thought that where such limestones are covered by clay their porosity is restricted.

All fissures and caves which were found filled with sand and clay during the excavations were cleaned out and filled with concrete. Tests with water pressure made on some of these sand-filled fissures proved that they were capable of sustaining high pressures; for example, a pressure of 16 kg on 5 cm^2 (50 lb. per in.^{2}) with unsupported sand in a fissure $2\frac{1}{2}$ m (8 ft.) in length.

Temporary diversion works

The alluvium in the dam foundation area had to be compacted in the region of the dam in order to sustain the pressure of the temporary diversion dams and to facilitate the digging of the foundations of the dam itself.

Upstream of the dam (*Figure 128*) sheet piling was driven into the alluvium for a distance of about 9 m (30 ft.), and below this the alluvium was compacted by injecting some 400 tons of clay and 30 tons of silicate, or 2 tons of clay and 0·02 tons of silicate per m^2 of sheeting.

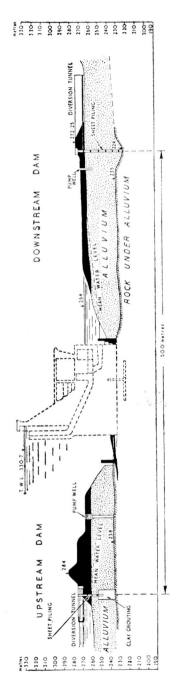

Figure 128. *Génissiat dam. Diversion works. To divert the Rhône into the large diversion tunnel the alluvium upstream was grouted with clay beneath some sheet piling and a long earthen bank (shown in black) constructed. Similarly, downstream an earthen dam was constructed to prevent the water flowing back into the main dam foundations and sheet piling was sunk through the alluvium to the Hauterivien rock. The sheet piling did not prove as effective as the clay grout used in the upstream dam*

can disappear and reappear and can *fill*, but need not necessarily leak out of Martel's 'vast kastic' caverns underground!

Proceeding eastwards from the Pont de Grèsin and the old course of the Rhône, at the southern end of a loop, the Rhône goes north over the Jurassic in a narrow gorge of the Fort d'Ecluse. Thereafter the width of the river increases in the flat moraine, Longery area and Quaternary to

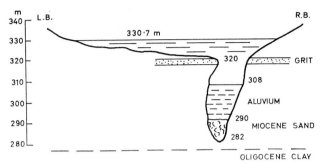

Figure 131 (c). Genissiat Reservoir. Near the Pont de Grèsin there is 8 m of Alluvium on which 18 m of water flows in a gorge about 10 m in width. Lugeon inferred that an additional 22 m head of water to 330·7 m O.D. would incur no loss

the frontier, even though the increase in depth from Génissiat is small it widens out. On the whole however the amount of additional land to be flooded by raising the river at Génissiat by 80 m is remarkably small.

In Lugeon's report (French) there is also much information on the anticipated amount of silting to be expected by increasing the depth in the 23 km stretch of river.

THE RHÔNE DEVELOPMENT (LOWER)

The Rhône flows out of Lake Geneva (Lac Léman) at Geneva and follows a tortuous course in a south-westerly direction via Bellegarde and Génissiat, to Lyons (*Figure 132*). It then flows almost due south from Lyons to Avignon and to the old Roman town of Arles, finally entering the Mediterranean 30 miles west of Marseilles.

From Lyons to the south, main roads and railways closely follow the present valley of the Rhône; there are many towns and villages along its course, and there is a large amount of water traffic and irrigation.

When the flow is low (in certain stretches of the river) there may not be enough water to maintain river traffic. When the river is high in flood time (in other stretches), the amount of water flowing may: (*a*) impede traffic proceeding upstream causing excessive expenditure in power*; (*b*) cause difficulties and danger for traffic going downstream on account of the high velocity of flow; (*c*) cause serious damage to crops and property by flooding.

The improvement of the river from the Swiss frontier to the sea with the

---

*M. R. Lefoulon says that for a 1,000 ton ship in the Rhine 850 horse power is necessary, whereas in the canal only 220 horse power or one-quarter of the power is necessary. (5th Int. Dam Congress Supplement Travaux No. 247 1955.)

three-fold object of generating electric power, improving navigation, and improving irrigation, is entrusted to a single Authority, namely the Compagnie Nationale du Rhône, which was formed in 1934.

The scheme for the general development of the French Rhône is divided into two sections, being known as the Upper Rhône between the Swiss frontier and Lyons, and the Lower Rhône between Lyons and the Mediterranean.

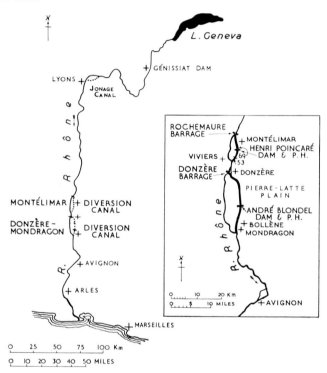

Figure 132. *Rhône development. The Rhône is being made navigable from the Mediterranean to Lake Geneva. In places this necessitates complete diversions and at the same time opportunity is being taken for generating electric power and improving irrigation. The major diversion canals (Donzère-Mondragon and Montélimar) were completed by 1958*

Twelve hydro-electric plants have been planned along the course of the lower Rhône, most of which have been provided with large diversion canals.

Two of the largest diversions have been completed, namely, the Donzère-Mondragon scheme (1947–53) and the Montélimar scheme (1954–58).

Basic principles

The principle of development is to build a low barrage across the existing river at a suitable place to raise the existing river level a few feet so that it can flow into the intake of the new diversion canal, built a few hundred yards above the barrage.

The canal will convey the water for several miles south to another suitable place for a dam, lock and power-house.

At the foot of the dam and power-house, the canal (with a suitable 'garage' for the boats taking their turn at the lock) goes on for several miles until it rejoins the existing river.

The principle of the scheme is similar to that of the development of the Rhine, the section between Basle and Vögelgrün having been completed. It has analogies with the improvements of the River Danube or with the St. Lawrence River development of America, to name only two.

The suitability of the sites where the barrage dams and canals are to be has been proved after many years of detailed hydrological and hydro-geological research of a most interesting nature.

Donzère-Mondragon Scheme (1947-1953)

The characteristics of the river at Donzère are as follows:

Area of gathering ground	26,530 sq. miles
Minimum flow	400 m³/s (14,000 cusecs)
Flow for 10 days in year	560 m³/s (19,600 cusecs)
Average flow for 6 months in year	1,400 m³/s (49,000 cusecs)
Powers allow the canal to divert up to	1,530 m³/s (53,600 cusecs)
Powers provide for a minimum flow to be left in the river of	60 m³/s (2,100 cusecs)

Donzère diversion barrage

At Donzère the barrage has been put across the Rhône and has the following characteristics.

Height	20 m
Length	798 ft.
Top water level	58·65 m O.D.

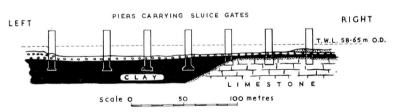

Figure 133. Donzère diversion barrage. Owing to the right bank of the Rhône being on Cretaceous limestone and the left bank on Pliocene clay, difficulties arose with unequal settlement of the piers which carry the massive gates of the barrage

The top water level of 58·65 m is some 6½ ft. lower than the highest known flood level and enables the filling of the diversion canal immediately upstream and ensures good navigation in the old Rhône up to the tail end of the Montélimar diversion canal, about 4 miles upstream (*Figure 132 (inset)*, *and Figure 133*).

The barrage, on the left side (and on a good deal of the centre part) of the valley, is founded on piles penetrating alluvium to a stratum of Pliocene clay. Under the right-hand end of the barrage, by contrast, the piles go through the alluvium and rest on compact Cretaceous limestone of Urgonian age which outcrops at the surface lower down the river.

Owing to the difference in bearing pressure sustainable by the clay on the left end and the limestone on the right, special construction joints had to be inserted in the foundations of the barrage, and the foundations of the piers on clay had to be enlarged. Variable settlements occurred in the piers according to the depths of water.

This type of barrage is used mainly to permit the flow to enter the canal and the difference of water level between upstream and downstream is not high, in this case 58·65–50·40 = 8·25 m.

The diversion dam comprises five passes of 31·5 m in width and a sixth pass of 45 m all equipped with sluice gates of 9·15 m in height.

The power-dam 'André Blondel' on the canal

Dam

Maximum height	43 m
Normal water level upstream	58 m O.D.
Low water downstream	32·45 m O.D.
Difference	25·55 m

Lock

The lock incorporated with the dam and power-house is 640 ft. in length, 40 ft. in width and 85 ft. in depth.

Power-house

Ultimate capacity 300,000 kVA ($2,100 \times 10^6$ kWh per annum)
Irrigation 7,500 acres

The site ultimately chosen for this very heavy structure (similar to that shown in *Figure 137*) is 11 miles south of Donzère in the 'Plain of Pierrelatte'. Here the Rhône valley is generally flat but flanked by hills of Urgonian limestones which dip underneath deposits of later age. The top of these buried limestones has an undulating surface and they sometimes rise to the surface (as at Pierrelatte), or to a height not far beneath the surface.

More than 300 borings were made in the Plain, as well as a geophysical survey, and this resulted in a 'hump' of sandstone being discovered near Bollène at a reasonable depth (30 ft.) below the surface, which provided a suitable site for the André Blondel dam, power-house and ship lock (*Figure 132 inset*).

The general sequence of strata in the Plain is as follows.

Variable silt, mud, slime, fine sand	10–23 ft.
Recent alluvium, sand	16–40 ft.
Old alluvium, sand, gravel	Variable thickness
Pliocene clay	Variable thickness
Cretaceous limestone	

Canal

When the site for the dam, lock and power-house near Bollène had been fixed, the route of the canal for 11 miles between Donzère and Bollène and for 7 miles from Bollène to Mondragon, where the canal rejoins the Rhône, could be investigated (*Figure 134*).

Figure 134. Donzère Mondragon scheme. Canal downstream of power-house and looking downstream near Bollène. The banks are formed of alluvial sands and gravel. Width 410 ft., depth 50 ft.

Some 11 miles of head-race canal 480 ft. in width, carrying 33 ft. depth of water, and 7 miles of tail-race canal 410 ft in width with 50 ft. depth of water had to be made in the variable silts, slimes, muds, clay, alluvium, sands (fine and coarse) and gravels outlined above.

Where the water level in the canal had to be above ground, the embankments made from such heterogeneous materials would have to retain the water. Where the water level in the canal had to be below ground level, care had to be taken not to have the cutting damaged by external underground water. Care had also to be taken not to penetrate into Cretaceous limestone. In this connection a mean velocity of $4\frac{1}{4}$ ft. per second was adopted.

The most hazardous part of the work was that of coping with the vagaries of underground water (*Figure 135*). Depending upon the degree of permeability of the bank and/or cutting the water level in the canal would be in communication with the natural underground water level outside, and any rise or lowering would affect the levels of wells owned by many people along the 18 mile stretch of the canal. These effects would occur both during construction and on completion, as well as on the filling and emptying of the canal after completion.

The work of investigation was described in detail by Delattre and by Henry, Director of the Company (Chargé des Services d'Etudes), and *Figure 135* gives an indication as to how the natural underground water levels were affected by the digging of the tail-race canal between Bollène and Mondragon.

That the preliminary investigations proved to be satisfactory is shown by the report of the Director General of the Company, Monsieur P. Delattre.

> The canal has been made in the alluvium of the great Rhône plain, alluvium carrying beds or lenses of varying permeability; argillaceous (clayey) slimes and muds (especially at the surface), beds of sand, gravel or pebbles, with pores now and then open and even visible to the naked eye.

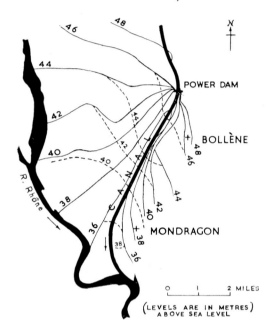

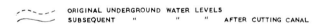

Figure 135. Mondragon. Underground water levels

The level of the underground water is often lower than that in the canal. To avoid losses which would reduce the flow of the canal and swell the underground water (which would modify the cultural condition of the Plain) one might have thought, at first sight, of covering the banks with some clayey mixture obtained from the excavation; but the placing and compacting of such an impermeable covering would necessitate a very complicated technique. Furthermore, whenever the canal is emptied and its water level is lowered below the ground water outside, there would be a risk of any such covering being destroyed by the upward water pressure from outside.

Therefore, the opposite procedure was decided upon, namely, not to attempt to make an impermeable covering but to make the lining of the canal with sands and gravels. These were well cleaned and uniformly graded, for it was found, experimentally, that this constituted an excellent medium for retaining the slimes of the turbid waters of the Rhône.

The mixing and placing of the clean alluvial materials was done (without special precautions or mechanical compacting) by ordinary earthwork-moving machinery.

The impermeability of the bank ought to be obtained in due course by the mud, carried by the percolating water, into the interior of the bank.

This method has so far succeeded perfectly, for on the first filling of the head-race canal, there were substantial losses which materially raised the underground water table.

After a certain time, notably after the storms of November and December 1952, clogging took place and the underground water level fell to its original level.

When and if the canal is emptied, an unclogging effect may be produced but any undue pressure behind the bank will be avoided and the bank should not disintegrate.

In brief, had the Donzère-Mondragon section been constructed near the present Rhône, high embankments (20 m) for many miles would have been necessary. Fortunately, by siting the canal to the east, as the Saint-Pierre rock came near the surface as an island, this determined the site of the André Blondel power-dam, near Bollène.

In addition, the banks of the canal, artificial above and below ground, were of gravel which, in due course, will silt up, thereby stabilizing the natural underground water conditions in the gravels on which or through which the canal has been constructed.

Montélimar Scheme (1954-1958)

The characteristics of the Rhône at Montélimar are not very different from those at Donzère, as the intake canal for the Montélimar diversion is only 10 miles upstream (*see* page 159).

Montélimar diversion barrage at Rochemaure

The diversion barrage consists of six openings, each of 26 m, founded on piles driven into the calcareous marl; the difficulties of unequal bearing pressure which occurred under the Donzère diversion barrage not being present.

Figure 136 shows a section of a pier, carrying the sluice gates, which is founded on piles over an area of 30 × 6 m.

The superstructure of the pier and gates is similar to the Donzère diversion barrage (*Figure 133*). The operating gates control the canal water levels as follows.

For opening, the top water level can be lowered by the top section of the gate (which is lowered), and the emptying of the canal can be effected by the lower section of the gate which is raised. Both these gates are operated by electric motors housed in the top of each pier.

The normal level of the Rhône raised to	77 m O.D.
The lowest downstream level	67·7 m O.D·
Difference	9·3 m

The maximum depth of water impounded (77 − 61) = 16 m

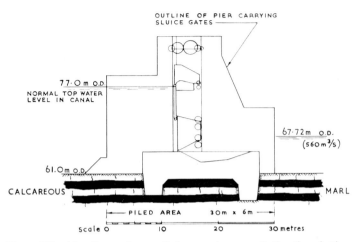

Figure 136. Montélimar scheme: Rochemaure barrage. Section through pier. In contrast to the foundations of the Donzère barrage, the calcareous marl of the Gargasian extended at a reasonable depth right across the River Rhône at Rochemaure (see Figure 133 on page 213)

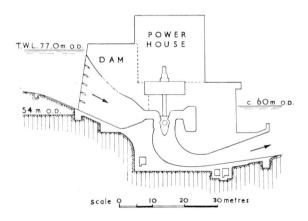

SECTION THROUGH POWER HOUSE
(HENRI POINCARÉ)

Figure 137. Montélimar scheme. Below the sandy plain, reasonable lacustrine sandstone was found at 54 m O.D.

Power-dam 'Henri Poincaré' on the canal

Like that for the Donzère-Mondragon diversion, the problem of searching in the middle of a dry sandy plain for a suitable site on which to build a power dam is somewhat novel.

The chief characteristics of the Montélimar power dam (named the Henri-Poincaré works) (*Figure 137*) are as follows.

Normal top water level	77 m O.D.
Maximum depth of foundations	43 m O.D.
Maximum height of dam	34 m
Original ground level	64 m O.D.
Rock level at power-dam, lacustrine sandstone	54 m O.D.
Depth of water impounded in the supply canal above the power-house is about	20 m
Operational head	19 m
Ultimate capacity . . . . 300,000 kVA (1,670 $\times$ 10^6 kWh per annum)	
Irrigation . . . . 15,000 acres	

Canal

The siting of the Montélimar canal, barrage and power-dam, like that of the Donzère-Mondragon scheme was the subject of considerable investigation, and to give some idea of the geological difficulties involved, the following is an extract from M. Henry's paper.

> To confirm that the lay-out of the scheme in detail should be sound, the Company asked Professor Jacob and Professor Denizot to proceed with a detailed geological study with several hundred bore holes, including a geophysical reconnaissance (resistivity or seismic as required).
>
> This study led to the conclusion that the geology of the plain is somewhat complicated, with the various limestones, grits and Pliocene clay, and these rocks were not deep anywhere which is a dangerous situation for the establishment of a canal.
>
> The diagram (*Figure 132 inset*), therefore, shows two levels of the top of the rock *in situ*; '69 m', the highest which one could consider placing the high level canal leading to the power-house, and '53 m', the highest level to be envisaged for the bed of the tail-race canal from the power-house.
>
> The levels of 69 m and 53 m are, therefore, the actual levels below which it is undesirable to site the head-race and tail-race canals respectively.
>
> At first sight it seems that the route should be near the Rhône, but the rock whether or not it exceeded 69 m in the Gournier district was proved to be Pliocene clay. As the Company had experience of the formation, by having to remove several million cubic metres of it in the Donzère-Mondragon section of the canal an alternative solution had to be considered.
>
> This was for driving the canal through the Pliocene of the Gournier plateau, then passing near the railway to avoid some Lacustrian limestone and rejoining the Rhône opposite Viviers. There were then two alternatives: (*a*) going west of Gournier; or (*b*) crossing the plateau. As the western route would have entailed high embankments and restricted the area open to flood water from the Rhône, the route was finally decided further to the east. Thus, the canal with its embankment on the right bank of 10 km and on the left bank 4 km as well as the barrage and tail race were sited.

The canal to the power-dam was designed to carry 1,900 m³ per second (66,500 cusecs) with a water speed of 1·3 m per second (4¼ ft. per second) for the head-race canal and 1·5 m per second (5 ft. per second) for the tail-race canal.

Owing to the existence of rock, the depth was often restricted, necessitating the widths being increased to 152 m and 194 m compared with the Donzère-Mondragon widths of 145 m and 127 m for the high level and tail race canals respectively . . . .

The high level canal went through, for the most part, Rhône gravel which is very permeable and the banks themselves were made of gravel; however, no attempt was made to render the canal watertight by any kind of facing (*Figure 138*).

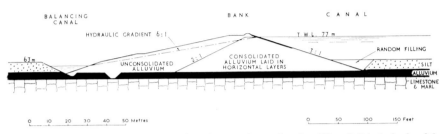

Figure 138. Montélimar canal. Where above the natural ground surface (63 m O.D.) the banks of the canal are built of permeable materials excavated near the site. The loss is estimated to be some ½ per cent of the dry weather flow of the Rhône at Montélimar

This method was adopted on the Donzère-Mondragon section and the experience there proved the method to be right.

The leakage of the high level canal has been much as anticipated, namely, 17 m³ per second or 1 m³ per second per km of canal for the scheme.

As the dry weather flow of the Rhône is 560 m³ per second, the loss of 17 m³ per second is only 3 per cent of the dry weather flow which is permissible having regard to the cost of a watertight facing.

At Donzère the losses were recorded as follows: (*a*) in winter, losses were halved because of the increase in the coefficient of viscosity of the water at low temperatures. (*b*) The warping effect is very noticeable and after 3 years the losses fell to one-quarter of the original value.

The Montélimar section is expected to operate in the same way with the exception of a section 900 m in length in the middle of the high level canal (near Grèzes south of Roubions) where the underground water level was found to be in free communication with the Rhône owing to the underlying fissured limestone.

The canal bed is, therefore, lined with a mixture of selected mud and gravel put in place with particular care, and the slopes being formed of gravel which would be subject to erosion by waves (caused naturally or by boats) have had placed upon them bituminous lining 6 cm thick to prevent this. I am especially indebted to M. Delattre for reading these notes on the Rhône development.

TIDAL POWER SCHEME IN THE RANCE ESTUARY, BRITTANY

The site of the power scheme for generating 544 GWh., is on the Estuary of the River Rance, between St. Malo and Dinard, Brittany.

There are two high tides in 24 hours 50 minutes (owing to the position of the sun and moon relative to the rotating earth) and the levels between high and low tides are:

> Highest, equinox, 13·5 m
> Average, full, 10·9 m
> Average, normal, 8·5 m
> Average, neap, 5·0 m

The exceptionally high range is due to the Brittany peninsular jutting out into the Atlantic.

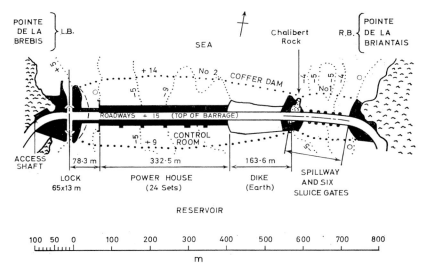

Figure 139 (a). *Rance Estuary Power Scheme. The bed of the estuary consists of Gneiss with bands of pre-Tertiary Dolerite running in a north-south direction like the river. On the Gneiss is a thin covering of sand and pebbles about a metre in thickness. The first work to be constructed was the shipping lock on the Gneiss of the left bank. This was followed by the construction of the spillway and gates section on the Gneiss of the right bank. The river was diverted through the whole of the gate control section when the main power-house and dike were constructed in the middle of the river between coffer dams, shown by dots enclosing an elliptical area. The power-house is shown in black under the roadway and the dike in white, joining the Gneiss outcrop*

The barrage across the river includes several units (*Figure 139a*) as follows:

(a) *A lock* sited on the left bank (the Dinard side of the estuary) for navigation between the sea and the river.

(b) *A power-house* which occupies a width of 332·5 m across the deepest part of the estuary. This is a very large concrete structure (to house twenty-four horizontal turbines) which was constructed in the dry within a large eliptical area enclosed by a coffer dam.

(c) *A dike.* As a large amount of excavation was necessary for the foundation of the power-house, the material was disposed of in an earthen embankment or dike, 163·6 m in length, one end of which abuts the power-house (*Figure 139b*).

(d) *A natural rocky island.* The other end of the dike abuts an island known as the Chalibert Rock.

(e) *Spillway.* The remaining width of the estuary is occupied by a spillway to dispose of the natural floods of the river, and

(f) *Sluices* which have to be opened and shut several times daily to let the water into the reservoir and to provide for the reversed flow operation for the turbines and to be closed during the time of their normal operation.

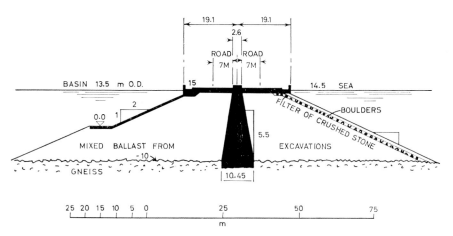

Figure 139 (b). Rance Estuary Power Scheme. *Showing the dike between power-house and Chalibert Rock Island constructed with gravel and sand material excavated for the power-house section. The clay core and embankment has to withstand rises and falls of tide up to $13\frac{1}{2}$ m twice in 24 hours 50 minutes in perpetuity*

The spillway and sluices, in concrete, adjoin the natural gneiss rock on the right bank on the St. Malo side of the Rance. The total length of the barrage is 750 m. There is a central carriageway of two lanes of 7 m each way with a central reservation of 2·6 m.

The Reservoir

The bed of the estuary consists of gneiss with bands of pre-Tertiary dolerite running as does the river in a north-south direction. On the gneiss is a layer of sand and pebbles about a metre in thickness. The level of the river bed is about 10 m O.D. at the barrage site. The top water level in the reservoir is $+ 13\cdot5$ m O.D.; the total depth of water is 13·5 m. The volume of the reservoir between highest and lowest tides, 0 and 13·5 m is 184×10^6 m³. The surface area of the water in the reservoir is 2,200 hectares (5,500 acres).

The construction methods for closing the estuary were as follows:

(1) The lock on the left bank was constructed in 1962.

(2) No. 1 coffer dam was erected between the right bank and the Chalibert Rock for constructing the overflow and sluice gates in 1963.

(3) Removing No. 1 coffer dam and erecting No. 2 coffer dam. All tidal water would go through the opened gates and river water in flood time and would flow over the spillway when No. 2 coffer dam is completed.

(4) When the No. 2 coffer dam was erected the foundations for the power plant and dike proceeded (1964–5).

(5) Before tidal water could go through the turbine conduits great stress was put upon the sluice-gate and spillway section; the tidal velocity reached at times two metres per second to go through the restricted gate openings (1,968).

In this respect a scale model was found to be most useful to foretell tidal conditions. The scale of the model was 1/150, the corresponding scale for *time* being $(1/150)^{\frac{1}{2}} = 1/12 \cdot 25$. As the period of the real tide is 12·25 hours, the model gives the probable conditions during the whole of one tide in one hour.

BIBLIOGRAPHY AND REFERENCES

Chambon dam
  CANEL, J. Le barrage-réservoir du Chambon. *Trav. publ.*, 82e Année, No. 892 (1936).
  GIGNOUX, M., and BARBIER, R. *Géologie des Barrages* Masson et Cie (1955).
  HAEGELEN, A., and BOURGIN, A. Le barrage-réservoir du Chambon. *Génie civ.*, 108 (1936).

Tignes dam
  GIGNOUX, M., and BARBIER, R. *Géologie des Barrages* Masson et Cie (1955).
  GIGUET, R. The Tignes hydro-electric scheme. *Soc. Ing. civ. Fr. Brit. Sect.* Extract from *Struct. Engr*, 376 (1955).
  PELLETIER, J. Construction of Tignes dam and Malgovert tunnel. *J.I.C.E.*, **2**, 480 (1953).

Sautet dam and reservoir
  DUSAUGEY, E. Le barrage-réservoir du Sautet, sur le Drac (Isére). *Génie civ.*, (25 July 1925).
  GIGNOUX, M., and BARBIER, R. *Géologie des Barrages* Masson et Cie (1955).
  THOMAS, J. L'amenagement du barrage-réservoir et de la centrale du Sautet. *Génie civ.*, **106** 625 (1935).

St. Pierre-Cognet reservoir and dam
  AILLERET, P. Mésure des fuites de contournement d'un barrage avant la construction de la barrage. *Ann. Ponts Chauss.*, Sept.–Oct. (1943).
  — *J.I.W.E.*, **5** 98 (1951).
  SITE PAMPHLET. Electricité de France. Chute de St. Pierre-Cognet (1956).
  VOLUMARD, P., and DUBOST, R. *Travaux* (Special Number) 256 (1958).

Monteynard dam
  d'ARCIER, D., FAIVRE, G. and CONTE, J. 'La consolidation des appuis du barrage de Monteynard'. *I.C.O.L.D.*, **1** 377 (1964).

Castillon dam
  EHRMANN, P., and SUTER, E. Barrage de Castillon. *Tech. d. Trav.*, 7–8 221 (1948).
  GIGNOUX, M., and BARBIER, R. *Géologie des Barrages* Masson et Cie (1955).

Chaudanne dam
  HAFFEN, M. Barrage de la Chaudanne. 'Travaux de consolidation et d'étanchment'. *I.C.O.L.D.*, **IV** 1105 (1955).

Chastang, Aigle and Marèges dams
  AUROY, M. Hydraulique et Electricité Françaises (Chastang) *Houille blanche* (Special Number) 68 (1950).
  COYNE, A. *J. Instn struct. Engrs*, **15** 70 (1937).
  — *Soc. Ing. civ. Fr. Brit. Sect.* (1947).
  GIGNOUX, M., and BARBIER, R. Géologie des Barrages Masson et Cie (1955).

Bort dam and reservoir
  GIGNOUX, M., and BARBIER, R. Géologie des Barrages Masson et Cie (1955).
  MARY, M. Le barrage de Bort. *Ann. Ponts Chauss.*, No. 2 265 (1949).
  SEVIN, M. Hydraulique et Electricité Françaises. *Houille blanche* (Special Number) 57 (1950).
  SPECIAL NUMBER OF *Houille blanche* (1953).

Génissiat dam
  DELATTRE, P. *J.I.W.E.*, **5** 641 (1951).
  — Le barrage et l'usine de Génissiat. *Génie civ.*, (1948).
  GIGNOUX, M., and BARBIER, R. *Géologie des Barrages* Masson et Cie (1955).
  LUGEON, M. Etude géologique sur le project de barrage du haut Rhône Française à Génissiat. *Mém. Soc. géol. Fr.*, 4e Série., t.2., Mem. No. 8 (1922).
  SPECIAL NUMBER OF *Houille blanche* (1950).

Rhône development
  DELATTRE, P. in *Géologie des Barrages* by Gignoux and Barbier Masson et Cie (1955).
  — and HENRY, M. Hydraulique et Électricité Française. *Houille blanche* (Special Number) 105 (1951).
  DELATTRE, P. The Rhône Development (Montelimar) Travaux. Page 125 (Aug. 1958).
  HENRY, M. La chute de Montélimar. *Bull. Gr. Sud-est.* Soc. Française Elect. (1955).

Tidal power scheme in the Rance estuary
  ALLARY, RENÉ. 'La technique des travaux'. **42** (1966). Translated by M. K. PERL, Cement and Concrete Association, (1968).

Hydro-electric power in France
  SANDOVER, J. A., *I.C.E.*, **26** 51 (1963).

19

GERMAN DAMS

The location of German dams is shown on the map of Austrian and German dams, *Figure 76* on page 143.

Rosshaupten Dam (Füssen, Bavaria)*

The Rosshaupten dam (*see Figure 140*) on the River Lech (53 miles S.W. of Munich) has a special interest as it simulates English conditions in many ways.

The chief particulars are as follows.

Type	Earthen embankment with reinforced concrete trench, Boulder Clay core and moraine gravel fill
Constructed . . .	June 1952 to September 1953 (15 months)
Top water level . .	781 m O.D. (flood level 782 m, top of bank 785 m)
Maximum height . .	41 m
Maximum depth of trench .	60 m
Length of crest . . .	about 250 m
Capacity of reservoir . .	165×10^6 m³ at 782 m O.D.

Geology of the site

The dam is founded on a very bad geological site; it is reported that alternative sites upstream and downstream were even worse, being ruled out for tectonic and morphological reasons; that is, geological reasons.

The dam site consists of soft layers of sandstone, clay marl with subordinate conglomerate, pitch-coal parting seams, known as Lower Marine Molasse of the Oligocene formation. These strata are almost vertical in dip and the strike is at right angles to the axis of the dam and parallel with the river, that is, west to east. As the strike of the vertical permeable seams is at right angles to any cut-off trench and grout screen, there is not any impermeable strata with which to make any contact and reliance must be placed on 'wire-drawing' to exclude leakage.

The proximity of the Bavarian (Lectaler) Alps, 3–4 miles to the south, is no doubt responsible for the vertical Oligocene at the deeper horizons

---

*I am very much indebted to Dr. Ing. Fredrich Treiber in Bayerische Wasserkraft Works, Munich, for reading these notes on the Rosshaupten dam and giving additional information.

in the trench being found to be well consolidated. The trench was therefore taken only 20 m into the Molasse rock with a suitable grout screen below (*Figure 140*).

The ground surface, top of rock, depth of trench, extent of grout screen) the dip and strike of the strata and the positions of 17 of the 19 (20–45 degrees, inclined grout holes to cut the vertical strata are shown in *Figure 140*.

Isotopes

There are two inspection galleries and originally there was a leakage from the drainage system of 9 litre/sec., which was reduced to 7 litre/sec. After 5 years, in 1957 the leakage fell off to 4·76 litre/sec (411 m³ per day).

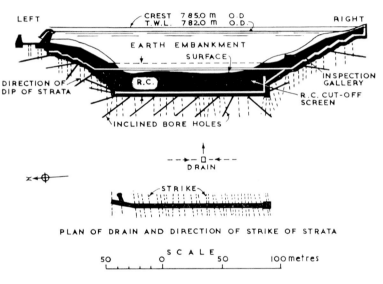

Figure 140. Rosshaupten dam. Longitudinal section of earthen embankment 41 m in height founded on vertical Oligocene sandstone, 'Molasse'. The strike of the beds are parallel to the river, the vertical dip of the beds being shown into which grout was pumped by inclined bore holes

From a determination of seepage by radio-active isotopes, a mean velocity of 0·32 m per day was ascertained from the mean flow in 4 test bore holes, and as the cross-section of the area of leakage was estimated to be 3,200 m², that is, 155 m in length and 19 m in depth, the leakage would be 0·32 × 3,200 = 1,024 m³ per day.

As the drainage adits show only 411 m³ per day, it is concluded that a good deal of the leakage does not enter the drains but emerges without doing any damage far beyond the dam toe.

The velocity of the water is simply measured by a geiger counter in each bore hole; the decrease in activity registered is directly proportional to the velocity across the bore hole. The activity drop per unit of time, therefore, is all that it is necessary to measure.

Geology of the dam

The soft material at the site and the soft sandstone 'Molasse' foundation suggested that the dam should be made of earth.

The impervious clay core and horizontal apron under the upstream slope (*Figure 141*) was obtained from local Boulder Clay which contained grains 50 per cent under 2 mm in diameter (*Figure 142*).

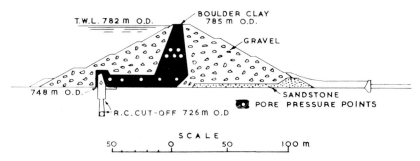

Figure 141. Rosshaupten dam. Section of embankment of moraine gravel filling with a sandstone drainage carpet. There is a Boulder Clay core and apron leading upstream to the trench which is filled with reinforced concrete. The pore pressure points which dissipated by about 20 per cent in 4 years are indicated. The embankment is on vertical strata, the strike of which is parallel to the valley

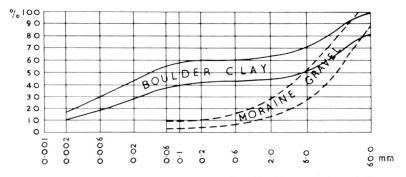

Figure 142. Rosshaupten dam. Composition of Boulder Clay core and gravel fill

The embankment is of granular gravel; the downstream slope is placed on a sandy mattress resting on a cleaned broad foundation, and the upstream slope is on the moraine Boulder Clay extending from the core up to the trench and thence on a mattress of stone on a cleaned foundation. The composition of the gravel is shown in *Figure 142*, 80 per cent by weight being over 2 mm in diameter.

The reinforced concrete trench is carried upward for 20 m into the gravel embankment of the upstream toe.

The gravel of the embankment was compacted in layers up to 120 cm in thickness with 8 ton vibrators with varying amounts of water. The Boulder Clay core was compacted by 8 passes of 18 ton tamping rollers, designed with shoes to lift vertically and not to disturb the compaction.

Pore-pressures were recorded on the horizontal clay apron and base of core at the level of 749·80 m and also at the higher levels, 765·75 m and 768·30 m.

At the completion of the dam in 1953, in April 1957, and subsequently in March 1960, the pore-water pressures percentages of superimposed loads were recorded at several levels, as follows.

Level (m)	Percentage of superimposed load			
	September, 1953	March, 1954	April, 1957	March, 1960
768·30	51·5	39	20	25
765·75	36–44	31–38	20–30	17–27
749·80	65	43·1	27	26

Sylvenstein Dam (near Bad Tölz)*

The Sylvenstein is a new dam built on the River Isar which flows northwards from the Bavarian Alps past Bad Tölz to Munich and thence into the Danube. The floods from the Tyrol are severe, and typical of the damage

Figure 143. Salzburg autobahn. Showing the damage caused by the floods from the Tyrol in August 1959

done by them was that on the River Salz, a few miles to the east, when a new autobahn bridge near Salzburg was wrecked in August 1959 (*Figure 143*). This bridge alone probably cost nearly as much as the Sylvenstein dam, the purpose of which is primarily for flood prevention.

*Very many thanks are due to the Ministerial dirigent, Dipl.-Ing. Joseph Krauss for reading the foregoing notes which are based on the papers by him and Dr. Ing. W. Lorenz in Bautechnik. Thanks are also due to Mr. E. D. Lawrence for help in the preparation of these notes.

The Sylvenstein reservoir has a capacity of 60×10^6 m³ (T.W.L = 764 m O.D.; reservoir area = 3·7 km²) and in *Figure 144* it is seen that in 1954 the catastrophic flood at Tölz, 20 km downstream of the dam, would have been reduced by one-half, from 780 m³ per second to 420 m³ per second; and at Munich, 60 km downstream, by one-third, from about 1,130 m³ per second to 710 m³ per second.

At the dam site the maximum flood is 600 m³ per second based on 60 years records, the gathering ground for which is 9,211 m². The average annual flow is 940×10^6 m³.

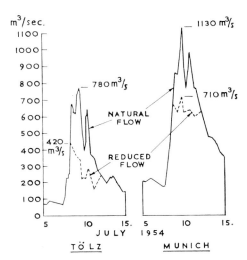

Figure 144. *Sylvenstein reservoir. The diagram shows how a great flood which occurred in 1954 would have been reduced had the reservoir been built*

Figure 145 shows a plan of the dam giving positions of the 'scour' tunnel on the left bank (capacity 370 m³ per second) and the 'power' tunnel on the right bank (capacity 225 m³ per second). There is also an automatic gate which can discharge over 200 m³ per second.

Figure 146 shows the conditions during the great flood in August 1959, when the Salzburg bridge (*Figure 143*) was destroyed.

The dam foundations are of great interest: the dam is built in a Dolomite valley of the Upper Trias filled to a width of about 140 m at the surface and to a depth of about 100 m below the surface with glacial moraine silts, sands, gravels and scree from the Bavarian Alps (*Figure 147*).

A rockfill dam (crest 185 m in length, maximum height 41 m) had to be adopted for such a soft foundation.

The cut-off for grouting such heterogeneous glacial deposits consisted of a thixotropic clay-cement grout, with sodium silicate, applied for a width of 6–21 m under the central clay core and gravel filter of the embankment (*Figure 148*). The physical analyses of these heterogeneous materials, as well as the composition of the grout injected, are shown in *Figure 149a*. The composition of the dam is shown in *Figure 149b*.

The underground impermeable block was first confined by the two thixotropic clay cement screens (18–21 m apart) for a depth of 15–20 m

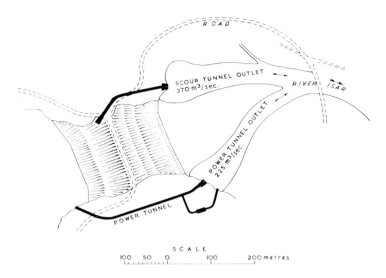

Figure 145. Sylvenstein dam. This is a rockfill dam built on a depth of about 100 m of glacial material in a Dolomite valley. It was built to control the floods from the Bavarian Alps in the River Isar. The capacity of the reservoir is 60×10^6 m^3 and there are two outlet tunnels as well as a spillway

Figure 146. Sylvenstein dam. The Isar in full flood below the dam in August 1959. Two streams of water are coming from the dam, one from the lower tunnel on the left, and one from the scour tunnel on the right. The dam is between two cliffs of Dolomite between which there is up to 100 m of permeable moraine

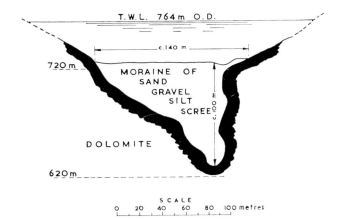

Figure 147. Sylvenstein dam. The Dolomite gorge is filled with soft material from the Bavarian Alps and this material was capable of sustaining the weight of a rockfill dam of a maximum height of 41 m. The cut-off was made by grouting a section of the glacial filling under the centre of the dam. The composition of the typical morainic material is given in Figure 125a

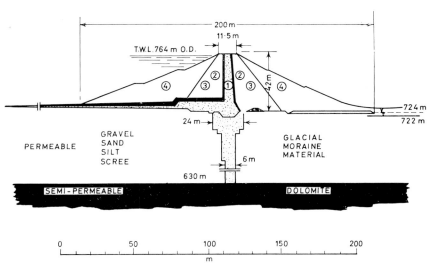

Figure 148. Sylvenstein dam. A thixotropic clay cement grout with sodium silicate was injected to form two curtains 18–21 m apart within the 24 m section and about 15 m in depth. Between these walls the whole of the moraine was grouted down to the Dolomite to form an impermeable block having a volume of some 70,000 m³. A grout curtain to a depth of 25 m was also formed in the Dolomite for which only a little grout was accepted. (1) Clay core; (2) sand filter; (3) selected fill; (4) rockfill. See Figure 149b for composition

which went all the way across the valley. Between these, pressure grouting was applied right down to the bottom of the valley, as shown in *Figure 148*.

The grout curtain

The raw product contained 35 to 40 per cent of clay with a plasticity index of 30 to 40. By adding sodium silicate the suspension becomes thixotropic and was selected because it would resist water currents, and the

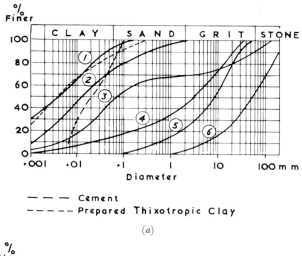

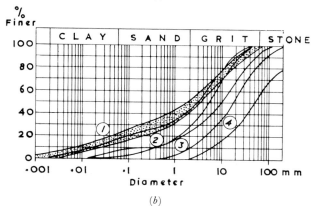

Figure 149. Sylvenstein dam. (a) Analysis of 6 morainic materials; silts, sands, grits, and stone on which the dam is built with the composition of the grout, thixotropic clay and cement, injected into them. (b) Analysis of materials used in the construction of the embankment, (1) core, (2) filter, (3) selected fill, and (4) rockfill. These reference numbers refer only to diagrams 148 and 149b

silicate/cement additive would resist erosion. The amount of water was adjusted for a specific gravity of 1·13 to 1·2 tons/m³ according to soil conditions.

The dimensions of the curtain, which took $2\frac{1}{2}$ years to construct are:

Area	5,200 m²
Thixotropic clay-cement suspension	37,000 m³
Volume of rock treated	70,000 m³
Length of boreholes	9,400 m

Below the grouted block, also at the sides of the valley, some cement grout was injected, under pressure, to a depth of 25 m, but little cement was accepted as the Dolomite is very impermeable being very free from joints and fissures.

Percolation tests

By 1967, after six years many records of percolation through the curtain were computed and a typical example for an impounding head of 30 m is assessed at 17 litres per second on the following lines:

Area of curtain, 5,230 m²
Width of curtain, 11·8 m
Area of valley, 8,500 m²
Length upstream of curtain, 200 m

From involved and detailed pressure-measurements, the coefficient, (k_1), of percolation in the natural moraine was assessed at 5×10^{-3} m³/sec. and from a ratio of $k_2/k_1 = 2 \cdot 5 \times 10^{-4}$ the coefficient of the curtain (k_2) is assessed at $1 \cdot 25 \times 10^{-6}$.

The percolation through the curtain would therefore be

$$1 \cdot 25 \times 10^{-6} \times \frac{30}{11 \cdot 8} \times 5,230 = 0 \cdot 017 \text{ m}^3/\text{sec.} = 17 \text{ litre/sec.}$$

This flow has remained stable for six years. The average estimate of ground water flow through the Alluvium, before the grout curtain was made, was 150 litre/sec.

BIBLIOGRAPHY

Rosshaupten dam
MOSER, H., and NEUMAIER, F. Determination of seepage flow under Rosshaupten dam by means of radio-active isotopes. *I.C.O.L.D.*, **4** 143 (1958).
TREIBER, F. Compaction methods, etc. for the Rosshaupten dam. *I.C.O.L.D.*, **3** 123 (1958).
— Measurements and observations on Rosshaupten dam. *Ibid.*, R5, **2** 215 (1958).

Sylvenstein dam
KRAUSS, J. Der Hochwasserspeicher am Sylvenstein. *Bautechnik* **6** 201 (1958).
LORENZ, W. Der Staudamm am Sylvenstein mit Dichtungsschürze. *Bautechnik*, 8 309 (1958).
LORENZ, W. *The grout curtain of Sylvenstein Dams. I.C.O.L.D.* **1** 19 (1967).

20

HOLLAND

Zuider Zee Dyke

In the thirteenth century the sea broke through the sand dunes which connected a chain of islands from the north coast of Holland (at Heder) to Germany. It was not considered practicable to join the islands together again owing to the deep channels which had been formed between them,

(a)

(b)

Figure 150. Zuider Zee dyke. (a) *North Sea side, 18½ miles in length, 24 ft. above mean sea level and 57 ft. maximum height. Basalt pitching on Boulder Clay embankment.* (b) *Lake Ijsselmeer side, stone pitched Boulder Clay embankment with road, cycle track, footpath, and space for railway by fence on the berm 13 ft. above mean sea level*

so the Zuider Zee dyke, 18½ miles in length, was the alternative adopted, and this was completed in 1932.
The main dimensions of the dyke are as follows.

Maximum depth below mean sea level	33 ft.
Normal depth below mean sea level	16 ft.
Height of bank above mean sea level	24 ft. (on North Sea side)
Height of berm above mean sea level	13 ft. (on Lake side)
Width of bank at mean sea level	285 ft.

The bed of the sea under the dam is Boulder Clay and other glacial material on to which willow mattresses of brushwood (350 × 80 ft.) were lowered and weighted with stones. Two parallel dams of Boulder Clay deposited by hopper-bottom barges with sand pumped between them formed the underwater foundations for the dyke. The Boulder Clay was obtained from the lake. The North Sea embankment was pitched with basalt in the tide range of wind and water and the toe was protected by brushwood held down by stones (*Figure 150a*). The lake-side embankment was not so heavily pitched (*Figure 150b*).

The final closing gap was made on May 28th, 1932, when a difference of head of 16 in. between the sea and lake had to be closed against a velocity of 9 ft. per second.

The Zuider Zee dyke enables: (*a*) polders (land reclaimed) to be formed by smaller dams of less height than the dyke; (*b*) maintenance to be saved by creating a quiet fresh water lake (270,000 acres); (*c*) provides a fresh water supply for irrigation; (*d*) suitable drainage for existing rivers such as the Ijsselmeer; (*e*) water transport; and (*f*) good road access between North Holland and Friesland.

21

ITALIAN DAMS

In Italy since 1930 there are 413 dams, of which 178 have been built since 1950; they are as follows.

Type of dam	Dams constructed up to the year	
	1950	Oct. 15th, 1966
Gravity	43	107
Buttress Hollow	16	23
Arch and dome	35	68
Gravity arch	21	22
Multiple arch or slab	0	8
Rockfill Dry Masonry	9	25
Earth	36	45
Masonry Weirs	3	84
Other types	21	31
Totals	178	413

Of these 413 dams, 53 are between 15 m and 50 m, 33 between 50 m and 100 m and 9 over 100 m in height.

As the cost of materials is high, it is important that concrete dams should be as thin as possible, particularly as labour and skilled labour are available at less cost than in many other countries.

For economy, but mainly for technical reasons, cement is often mixed with natural pozzuolana found in many places in Italy, especially near Rome and Naples.

Geology plays an important part; in the western central Alps and in the Apennines, in crystalline and metamorphic regions, sites are wide, necessitating gravity dams—solid or hollow.

Whereas in the north-eastern Alps, both in the Triassic Dolomites V gorges and U valleys, high cupola and arch dams (built largely to designs tested by models in the Bergamo laboratories) have become some of the most remarkable dams in Europe.

As so many sites for hydro-electric power have been developed, thermal stations may become more economical. Water supply and irrigation dams in low level areas and in the south are becoming more numerous. Earth/

I am very much indebted to Dott. Ing. Carlo Semenza for reading all the notes on the Italian Dams of the S.A.D.E.

rockfill dams built during the 15 years, 1950–1966 number 70 compared with 5 for the previous 20 years 1930–1950.

Some of these are described in the following pages; the localities are shown in *Figure 76* on page 143.

Piave di Cadore Dam

The Piave di Cadore dam, constructed 1947–50, is of geological interest as it is built partly on a 'plug' of concrete in a gorge and partly on a rocky

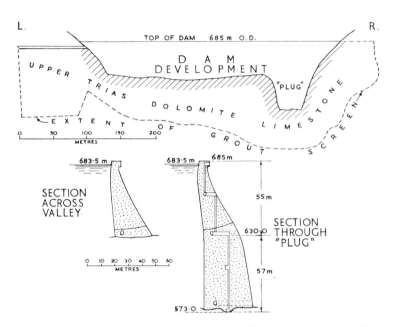

Figure 151. Piave di Cadore dam. In the valley of the Piave there is a gorge over 50 m in depth which was filled with concrete (known as a 'plug') to enable a curved, thick, arch dam to be constructed, typical sections of which are shown. An extensive grout curtain in the limestone was necessary

plateau aside the gorge, all on Upper Trias Dolomite limestone in the valley of the River Piave. The site of the dam is near Cortina in the heart of the Eastern Alpine Dolomites.

The gorge which had to be plugged is some 57 m in depth and 55 m in width (*Figure 151*), and the dam built on top of the plug is another 55 m in height, or 112 m above the bottom of the plug.

Figure 152a and *b* shows a view of the dam with the gorge on the left, and *Figure 152c* is taken from the top of the dam looking down into the plug and showing the road bridge over the gorge, a little way downstream.

The first suggestion was to build a gravity dam on the plateau (Pian delle Ere) with an arch dam in the gorge, but the tension stresses at the base of the gravity section in the neighbourhood of the abutment were too high.

(a)

(b)

(c)

Figure 152. Piave di Cadore dam. (a) The dam is put upon a fairly wide flat valley having a deep gorge at the right abutment end (left in photo). The gorge was filled with a so-called 'plug' of concrete, the top of which can be seen. (b) Left abutment showing the long length of curved dam on the flat part of the valley in continuation with (a). (c) The gorge looking down from the top of the dam. The bridge can also be located in (a)

The second suggestion was to have a gravity dam resting on a plug in the gorge, but the plug would have had to be large and the stress conditions between the base of the gravity dam and the plug were uncertain.

The third suggestion (which was adopted) was a thick arch dam on the rocky plateau and a smaller plug.

Radius of arch	160·9 m
Length of crest	390 m
Thickness of base of arch	26 m
(This is 65 per cent of the thickness of a gravity dam. Young's modulus of the rock was taken at 100,000 kg per cm^2 (or $1·422 \times 10^6$ lb./sq.in.))	
Chord of dam at crest	305 m
Height of dam above plug	55 m
Chord-height ratio	5·5 m

The thickness of the dam being substantially less than a gravity dam is of special interest for such a chord-height ratio of 5·5.

The reservoir has a useful capacity of $64·3 \times 10^6$ m^3 between the top water level of 683·5 m and 625·5 m O.D.

Grout

The limits of the grout screen are shown in *Figure 151*, and the quantities of grout and bentonite used and length of bore holes drilled (reported 1956) are as follows.

Bore holes	64,267 m	6,179	tons of cement
Bore holes	9,125 m	85	tons of bentonite
Totals	73,392 m	6,264	tons of chemical products

Area of grout screen	37,000 m^2
Amount of chemical products per m of bore hole	85·5 kg per m
Amount of chemical products per m^2 of screen	170 per kg m^2

Dott. Ing. Carlo Semenza reported that some areas on the left abutment with maximum hydrostatic head, had some leakage of a few litres per second which was harmless since the nature of the limestone is not affected by the presence of water. However, to reduce this leakage supplementary holes and grouting were carried out on the left shoulder, and by the end of September, 1950, a total length of 62,000 m had been drilled and 5,700 tons of cement injected. The injections had been carried to a depth of more than 135 m, that is, below maximum flood level.

Valle di Cadore Dam

The Valle di Cadore dam is 4 miles south-west from the Piave di Cadore dam and was constructed about the same time (1948–50).

The reservoir capacity between the top water level of 706·5 m O.D. and the level of 673·7 m O.D., is $4·3 \times 10^6$ m^3 (840×10^6 million gal.) (*Figure 150*).

The dam is in a gorge in the valley of the River Boite, a tributary of the Piave, which flows in the Middle Triassic Dolomite, Ladinic period (below the Noric Stage). This horizon includes limestone, marly clay and flysch.

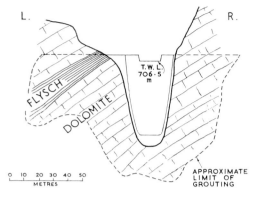

Figure 153. Valle di Cadore dam. Section of dam (developed). Plugged gorge and extent of grouting in Middle Trias Dolomite. A cupola dam is appropriate in such a valley for a dam twice as high as its crest width

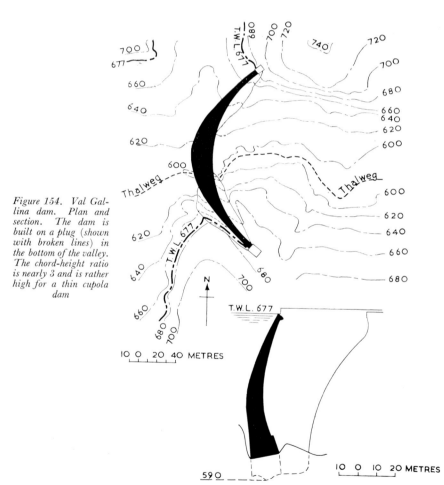

Figure 154. Val Gallina dam. Plan and section. The dam is built on a plug (shown with broken lines) in the bottom of the valley. The chord-height ratio is nearly 3 and is rather high for a thin cupola dam

The general dip is downstream but at the dam site it is across the thalweg.

Type	Cupola
Height of dam	58 m
Length of chord (approximately)	25 m
Chord-height ratio	0·425

Grout

Length of bore holes	4,080 m
Area of screen	6,500 m^2
Cement	1,500 tons
Cement per m of bore holes	375 kg per m
Cement per m^2 of screen	230 kg per m^2

Val Gallina Dam

Like the Vaiont valley, 5 miles to the north, the Gallina is a tributary of the Piave, and the dam is about 2 miles east of the Piave up the Gallina valley.

The Val Gallina dam, constructed in 1950–51, forms a large reservoir of $5\cdot84 \times 10^6$ m^3 at a top water level of 677 m O.D. The gathering ground is small but the reservoir acts as a balancing reservoir for the comprehensive hydro-electric system S.A.D.E.

The dam site is in an anticline of Upper Trias Dolomite and is comparatively wide above the plugged gorge (*Figure 154*), whereas the Vaiont dam is in a syncline in a narrow Jurassic gorge with almost vertical sides.

The maximum height is 92·4 m and the gorge in Upper Trias limestone (Noric stage) has been plugged with concrete upon which the curved arch (cupola) dam (99,000 m^3 concrete) has been built.

The chord-height ratios at the crest are 2·71 inside, and 2·15 including the abutments, rather high for a thin arch dam, according to Dott. Ing. Carlo Semenza. It is about 5 m in thickness at the top end, and 15 m in thickness above the plug, the latter being slightly thicker.

The right and left abutments and the nature of the limestones are shown in *Figure 155a, b and c*.

A large amount of grout was used, fairly well distributed on both sides of the dam in the Upper Trias limestone (*Figure 156*), the details of which are as follows.

Length of bore holes	30,000 m
Cement used	28,647 tons
Area of grout curtain (approximate)	20,000 m^2
Grout used per m of bore hole	950 kg per m
Grout used per m^2 of screen	1,432 kg per m^2

(Five times that of the Middle Trias limestone of Valle di Cadore, and nine times that of the Piave di Cadore dam also in Upper Trias limestone.)

(a)

(b)

(c)

Figure 155. Val Gallina dam. (a) The right-hand abutment and top of plug is seen resting in Upper Trias (Noric) limestone. (b) Left-hand abutment in Upper Trias limestone, base of dam resting on plugged gorge. (c) The gathering ground is composed mainly of Triassic limestone mountains with regular slopes. As the valley is in a broad anticline, the slip on both sides of the valley, is into the hillsides

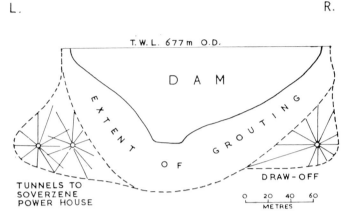

Figure 156. Val Gallina dam. Extent of grouting in a wide valley of Upper Trias limestone (Noric Stage). Nearly 1,000 kg of cement per lineal metre of bore hole were consumed or over 1,400 kg. of cement per m² of curtain to secure watertightness

Vaiont Dam

The Vaiont dam (under construction in 1957–59) is a remarkable dam in a deep Jurassic gorge, 10 miles north-east of Belluno, about a mile from the main valley of the River Piave on the Vaiont tributary.

Type	Cupola
Top of dam	725·5 m O.D.
Normal top water level	722·5 m O.D.
Bottom of foundation	460 m O.D.
Maximum height	265·5 m
Useful storage	150×10^6 m³

The limestone in that part of the gorge in which the dam is founded is of Middle Jurassic age having, in general, a fairly steep dip upstream, but affected by orogenic action. The limestones and dolomitized limestone layers known as the 'Dogger' are generally oolitic and massive, hence it is possible to submit it not only to a water pressure nearly 900 ft. in depth (25 tons per ft.²), but also to the thrust of the 'arch' abutments, calculated not to exceed 68 kg per cm² or 60 tons per ft.²

The gathering ground is basically the Middle Jurassic limestone and Upper Jurassic marls, etc., and Lower to Upper Cretaceous (Senonian), red marls, Scaglia, etc.

Figure 157a, b and c shows the excavation from top to bottom of one abutment; men can be seen at work in (c) which gives an idea of the scale. The dark lines are gangways cut in the rock connected with tunnels for access,

(a)

(b)

(c)

Figure 157. Vaiont dam. (a) Excavation in Jurassic Dolomite for the right-hand abutment. The horizontal dark bands are the walkways for the men. (b) A little lower down than (a) there are ladders and stages going vertically down the excavation. Some of the men can be seen at the bottom of the black vertical marking. (c) The base of the 875 ft. excavation showing the vertical black marking access ladders and men just below in the top left-hand corner. At the base of the abutment the dip of the strata upstream is well seen

access in detail being given by ladders to each stage of the work as it proceeds with rope ladders and ladders with protective hooping (*Figure 158*).

Figure 159 shows the suspension bridge (nearest the camera), the 'pipe-bridge' (further away) which conveys water from the higher Piave di Cadore reservoir to the large Soverzene power station, 5 miles distant, and beyond

Figure 158. Vaiont dam. Near the top of the left-hand abutment with access protective cage

Figure 159. Vaiont dam. Looking up the valley. The works suspension bridge at the top is nearest the camera. The pipe bridge conveying water from the existing reservoir to the turbine house is further off; beyond, on the left, can be seen the excavation for the right-hand abutment of the dam with the black horizontal streaks indicating the access paths shown on Figure 157

on the left, the smoothed right-hand abutment (indicated by the access paths shown in *Figure 157a* and *b*.

Although the limestone in *Figure 159* seems as if it would be extremely permeable, it becomes much denser when protected from weathering, that is, at a depth of a few feet behind its exposed face.

The rocks have to be treated with great respect as minor faults and joints cause chunks to break away. There are hard nodules of radiolaria, 'almonds', which cause difficulties in drilling and parting seams of thin clay which may cause slipping.

After a good deal of investigation, the type of dam adopted was a thin dome or cupola type having the following characteristics.

Crest length	190·5 m
Height of dam (maximum)	265·5 m
Chord	168·6 m
Chord-height ratio	168·6/265·5 = 0·7 approximately
Volume of concrete in dam	355,000 m³

Figure 160. Vaiont dam. A view looking upstream showing the valley taken from the suspension bridge in Figure 159. The top right-hand corner shows the area of the disastrous slide of the weak Cretaceous on the Jurassic limestone which is exposed on the left

The dam is built on a cylindrical plug 50 m in height, 22·71 m in thickness at 460 O.D., and 19·71 m in thickness at 510 O.D., the upstream face radius being 46·5 m and the downstream radius 23·9 m (Figure 161).

The dome rests on the plug (which is slightly wider) and has the following dimensions.

At 550 O.D. thickness 16·5 m; horizontal radii 49·5 and 33 m (and 27 m).
At 720 O.D. thickness 3·4 m; horizontal radii 109·35 and 105·94 m (and 96·04 m).

The Vaiont Reservoir area

The Vaiont reservoir area has been referred to on page 71 and *Figure 160* is a birdseye view of the valley looking upstream from the suspension bridge shown in *Figure 159*.

Figure 162a of the reservoir area is adapted from Professor Edoardo Semenza's remarkable geological synthesis, carried out mainly in collaboration with F. Giudici in 1959 and 1960 and with D. Rossi in 1963–1965 after the disaster on 9 October 1963.

The basic formation upon which the dam is founded is Middle Jurassic limestone with massive dolomitized limestone, 'Doggers'. The submerged

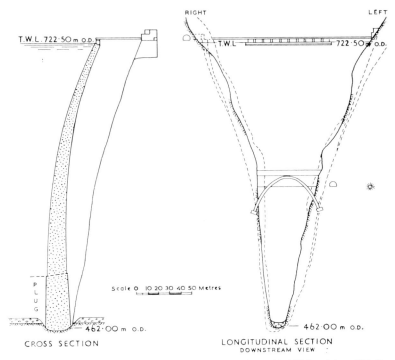

Figure 161. Vaiont dam. Section of the highest dam in the world, 265·5 m (about 875 ft) in height, impounding 860 ft of water. The height of the dam is half as high again as the width of the valley at the top, the chord-height ratio being 0·7. A thin cupola type of dam has been adopted as the massive dolomite Jurassic limestones are well able to take the thrust of the abutments. There is a road tunnel at the top of the dam on the right side of the valley, and a road tunnel on the left side of the valley about half way up the valley at the level of the bridge, which carries a large pipe

area consists of Upper Jurassic marls, clays and limestones underlying the Cretaceous. The top strata of the Jurassic are known as 'Malm' which is a calcareous Marl.

The Lower Cretaceous strata are very disturbed and are in juxta-position with large areas, for want of a better term, undifferentiated disturbed detritus, 'Detrito prevalente'. (See 'Alluvium, etc.' of *Figure 162 (b)*.) Owing to further research by C. Loriga and M. G. Mantovani (1961–1965) some some further identification of the Cretaceous strata was revealed and corrected through micro-palaeontological investigation.

Figure 162b shows Professor Edoardo Semenza's suggested slip planes distant 500 m and 800 m from the drop on the Malm/Cretaceous junction. *Figure 163* is an example of the complexity of the Alps, although the causes of the complexity remain unsolved. It includes the Gallina and Vaiont valleys, the one in an anticline the other in a syncline, from Professor Piero Leonardi's book, Le Dolomoti published in 1967, (Vol. II, p. 44). On the northern flank of Mount Toc, the complete series of Jurassic limestones, Dogger, Malm, Lower and Upper Cretaceous to the Eocene form a syncline whose axis N. 80° E. to S. 80° W., approximately along the Dogger-Malm

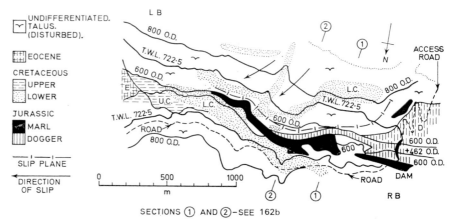

Figure 162 (a). Vaiont Reservoir. The dam is on the right, the bottom water level is 462 m O.D. and the top water level 722·5 m O.D. The level of the water was about 700 m O.D. when the slip occurred and the depth of water into which the mass of 600 million tons fell in 6 seconds would be of the order of 200 m. The contours show the level of the valley before the slip and the geology is approximate; a good deal of the Lower Cretaceous area of the slip, shown by the two arrows is undifferentiated. The lower limit of the slip plane appears to be approximately parallel with the 600 O.D. contour

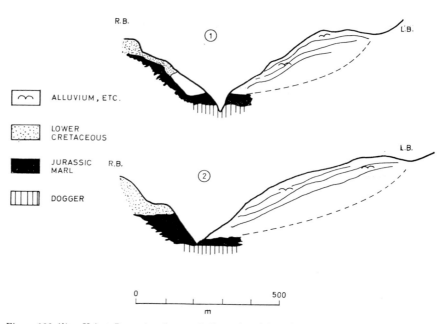

Figure 162 (b). Vaiont Reservoir. Suggested slip sections (1) and (2) of Figure 162(a) of the slide of the Cretaceous on the Jurassic Malm. (Figures 162 (a) and (b) are taken from the report Giudici-Semenza of June 1960 and are not exact in all details. For mor eexact details the reader is referred to the large-scale (1/5,000) maps of Rossi-Semenza, 1965)

outcrop in the bottom of the valley shown in *Figure 162* (*a*). The slip movement was from north to south in the valley bottom, the momentum drove the slide up the right bank and, on the left bank, marly limestones of Malm age and crystallized limestones of Dogger age were exposed in thick massive beds. Notwithstanding the complexities of detail in the formation, faults and thrust faults, folds, talus, joints, and tumbling effects, the whole of the solid mass has preserved its original shape. *Figure 164* gives a general impression of the shape of the slip from a paper by R. Selli and R. Trevisin.

The following is a Report on the Vaiont dam and reservoir by the Italian National Committee on Large Dams and Ministry of Public Works Dam Service (*I.C.O.L.D.* **IV** 730 (1967)).

'The Vaiont disaster was the severest catastrophe involving hydro-electric reservoirs that ever occurred in Italy, and it was caused by the sudden breaking loose of a large portion of the Mount Toc hillside, on the left side of the valley, into the reservoir.

The dam was undamaged, notwithstanding the tremendous impact to which it was subjected, thanks to its excellent design and construction.

The Vaiont tributary which flows in a deep ravine eroded through a series of Cretaceous and Jurassic calcerous formations was barred by an arch dam, 261 m high, across a gorge located approximately 1400 m upstream from the River Piave.

The dam, built in 1956–1960, created a reservoir having a useful capacity of 150 million m^3 and a maximum storage level of 722·50 m above sea level.

On October 9th 1963, a little upstream of the dam, an enormous mass of rock, 200 m high and 1,800 m long with an estimated volume of 250–300 million m^3 broke off in one piece from Mount Toc and started sliding rapidly. The front of this mass travelled approximately 400 m, and reached the opposite bank in a few seconds and slid up into it.

At that time the water-level in the lake was 25 m below the dam crest. The rocky mass retained its continuity to a surprising degree in the slide and it filled most of the valley above the dam crest.

The moving mass caused a terrific air-blast which thrust the water, 200 m up the right bank. In falling back on to the landslide some water flowed upstream and the rest rushed downstream producing a disastrous flow over the dam. The volume of the overflow was 25–30 millions/m^3.

The sloping "blade" of water reached a height of about 200 m on the right bank and 100 m on the left bank. After having travelled the 1400 m from the dam to the Piave river in less than a minute, the huge wave, with a front at least 70 m high, rushed across the 400 m of the Piave river bed and crashed first into the town of Longarone and then into other inhabitated areas both upstream and downstream.

The stability of the reservoir banks had been checked thoroughly during both the design and construction stages. Later on, while water was being impounded for testing purposes, a fall on a slope occured and special attention was directed to a particular area on the left bank where, in the autumn of 1960, a thin crack had appeared.

A careful check was kept by means of a complete series of observations and investigations. These included boreholes, wells for water-level observation, exploratory tunnels and a network of ground levels with the setting up of many bench marks. A seismographic station was also installed to record local earth movements.

It was found that the movements of the part of the hillside delimited by the crack, always very slow, showed no tendency to be related to the storage level.

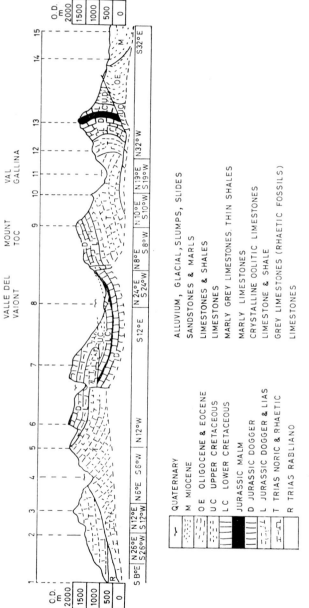

Figure 163. The Gallina Dam lies in an anticline of Upper Trias limestone (Noric and Rhaetic) where both right and left banks slope into the valley sides. (The Rhaetic limestone is hundreds of metres in thickness in the Dolomites, compared with 25 ft. in Britain.) The Vaiont Dam is in a syncline of Upper and Lower Cretaceous (limestone and shales) on Jurassic Malm (shale and limestone) and Dogger (hard crystalline limestone). On the right bank these strata have a gentle slope dipping into the bank and are capped by a thrust horizontal fault with hard Dogger Limestone. On the left bank, the Lower Cretaceous Malm, and Dogger, dip steeply into the valley (shown with a dotted line before the slip) on the lower slopes of Mount Toc. Mount Toc itself appears to be well stabilised by a fault pushing up the Dogger, Lias and Trias Dolomites

1. T. Valmontina
2. M. Val della cima (1,567 m)
3. Sasso di Mezzodi (2,034 m)
4. V. Pagnac di dentro
5. M. Citt (2,190 m)
6. M. la Palazza (2,298 m)
7. M. Borga
8. Valle del Vaiont
9. M. Toc (1,921 m)
10. Col Dcon (1,303 m)
11. Val Gallina
12. Basc pieta
13. Col Mat
14. F. Tesa
15. T. Tessina
} Alpago

250

Nothing in the geological, geotechnical and hydrological information on the area, or in its past history, gave any indication that there would be such an abrupt change in the characteristics and trend of the landslide. All previous observations indicated, as said above, a very slow movement, even in the last 24 hours preceeding the disaster, and there was no reason to predict that the whole mass would slide in one piece so suddenly and so quickly.

This is the story of the tragic number of casualties suffered including the engineers for the dam.

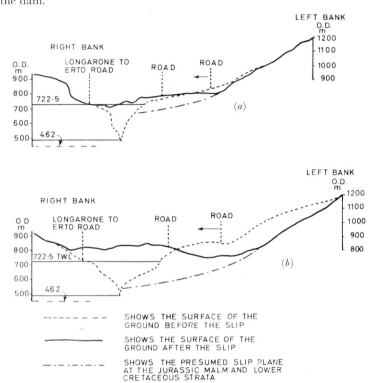

Figure 164. Section near the dam before and after the slip. The top water level is 722·5 m O.D. and at the time of slip was about 600 m O.D. Datum line taken as 462 m O.D. which is the base of the foundation of the dam. (a) shows the conditions before and after the slip near the dam, and (b) a short distance upstream.

With regard to the dam it should be pointed out that it did not suffer any damage nor show any anomaly in its elastic behaviour not even from the new stress condition imposed on it by solid material replacing water or the dynamic thrust it had to withstand.'

Pontesei Dam (River Maè)

The Pontesei dam, which was built in 1956–57 and is part of the system of the Società Adriatica di Elettricità (S.A.D.E.), is one of many in the Italian Dolomites for impounding water for generating electric power.

The gathering ground is on a tributary of the River Piave known as the River Maè, and commands about 150 km² of the Permian to Upper Triassic

(a) (b)

Figure 165. Pontesei dam. (a) Right end abutment on Upper Trias Dolomite. The outlet of the tunnel from the morning glory spillway can be seen near the left bottom corner below which the water cascades down the white part of the cliff. (b) Left end abutment on Upper Trias limestone. The slipped area is above the top of the dam on the left

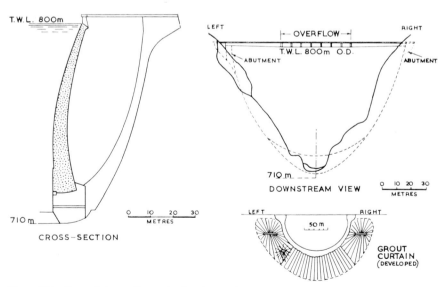

Figure 166. Pontesei dam. Shape of valley for which the chord-height ratio is 1·66. Grout curtain which consumed 42·5 kg of cement per m^2. Section of dam, a cupola type well suited to the shape of the valley which is sound limestone capable of sustaining 60–70 tons per $ft.^2$ without deformation

formations, giving rise to a hard water. The site of the reservoir lies between Belluno and Cortina near Forno di Zoldo, about 65 miles from Venice.

The dam is founded on Upper Triassic limestones, the Noric stage (in age, lying somewhere between our English Keuper marl and Rhaetic limestone). Superficially this great outcrop of Italian limestone resembles something between our Carboniferous and Jurassic limestones. However, from the building point of view, the limestones have strength, and although they are capable of sustaining high pressures they are much jointed and fissured and blocks may become detached; but they are not karstic, that is, cavernous, and for this reason Dott. Ing. Carlo Semenza, chief engineer of the S.A.D.E., says they are 'honest' limestones.

The Pontesei dam has the following characteristics.

Type	Cupola
Top of dam	803 m O.D.
Maximum top water level	800 m O.D.
Bottom of foundation	710 m O.D.
Maximum depth of water	90 m
Capacity (useful)	$9 \cdot 1 \times 10^6$ m^3

The right and left banks are both in solid Upper Triassic limestone of Noric Dolomite dipping generally at 30 degrees upstream.

The type of dam adopted for this U-shaped limestone valley is a symmetrical thin dome, or cupola, involving a maximum thrust of 67·5 kg per cm^2 under hydraulic pressure, temperature variation, and shrinkage.

Crest length	150·2 m
Maximum height of dam	93 m
Chord	115·3 m
Chord-height ratio	1·66
Thickness of crest at centre	2·64 m
Thickness of bottom at centre	12·70 m

Figure 165a and b shows the right and left abutments respectively, and *Figure 166* the shape of the valley.

The grout curtain consists of a comprehensive system of inclined bore holes drilled from the surface and from galleries.

About 18,000 m of bore holes were drilled into which 700 tons of grout were injected, equivalent to about 39 kg per m.

Area of screen (approximate)	16,500 m^2 (*Figure 166*)
Amount of grout per m^2 of screen	42·5 kg per m^2

The flood control devices included the following.

(1) A morning glory spillway having a lip of 20 m in diameter or a length of 62·8 m, capable of discharging 350 m^3 per second (*Figure 167*). The bellmouth leads the water into a vertical shaft 5·6 m in diameter. The water drops 40 m and flows into a horizontal tunnel 5 m in diameter and 110 m in length cut in the right bank of the valley. The open end of this tunnel is some 50 m above the stream and is visible in *Figure 165*, and any flood water cascades down the hard limestone into the bottom of the valley.

Figure 167. Pontesei dam. Morning glory bell-mouth overflow with a 20m diameter lip and with wings to prevent vortex which would lessen the flow.

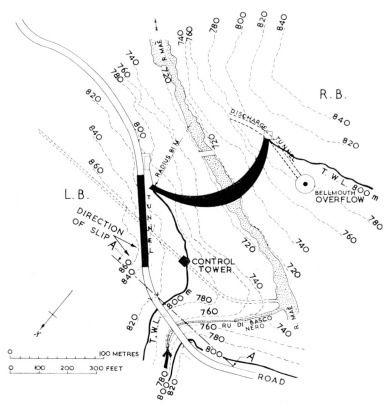

Figure 168. Pontesei dam. The contours by the control tower (below the normal top water level of 800 m) are steep, and the south-east to north-west dip of the beds and the tributary Rio di Bosco Nero entering the Maè at right-angles favours instability. Some slipping of the limestone cliff has occurred between the tunnel and control tower, but the dam is not threatened. The positions of Figures 165–171, can be located from this diagram. (See Figure 172 for Section A–A.)

(2) An auxilliary tunnel in connection with the above for 57 m³ per second.

(3) An outlet at the bottom for 175 m³ per second.

(4) A natural overflow on the crest of the dam having a length of 47·6 m for discharging 270 m³ per second.

The total overflow provided is, therefore, 852 m³ per second (28,820 cusecs) for the gathering ground of 150 km² (37,500 acres). This compares with the 11,250 cusecs estimated by the I.C.E. Floods Committee, ignoring any balancing effect of the reservoir.

Figure 169. Pontesei dam. Road tunnel under threatened slip of spur of Upper Triassic limestone. Open joints are visible on the right of the tunnel, the concrete lining of which has developed cracks. Remedial measures commenced in 1957 to stabilize the spur (see Figure 172)

Figure 168 shows the contours of the limestone hills overlooking the reservoir, and on the left bank there is a main road in a new tunnel under the hill.

Figure 169 shows the mouth of the tunnel and the limestone which has a dip of about 30 degrees towards us. To the right of the tunnel there are vertical joints opening out and large masses threaten to become detached.

The tributary stream Rio di Bosco Nero, by entering at right-angles to the main river, has left a pinnacle of rock. As the control tower (*Figures 170 and 171*) had moved some 5 cm (August, 1957), and the tunnel lining had cracked in several places, remedial measures were taken. A concrete support was built at the foot of the rocky pinnacle, as shown in *Figure 172*. The behaviour of the support is observed with strain-gauges. It is apparent from the contours shown in *Figure 168* that further movement is possible, although it might not endanger the dam it would certainly be very disquieting.

Figure 170. Pontesei dam. Road bridge over mouth of Rio di Bosco Nero stream which enters the reservoir at right-angles. The salient of rock, through which runs a tunnel in continuation of the bridge, threatens to slip and the control tower is slightly affected

Figure 171. Pontesei dam. Dip of Upper Trias limestone (Noric stage), slipped area with roof of control tower in middle distance and junction of Rio di Bosco Nero stream with reservoir on left

At a later date (1.1.60) Dr. Semenza wrote:

On the 22nd March 1959, a huge landslide of loose material (about 3 million m³ of gravel, sand blocks, clay, and so on) fell into the reservoir from its left side, from an area noticeably upstream (about 500 m) from Rio di Bosco Nero. No movement was apparent at the rocky pinnacle of Rio di Bosco Nero.

The level of water in the reservoir on that day was 13 m lower than the maximum.

The slide gave way to a sudden wave which overflowed the dam not only through the spillway but up to a few metres above the parapet.

The stresses in the dam probably became twice as high as under normal static conditions, but its behaviour was excellent. The control instruments, with the exception of only one pendulum, showed no movement at all. This is considered a further proof—not necessary but interesting—of the capacity of the curved structures to face exceptional stresses. Should the event come on a straight gravity (solid or hollow) dam something more serious would perhaps happen.

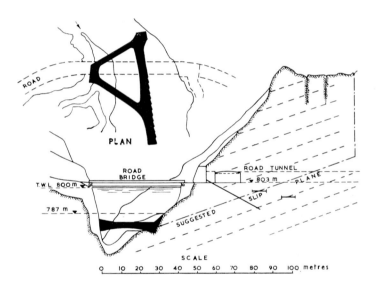

Figure 172. Pontesei dam. In 1957 cracks were noticed in the road tunnel through the Upper Triassic limestone in the area shown in Figures 168–177, and the remedial method consisted of a large concrete buttress (in black) supported by the rock on the other side of a ravine (Rio di Bosco Nero). In 1959 a large landslide occurred in the reservoir 500 m upstream of this site which made a great wave over the dam. The tunnel is behind the section A–A in Figure 168

Fedaia Reservoir

The reservoir is just below the highest mountain in the Dolomites, Marmolada, (3,342 m O.D.) in the midst of a climbing centre, complete with hotel, meteorological and cosmic ray observatories. The dams were built in 1953–56.

A reservoir which has a dam downstream to keep the water in and another dam upstream is always interesting.

Details of the reservoir are as follows.

Top water level	2,053 m O.D.
Area of gathering ground (natural, plus catchments)	$8·2 + 4·4 + 7·1 = 19·7$ km²
Capacity of reservoir	16×10^6 m³

The very high elevation of the reservoir above the power-house of Malga Capela (1,468 m O.D.) commands the useful head of 585 m, and downstream of Malga Capela a further series of heads are utilized in 8 power stations almost down to the Plain.

The details of the dams are as follows.

Main dam, maximum height	64·5 m
Secondary dam, maximum height	18 m
Type and length of crests	
Main dam; gravity (280 m), buttress (342 m)	total length = 622 m
Secondary dam; earth	length about 300 m

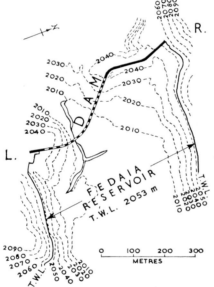

Figure 173. Fedaia reservoir. Primary dam. For this dam, a mixture of buttress and gravity sections, advantage is taken of the high level of the middle Trias limestone on the right bank. This forms a ridge at about 2,040 m O.D. The two ends of the dam are in gravity section, the middle part buttress

The great geological interest of the site is:

(a) the contour of the rock (as shown in *Figure 173*) which gives rise to the 'S' shape of the main dam and to a mixture of gravity and buttress types; and

(b) the leakage under the secondary earthen dam.

The 'S' shape

As will be seen from *Figure 173*, for the right-hand end, advantage is taken of the 2,040 m contour; the rock, on which a length of 233 m of gravity dam is placed, is hard Middle Trias limestone. The bottom part of the 'S' points downstream.

Where the dam is over 10 m in height, for economy the middle section is constructed with buttresses for a length of 342 m and up to a maximum

Figure 174. Fedaia reservoir. Primary dam. Detail of gravity section on rock outcrop at 2,040 m O.D. The high-lying rock is in the foreground and the dam points downstream. As the top of the rock is narrow the upstream slope of the dam was reduced in places

Figure 175. Fedaia reservoir. Primary dam. Detail of buttress section (342 m in length and up to 64·5 m in height) in centre part of the dam

height of 64·5 m, also founded on limestone. The middle part of the 'S' points upstream.

At the left end, for a length of 47 m, and where the depth is under 10 m, the gravity section is adopted, again for economy.

Figure 176a. Fedaia dam in the Dolomites below Mount Marmlada

Examples of the site influencing the shape of the dam, on plan, are not very common and, certainly, the pointing of curves both downstream and upstream in the same dam across a valley is rare.

Storage has been substantially increased at Fedaia by the method adopted. *Figures 174 and 176* (*a*) give a general view of the dam and show the gravity part of the dam on the rock outcrop. *Figure 175* shows the buttress portion and left side of the valley.

Leakage under the secondary dam

Figure 176 (b) shows the plan of the earthen dam and the contours of the ground in the area of leakage, while *Figure 177* shows the section and the position of the grout screen at the toe. *Figure 178* shows the extent of the grout screen (of bentonite) across the valley through the glacial and other debris.

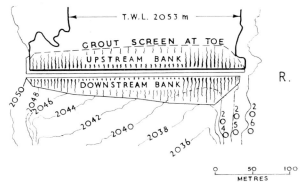

Figure 176b. Fedaia reservoir. Secondary dam. The surface contours fall away, as shown on the right, and there is a considerable amount of glacial material filling up an old lake or 'cirque'

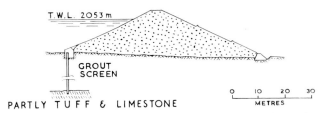

Figure 177. Fedaia reservoir. Section through earthen embankment of secondary dam showing position of grout screen

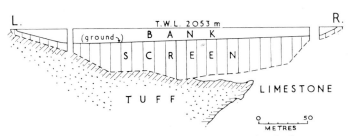

Figure 178. Fedaia reservoir. Longitudinal section (developed) of grout screen across valley and junction with tuff (volcanic debris) and Middle Triassic limestone

It will be seen that on the left side the grouting was taken down to the tuff, that on the right had to end in the limestone.

As leakage is excessive and fluorescein has not indicated its whereabouts, tracers using isotopes from Harwell were used, but without result.

(a)

(b)

(c)

(d)

Figure 179. Barcis dam. (a) *Right abutment in Upper Cretaceous limestone dipping upstream.* (b) *Right-hand side of dam showing dip of limestone top left, and head-gear for circular overflow shaft top right and overflow weir. The plug is at the bottom right-hand corner.* (c) *A circular overflow weir 13m in diameter the top water level of which can be controlled by mechanical means for lowering the water in the reservoir.* (d) *The plug in the gorge being concreted*

(Photo by J. B. K. Ley)

Barcis Dam

The Barcis dam and reservoir were constructed between 1952 and 1953 across the River Cellina about 2 miles east of Barcis (6 miles north-northwest of Maniago).

Dam type	Concrete cupola
Top of dam	405·15 m O.D.
Top water level	402 m O.D.
Bottom of foundation	355 m O.D.
Maximum depth of water	47 m
Storage	22 × 10⁶ m³ (useful) 20 × 10⁶ m³

(a) (b)

Figure 180. Barcis dam. (a) Left abutment in Upper Cretaceous limestone dipping upstream. The broken nature of this limestone is shown on the top half of the photograph. (b) Left abutment top left corner (downstream of dam) with a large bedding plane on right above river

The gathering ground is a large one (392 km²) consisting mainly of Upper Trias limestone (Noric stage). Nearer the dam, however, the scenery becomes tamer, almost English in fact, and consists of Upper Cretaceous limestone and Eocene sandy marl, flysch and broken material with breccia.

The dam is on the Upper Cretaceous limestone dipping generally upstream, as shown in *Figure 179a and b* and *Figure 180a and b*.

On the right bank the limestone is sound enough to take the thrust of the abutment, but has a flatter slope at the higher part of the dam above 375 m O.D. (*Figure 181c*).

On the other hand, the left bank, although steeper, is very much broken not only near the abutment (which can be inferred from *Figure 180a and b*), but also for a distance of over 200 m which is in contact with the water in the reservoir, as shown in *Figure 181a, b and c*.

Doubts and *difficulties*

Before a decision could be made on the type of dam, the feasibility of constructing any reservoir at all was very much in doubt and formed the subject of several geological reports. These reports called attention to the following.

(1) The apparent favourable dip of the Upper Cretaceous limestone upstream at the dam site.

(2) The soundness of the limestone for the right bank abutment.

(3) The weakness of the limestone for the left bank abutment.

(4) The prospect of abnormal leakage on the left bank for possibly several hundred metres until the red sandy clay (flysch), and reddish Cretaceous limestone, which are impermeable, could be reached.

Signor C. Corrado reported as follows.

> On the right, the gigantic limestone mass reduces the possibility of great leakage, but on the left the diaclastic (broken) limestone causes grave doubts that the reservoir would be watertight, particularly between the River Cellina and the tributary stream, the Dint, downstream from the dam site. Numerous exploratory bore holes and tests with coloured water confirmed these doubts.

He also says that the 'Barcis limestone is a good example of how Italian limestones of northern Italy are more permeable across rather than down a valley', thus supporting Dott. Ing. Carlo Semenza's statement (which provoked a good deal of discussion on his paper at the Institution of Civil Engineers in 1952).

It was decided, however, that if a comprehensive grouting programme were envisaged, in principle a reasonably watertight reservoir should be feasible. The actual grouting programme which was necessary is described in detail below.

Having surmounted the geological difficulties, therefore, the first step was to determine the type of dam. Concrete seemed obvious, and owing to the configuration of the valley, a thin curved dam was indicated; but the question arose as to what kind of curved dam was best suited for a valley with a narrow bottom (roughly 20 m in width and 20 m in depth), and one side, the left, sloping much more than the other (*Figure 181c*).

Three types were therefore considered.

(1) A symmetrical arch dam extending into the plug utilizing the flat slope of the right bank by constructing a reinforced concrete abutment incorporating flood gates.

(2) A symmetrical dome dam built over a plug in the gorge.

(3) An asymmetrical dome dam built over a plug in the gorge.

The asymmetrical thin dome or cupola type was adopted as it entailed less concrete and excavation.

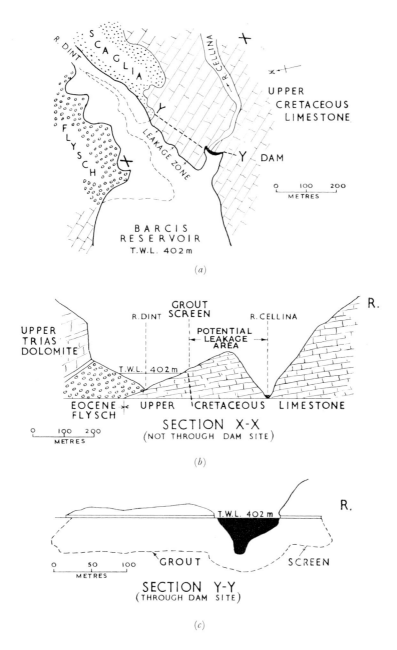

Figure 181. Barcis reservoir. (a) Geological map of Barcis reservoir showing position of dam and grout screen in the upper Cretaceous limestone. On the left bank this limestone is considerably diaclastic and the screen had to be taken for 200 m to the impervious scaglia and flysch. (b) The potential leakage area in the rotten limestone. (c) A section of the extent and depth of the grout screen

266　　　　　　　　　　　　　ITALIAN DAMS

Maximum height	50·15 m
Chord (at crest)	62·15 m
Chord-height ratio	1·25 m
Crest	83·7 m
Radius at crown, extrados	39·975 m
Radius at crown, intrados	38·125 m
Thickness at 400 O.D. (top)	2·0 m
Thickness at base	4·50 m

There are 5 vertical construction joints with copper strips, and both faces of the dam are lightly reinforced.

The grouting programme

As there were many vertical as well as horizontal joint planes to seal, many inclined bore holes as well as vertical bore holes had to be drilled.

The bore holes were spaced from 2·5 to 5 m apart, with pressures up to 20 atmospheres and up to 50–60 m in depth, the extent of the grout screen being indicated in *Figure 181c*.

The following quantities were used for the work.

Bore holes, diameter	45 mm
Bore holes, total length	14,413 m
Cement used	982 tons
Cement per m of bore hole	68·2 kg per m
Sodium silicate	44 tons
Sodium bicarbonate	1·8 tons
Area of screen	22,692 m²
Cement per m² of screen	43·3 kg per m²
Bore per m² of screen	0·638 m

Floods

The gathering ground of 392 km² (98,000 acres) consisting mainly of Triassic and Cretaceous limestones is subject to rather 'large and sudden' floods owing to the immediate influence of the sea.

The maximum flood anticipated is 1,400 m³ per second (49,000 cusecs).

The corresponding flood recommended by the Institution of Civil Engineers Flood Committee for an 'upland' area in Great Britain would be 19,600 cusecs.

The works, to take care of the maximum (1,400 m³) flood anticipated are as follows.

(1) A circular shaft 13 m in diameter which can be raised by mechanical means, therefore lowering the top water level to give a discharge of 970 m³ per second (*Figure 179c*).

(2) A bottom outlet having a capacity of 212 m³ per second.

(3) A weir (44 m in length) in the central part of the dam (8 No. 5·5 m openings). This has a capacity of 248 m³ per second with 2 ft. head.

Casoli Dam

This is situated 130 km east of Rome and was constructed during the period 1955–58. Maximum height, 54 m; crest length, 193 m.

Figure 182 (a). Casoli dam. Steep right bank with Miocene limestone regular bedding dipping downstream. The gravity section buttresses follow the limestone foundation by curving downstream on plan, seepage pump at bottom left hand corner. (b) Gently sloping left bank with disturbed Miocene at gravity abutment. (c) Drainage between sections X and XI. Seepage from each pair of buttresses are collected as well as from vertical drains and led outside to the toe of the buttresses

The Casoli dam is a buttress dam with ten gravity buttresses and two gravity ends sited across a gorge of the Miocene formation which attains a thickness of 200 m here.

These strata of alternating limestones breccias and shales dip steeply near to verticality downstream. *Figure 182a* shows evenly bedded strata dipping downstream on the right bank (*Figure 182b*); very broken strata at the gravity section on the left bank (*Figure 182c*), drainage coming between sections X and XI, i.e. between buttresses Nos. 2 and 3 from the left.

The variable nature of the Miocene, dipping almost vertically and the curvature of the gorge, ruled out an arch dam and the buttress dam was decided upon. Even so the buttresses had to follow the best line of a suitable stratum which necessitated an irregular curvature pointing downstream. The weak strata under the buttresses were also reinforced and inclined steel bars, 24–30 mm diameter, inserted in 10 m deep boreholes, spaced at 1 m. There are some 1,335 bores aggregating 12,149 m filled with 78,000 kg of steel bars and 3,525 quintals (352·5 tons) of cement.

Figure 183. Bomba dam. Two 25 m diameter bellmouths with Creager lips and struts dividing them internally into quarters to restrict vortex flow. Miocene cliff on left top corner and hills beyond, between which, behind the building, is the rockfill dam

Bomba Dam

The Bomba dam (1958–61) is 7½ miles distant from the Casoli dam. Also on Miocene limestone it is a rockfill dam in contrast to Casoli as it is sited in a tame, open, flat, valley although there are imposing hills in the vicinity as shown in *Figure 183*. In the foreground are two 25 m diameter bellmouth overflow weirs. The maximum height of the dam is 60 m and the crest 681 m in length. The trench is 20–30 m deep, excavated through alluvium to Scagliosa, Plia-Miocene clay.

The Casoli and Bomba reservoirs are on different rivers and supply power to a common station.

The Cecita Dam

(20 km north-east of Cosenza: 240 km south-east of Naples.)

The Cecita dam is an attractive curved gravity arch dam built in 1949–1951 on adamellite, a rock half way between granite and quartz-diorite (*Figures 184a and b*).

Figure 184 (a). Cecita dam. This dam which is located near Cosenza (S. Italy) is a gravity symmetrical arch dam 55 m in height. An entrance to an inspection gallery and seepage apparatus are shown on the left bank. The dam at the top right hand corner and the trees and grass below are in sunshine, most of the dam is in shadow. The dam is founded on partly decayed adamellite, a granite quartz diorite, an igneous rock of pre-Triassic age found in Scotland. (b) The right bank with another entrance to the same gallery as shown on the left bank

Maxium height, 55 m with a base-width of 19 m.
Crest length, 166·15 m (chord 143·5 m at crest level)
Volume of concrete, 60,000 m^3

The dam which is entirely symetrical, is partly on decayed granite; 1,500 tons of cement grout was absorbed in strengthening the foundation which accounts for its being a gravity arch rather than a true arch. (The chord/height ratio is about $\frac{1}{3}$). It was constructed in fourteen vertical sections which were later grouted. The concrete was air-entrained and the cement mix 275 kg/m^3. The aggregate was taken from an alluvial quarry in the valley, fine sand being substituted by ground limestone.

Figure 185a. Ancipa dam. The Miocene Eocene formations are much disturbed in Sicily and the right bank is reinforced heavily downstream of the abutment of a buttress dam

Figure 185b. Ancipa dam, Sicily. The upstream weak right bank end showing limestones above gravity abutment at end of dam and outlines of seven of the nine hollow segments, each 22 m in width consisting of isosceles triangles with buttress slopes of 20 vertical and 9 horizontal. The soft sandstone under the buttresses was consolidated up to a depth of 100 m while the grout curtain was extended for a distance of 300 m, at the right abutment

SICILY

Ancipa Dam

There are several types of dams in Sicily, the highest is the Ancipa 1949–1952, 111·5 m high, 13 miles west of Bronte (west of Etna). It is set on stratified Eocene sandstone with marl intrusions intersected with faults and fissures overlying Eocene clay, and on this account some of the right bank is reinforced with massive concrete (*Figure 185a*).

The dam consists of nine hollow triangular segments each 22 m in length, comparable with the Haweswater dam (page 97) completed in 1942. At each end of the dam there is an abutment of gravity section. The maximum height of the dam is 111·5 m and the length of the crest 253 m.

Figure 185b shows the right bank junction with Miocene formation and solid gravity section beyond the seven of the nine triangular hollow segments on the upstream face, in the foreground. It is claimed that the hollow section affords a saving of 30% in cement compared with a solid gravity structure. The contact grouting was carried down to a maximum depth of 100 m while the grout curtain at the weak right abutment went a distance of 300 m in length.

Pozzillo Dam (1956–8)

At 13 miles west of Adrano also west of Etna on the edge of the Basalt complex is the Pozzillo dam where the Miocene is much thrown about (*Figure 186a*). The reservoir area is on Eocene clay, and the overlying Miocene sandstone, and sandstone alternating with quartzite and marl appears in a gorge which is much disturbed.

As the maximum height of the dam is 59 m above the river bed, it was decided to build a concrete block dam for the left bank on the sandstone and sandstone quartzite, Miocene beds, which necessitated curving the dam downstream to avoid the clay and for the rest a gravity section of no great height on the sandstone towards the right bank. The result is an S-shaped dam some 407 m in length.

The small sections 1-1, 2-2, 3-3, on the right-hand side of *Figure 186a* show the alternating sandstones, quartzites and marls to be nearly vertical and hence some leakage problems may occur although the strike is parallel to the valley. On the upstream face however there is a steel lining 6 mm in thickness) of welded sheets.

The grout curtain at the upstream toe included 100 m holes 5 to 20 m apart and 50 m holes $2\frac{1}{2}$ to 5 m apart and many inclined holes in addition.

Figure 186b shows, behind the bywash, on the left bank on the downstream face of the dam 'patented' block work adopted to take up any unequal settlement due to the mixed nature of the foundation. The block work, 59 m in height, curves downwards to follow the sandstone as indicated on *Figure 182a*. The blocks are made of lean concrete with $3 \times 4 \times 4$ metre faces. The block work aggregate is basalt from a source 20 km away and the cement purzolan ferric cement for the foundation blocks and purzolan cement for those above 150 kg/m³.

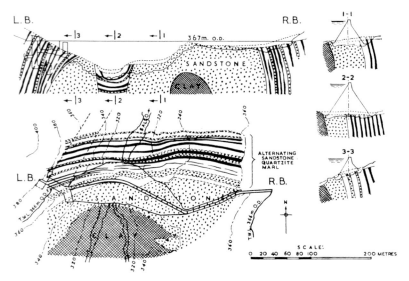

Figure 186a. Pozzillo dam. A solid concrete block dam was decided upon and designed to be constructed upon the Miocene sandstone and sandstone quartzite avoiding the Eocene clay. The resulting highest part of the dam curved downstream on the left and on the right curved upstream taking advantage of the sandstone outcrop thereby reducing height

Figure 186b. Pozzillo dam, Sicily. Showing (on left bank) the downstream curve of blocks to take up unequal settlement to suit sandstone foundation. The maximum depth of the blockwork is 59 m and ends at a natural outcrop of sandstone which has been deliberately utilized. It is seen between the block work and white concrete curves. Thereafter the dam continues in concrete, under 20 m in height, to the right bank

The block work ends at an outcrop of sandstone is always acceptable and beyond there is a long low concrete wall under 20 m in height leading to the right bank.

Figure 186c shows a typical contortion in the Miocene.

Two other dams with blocks have been built, one the Platani dam in Sicily (66 m) and the other Pian Palu in the Italian Alps (52·5 m). Both these dams were built on sites considered suitable only for rockfill, earth, or dry masonry and some flows during construction can be tolerated as well as possible overtopping.

Figure 186c. Pozzillo dam. In the left bank a contortion in the Miocene

The Eleuterio Reservoir (Scanzano and Rossella dams)

The Eleuterio reservoir scheme is for the water supply of Palermo. The reservoir (on Cretaceous Tertiary strata) is enclosed by two dams, the Scanzano and the Rossella, two tributaries of the River Eleuterio The gathering ground is 26 km². Storage 17×10^6 m³.

Minimum discharge is 20×10^6 m³/year. The scheme is designed to give 800,000 people, in Palermo in A.D. 2,000, 66 gals./head/day. Some irrigation and electric power will also be available from the scheme.

Earth dams with sloping cores	Scanzano	Rossella
Constructed	1961–1964	1958–1961
Maximum height, m	40	30
Crest length, m	530	350
Volume, cubic m	1,550,000	320,500
Crest O.D., m	529	529

* I am very grateful to Dott. Irrg. Clandia Marcello for reading the above notes on Sicilian dams.

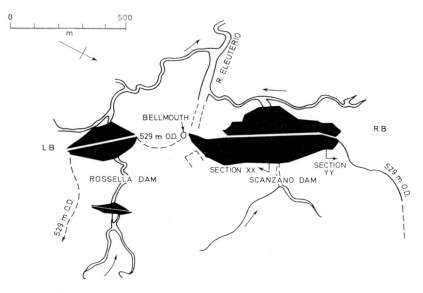

Figure 187a. Scanzano dam. The River Eleuterio reservoir is enclosed by the Scanzano and Rossella dams which forms the water supply for Palermo. The crest level is 529 m O.D. On the left end of the Scanzano dam is a bypass used during construction; the permanent draw off valve tower and pipe connected to which is the pipe from the bellmouth overflow. The natural ground between the dams is well above 529 m O.D. the crest of the dams.
    Both dams are on the soft tertiary cretaceous boundary
(Section XX denotes the direction of Figure 187c for core 4 and filter 2 Figure 187b). Sections YY shows the position of Figure 187d

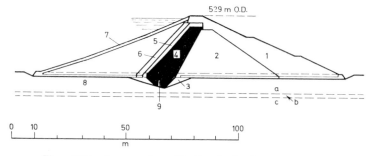

Figure 187b. Scanzano dam. The Rossella dam is very similar.

1. *Calcareous detritus supporting filters and core.*
2. *Sand for downstream filter.*
3. *Yellow clay surrounding toe of black clay core.*
4. *Clay core.*
5. *Filter upstream. Calcareous detritus with much clay next core.*
6. *Filter upstream of crushed sand.*
7. *Rip-rap.*
8. *Sandstone drainage.*
9. *Concrete trench sunk by I.C.O.S. process.*
a. *Altered clay.*
b. *Permeable alluvium.*
c. *Compact clay (Tertiary).*

Figure 187c. Scanzano dam. The white strip in the foreground is the downstream filter being rolled by a light roller. On the left of the filter is the core which has been rolled by a very heavy roller and on the right of the filter, calcareous detritus. On the left of the clay core there will be two filters and supporting calcareous material and rip-rap. In the distance at the abutment end of the dam is shown the valve tower, bridge, (and, behind, the bellmouth overflow). Owing to the width of each of the several layers, there is no problem if they are all brought up together in lifts of about a metre. See Section XX, Figure 187a

Figure 187d. Scanzano dam. Rockfill sloping core under construction July 1961. On the right is the clay-shale material being placed in position for rolling; in the middle there is a wide layer of sand to act as a filter, and at the extreme left rubble material will be placed for the downstream slope. The broad foundation and trench of the right bank is seen to be under construction beyond See Section YY, Figure 187a

Figure 187a shows part of the Eleuterio reservoir, about 20 km from Palermo enclosed by two earth dams, Scanzano and Rossella, and *Figure 187b* shows the section of the Scanzano dam, the Rossella dam is similar.

On *Figure 187a* is shown the positions of *Figures 187c and 187d* and of the construction in 1961 in the early stages of laying the clay core, sand filter and supporting calcareous material. *Figure 187c* shows the bridge to the valve tower at the level of the crest. The smooth white strip is part of the downstream sand filter, which is being lightly rolled to the left of which is the clay core having been compacted by a heavy roller. On the extreme right is the calcareous material and the natural ground for supporting the filter.

Figure 187d shows the clay for the core in the foreground and preparations for the reduced width of the core and filter as they rise to the right end abutment.

Where the ground rises is a bellmouth overflow at the left end of the Scanzano dam beyond which the natural rock level is well above top water level as far as the Rossella dam.

BIBLIOGRAPHY AND REFERENCES

Piave di Cadore dam
>PIAZ, G. DAL. Descrizione geologica del bacino del Piave in relazione agli impianti idroelettrici della S.A.D.E. In *Impianto Idroelettrico*. Piave-Boite-Maè-Vaiont. S.A.D.E., Venezia. (1956).
>SEMENZA, C. The most recent dams by the S.A.D.E. in the Eastern Alps. *J.I.C.E.*, **1** 508 (1952).

Valle di Cadore dam
>INDRI, E. La diga di Valle di Cadore. In *Impianto Idroelettrico*. Piave-Boite-Maè-Vaiont. S.A.D.E., Venezia. (1956).
>SEMENZA, C. The most recent dams by the S.A.D.E. in the Eastern Alps. *J.I.C.E.*, **1** 508 (1952).

Val Gallina dam
>CAPRA, V. Descrizione della diga di Val Gallina. In *Impianto Idroelettrico*. Piave-Boite-Maè-Vaiont, S.A.D.E., Venezia. (1956).
>SEMENZA, C. The most recent dams by the S.A.D.E. in the Eastern Alps. *J.I.C.E.*, **1** 508 (1952).

Vaiont dam
>*Impianto Idroelettrico*. Piave-Boite-Maè-Vaiont. S.A.D.E., Venezia. 1955. (Proposals.)
>— BIADENE, N. A., and PANCINI, M. Vaiont dam. *6th Congress on Large Dams.* C.23, **IV**, 359. (1958). (Larger dam as adopted.)
>GASKILL, G. The night the mountain fell. *Reader's Digest*, 87, 160–174 (1965).
>LEONARDI, PIERO. le Dolomiti. Propili Geologici. A cura del consiglio nationale delle reherche e della guinta provinciale di Trento, **1**, 512 (1967).
>LEONARDI, PIERO. le Dolomiti. Zona di Longarone. *Ibid.*, **II**, 925 (1967). (with summaries in English)
>LEONARDI, PIERO. le Dolomiti. la bassa valle del Vaiont e lo scivolamento gravitatrio del 9 Ottobre 1963. *Ibid.*, **II**, 937 (1967).
>Report of Italian Committee on Large Dams. Vaiont Dam *I.C.O.L.D.*, **IV**, 687 (730) (1967).
>ROSSI, D. and SEMENZA, E. Carte Geologich del versante settenronale del Monte Toc e Zone Limitrofe, prima e dopo il finomeno di scivolamento del 9 Ottobre 1963. Scala: 1/50,000 *1st Geol. Univ. Ferrara* (1965).
>SELLI, R. and TREVISAN, L. la frana del Vaionti. *Gion. di. Geol. Sez.* 2., **32**. Bologna.

SEMENZA, E. Sintesi degli studi geologici sulla frane del Vaiont dal 1959 al 1964. *Mem. Mus. St. Nat.*; *Trento AXXIX–XXX.* **XVI,** 1–51 (1966–67).
SEMENZA, C. The most recent dams by the S.A.D.E. in the Eastern Alps. *J.I.C.E.*, **1** 508 (1952).
LORIGA, C. and MANTOVAVI, C. le biofacies del Cretaccio della Valle de Vaiont. *Riv. Ital. Pal. Strat. Fasc.* 4. Milano.

Pontesei dam
  LINARI, C. Associazione Geotecnica Italiana Quarto Convegno di Geotecnica, Padova, Maggio (1959).
  SEMENZA, C., CAPRA, U., and INDRI, E. Pontesei dam on the River Maè. *6th Congress on Large Dams.* C.18, **IV,** 299 (1958).

Fedaia reservoir
  SERBATOIO DELLA FEDAIA (1956).
  MOSER H., and NEUMAIER, F. Determination of seepage flow under Rosshaupten dam (Germany) of radio-active isotopes. *6th Congress on Large Dams.* C. 10, **4** 143 (1958).

Barcis dam
  PIAZ, G. DAL. In *Impianto Idroelettrico*. Piave-Boite-Maè-Vaiont. S.A.D.E., Venezia. (1956).
  SEMENZA, C. The most recent dams by the S.A.D.E. in the Eastern Alps. *J.I.C.E.*, **1** 205 (1952).
  —FORLI, G., and INDRI E. Barcis dam on the River Cellina. *6th Congress on Large Dams.* C.19, **IV,** 305 (1958).

Casoli and Bomba dams
ARREDI, F. Technics for rock characteristics improvement at two dams in Central Apennines (Italy). *I.C.O.L.D.*, **I,** 551 (Aventino 560). (1964).
Site Pamphlets (Italian). L'impianto Idroelettrico di St. Angelo. Ministero dei lavori Pubblici. Grandi dighe Italiane, Casoli (247), Bomba (252). (Italian, English and French) *I.C.O.L.D. Conference.* Rome (1961).

Cecita dam
  Ministeri dei lavori Pubblici. Grandi dighe Italiane. Cecita (304) (Italian, English and French) *7th Cong. I.C.O.L.D.*, Rome (1961).
  Site pamphlet. Diga di Cecita (Italian) and Gli Impianti Idroelettrici sul Fiume Mucone, 1950–55. *Societa meridionale di elettricità.* Naples.

SICILY

Ancipa dam
  MARCELLO, C. Considérations sur les exemples réalisés ol'un type de barrages a eléments évidés. (Ancipa dam and others.) *I.C.O.L.D.*, **IV,** 1479 (1955).
  Ministero dei lavori Pubblici. Grandi dighe Italiane. (Italian, English and French) (Ancipa 325). *I.C.O.L.D. Conf. 1961.* Rome (1961).

Pozzillo dam
  MARCELLO, C. Un nonveau type de barrage covenant aux terrains de fondation fortement compressibles. *I.C.O.L.D.*, **IV,** 1517 (1955).
  Ministero dei lavari Pubblici. Grandi Dighe Italiane. (Italian, English and French). *I.C.O.L.D., Conf. 1961.* Pozzillo 330.
  Site Pamphlet. Ente Siciliano di Eleottricita.

Scanzano and Rosella dams
  la Cassa per il Mezzogiorno. Le Ente Acquedotti Siciliani. *Serbatio sul Fiume Eleuterio.* Palermo. *Site Pamphlet* (1961).

22

JUGOSLAVIAN DAMS

The potential water resources of Jugoslavia are comparable to those of France and Italy, but whereas 40 per cent of those in France and 60 per cent of those in Italy have been developed, only 4 per cent of those in Jugoslavia were in use in 1958; this means that there could be a reserve of power available not only for Jugoslavia but also for neighbouring countries.

Since 1958 a good deal of activity has occurred in utilizing the large amount of water available for electricity and recently for water-supply from 43,000 km of rivers (longer than 10 km). Thus by 1965, seventy-five

Figure 188a and b—Two concrete barrages across the River Sava

large dams were in use or under construction which included fourteen over 50 m and four over 100 m in height, namely, Gran Carevo (arch, 123 m); Tikves (rockfill, 113·5 m); Spilje (earth, 111 m); Rama (rockfill, 100 m).

Of these new fourteen dams, eight are rockfill and earth, four concrete gravity and two arch.

Some smaller sources have been developed in the north on the tributaries of the Danube; typical of these are two concrete barrages across the Sava

(*Figures 188a and b*) near the Austrian and Italian borders north-west of Ljubljana.

BIBLIOGRAPHY

Monveiller, E., Verčov, M., Kujundzić, B. Jugoslav National Committee on Large Dams. *I.C.O.L.D.* **IV,** 889 (1967).
Rajcevic, B., and Vercon, M. Various papers in IVe, Ve and VIe, Congrès des Grands Barrages (1951, 1955, 1958).
Jugoslav Water Resources as a European Power Reserve, Beograd (1958).

23

LUXEMBOURG

Esch-sur-sure Dam

An interesting thin arch dam (*Figure 189a and b*) situated 1 km west of Buderschide and 4 km north of Esch-sur-sure, nearing completion in June 1958 on the Devonian schists has the following characteristics.

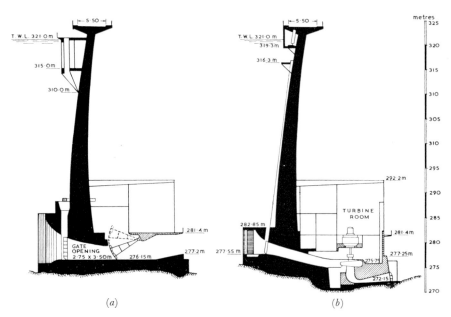

Figure 189a and b—*Esch-sur-sure dam. Sections through gate control and turbine room*

Top water level	321 m O.D.
Maximum height of dam	47 m
Length of crest	174 m
Thickness at crest	1·5 m
Thickness at base	4·4 m
Width of roadway	5·5 m
Width of roadway with paths	8 m

Figure 190. Esch-sur-sure dam. Upstream face of dam before filling, showing steep left-hand embankment of cleaved Devonian schists beyond

Figure 191. Esch-sur-sure dam. The left abutment of the thin arch curved dam on sound Devonian schists (seen at top right-hand corner). Due to the sun, curious shadow effects are produced on the curvature of the downstream face

The grout screen was taken to a depth of 30 m below the broad foundation with 1,200 m linear of bore holes, using 6 tons of cement per m.

The reservoir to be impounded commands a gathering ground of 428 km², the mean annual flow from which is 6 m³ per second.

Figure 192. (a) Esch-sur-sure dam. The cut-off trench is soft material at top of right-hand abutment under construction. The view is taken from the right looking across to the left side of the valley. The cut-off wall is in the foreground, the abutment is on the left of the wall, and the top of the downstream face of the dam shows above the top of the excavation on the right. (b) Top of the right-hand abutment with loose material on the right

Capacity of reservoir . . . . 62 × 10⁶ m³
Perimeter of reservoir . . . . 55·2 km
Surface: fields, 219 ha.; woods, 161 ha.; total 380 hectares

The Devonian rocks appear to dip upstream, but owing to cleavage the precise dip is uncertain.

The left bank of the valley is steep and undoubtedly provides a good foundation for the thrust of the arch (*Figures 190 and 191*).

The right side, however, is flat for about 100 yd. (comparable to the Piave di Cadore site) and the soft top material caused some difficulty so that a cut-off trench had to be excavated 5–10 m in depth and a large block of concrete formed to take the thrust of the top part of the arch, the lower part being in sound Devonian rock (*Figure 192a and b*).

Details of the Esch-sur-sure dam are taken from the site pamphlet by courtesy of the Administration of Ponts et Chaussées through the Director, Dr. Théo Sunnen of the Syndicat des Eaux du Sud à Koerich, Luxembourg.

24

PORTUGUESE DAMS

From a list of hydro-electric and irrigation reservoirs published by the Ministerio das Obras Publicas (1955), all types of dams are well represented of which the following are typical.

Name	Maximum height (metres)	Type
Paradela	112	Rockfill
Alto Rabagão	103	Buttress and rockfill (contemplated)
Venda Nova	94	Gravity arch
Salamonde	75	Thin arch
Caniçada	76	Thin arch
Vilar	49	Gravity
Guilhofrei	49	Gravity
Picote	95	Arch
Santa Luzia	72	Two thin arches
Idanha	53·5	Gravity
Cabril	132	Cupola
Bouçã	65	Cupola
Castello-do-Bode	115	Gravity arch (*Figure 161*)
Pracana	60	Buttress
Alvito	135	Cupola
Maranhão	50	Earth
Montargil	43	Earth
Pego do Altar	63	Rockfill
Vale de Gaio	51	Earth and rockfill
Campilhas	35	Earth
Silves	43	Earth

In the table the first five dams, together with another dam, Alto Cávado, form a comprehensive hydro-electric development of the River Cávado and its tributary, the Rabag˜o, in Northern Portugal. The Alto Rabagão dam was originally intended to be 103 m high, partly buttress and partly rockfill. This has been altered to a Cupola dam, 94 m in height and a gravity dam 59 m in height. The total length of the dam 1,970 m and the volume of concrete $1 \cdot 129 \times 10^6 \text{m}^3$.

Extensive site investigation, 1958–65, showed sound granite on the right bank with a modulus of 200,000 kg/cm² equal to concrete, but in the centre and left bank the granite contained feldspars and kaolinites giving moduli of 60,000–10,000 kg/cm². These lower values are $\frac{1}{3}$ to 1/20 of that of concrete, and even for an abutment of 40 m in height it would mean substantial reinforcement on the left bank. Nevertheless, this was considered preferable to the buttress-rockfill originally contemplated, especially with the reduced height of 94 m.

The buttress type was ruled out owing to unsuitable foundation downstream and the rockfill owing to absence of clay for the core, and a membrane type of dam, which would entail emptying the reservoir for repairs, was also ruled out.

This dam would seem to show that a cupola dam can occasionally be built on weaker or partly weaker strata than was hitherto acceptable.

Other dams in Portugal and one in Brazil have been recorded by J. L. Serafim and J. C. Rodigues namely:

The *Odeaxere dam* about 40 m in height, a cupola arch dam founded partly on rock with a modulus of elasticity varying between 10,000 and 2,000

Figure 193. Castello-do-Bode gravity arch dam
(Reproduced by courtesy of R. C. J. Walton)

kg/cm². The thrust was distributed by widening the upper part of the abutment and building a wide socket below in the lower part of the arch.

The *Cachi dam* on metamorphic rocks with a modulus of 15,000 kg/cm² had to be reinforced like Alto Rebagão.

The *Cambambe dam* and another dam on granite had extensive faults filled with clay which was removed and replaced by reinforced concrete.

The *Santa Eulalia dam* and *Funil dam* (in (Brazil) both have faults under the abutment reinforced by pre-stressed cables mainly at right angles to the faults.

The behaviour of these arch dams over the years should be interesting.

BIBLIOGRAPHY

ROCHA, M., da SILVEIRA, A. F., Cruz AZEVEDO, M. C., and LOPES, J. B. Influence on a very high deformability of the foundation on the conception and behaviour of an arch dam (Alto Rabagão) *I.C.O.L.D.* **I,** 441 (1967).

SERAFIM, J. L. and RODIGUES, J. C. Improvements on the safety of foundations of concrete dams. *I.C.O.L.D.* **I,** 935 (1967).

25

SPANISH DAMS

Some of the oldest dams in the world, and still in use, are to be found in Spain.

In an interesting paper by Sig. E. Becerril there is mentioned the Proserpina masonry dam near Merida, probably built in the time of Trajan in the first century A.D. The dam was restored in the eighteenth century and repaired in 1930.

Other ancient Spanish dams are Almansa (14 m) (1384?) raised to 21 m in 1586; Elche (23·2 m) ? Arabic origin; Arguis (24 m) raised to 27·3 m in 1926. These are described in 'Heritage of Spanish Dams' (see Bibliography).

The following table gives some details of a few dams over 100 ft. in height (with one exception) and the geological formation on which they are built.

Name of dam and river	Maximum height (metres)	Approximate chord-height ratio	Type	Formation	Remarks
Las Conchas R. Limia	44·5	—	Gravity (curved)	Fissured granite	
Los Peares R. Mino	110	3·5	Gravity (straight)	Granite intrusion	Built in two stages
San Esteban R. Sil	96	3	Gravity arch	Granite	
La Coruna R. Tambre	48	3·5	Gravity	Granite	
Doiras R. Navia	95	—	Gravity arch	Thick alluvium on Silurian	
Salime R. Navia	127	—	Gravity arch	Alluvium on schist	Built in two stages
Puenta Alta R. Frio	40	3·8	Diamond buttresses at 10 m C. to C.	Fissured decomposed limestone	
Ricobayo R. Esla	92	—	Gravity arch	Granite	Two dams with a verrou
Villalcampo R. Duero	45	—	Gravity arch	Schist and granite	Bores, 3,450 m cement, 1,375 tons
Castro R. Duero	56 42	3 3	Gravity (curved) do (straight)	Micaceous Pizarra	Bores, 5,600 m cement, 930 tons Two dams with a verrou
La Sarra R. Aguas Limpias	32	4·5	Earth	13 m of boulder clay in limestone	Bores, 795 m Activated clay 670 m³

Name of dam and river	Maximum height (metres)	Approximate chord-height ratio	Type	Formation	Remarks
Respomuso R. Aguas Limpias	48	—	Buttress	Granite	—
Escales R. Noguerra-Ribagorzana	110	1·5	Gravity	Shale and limestone	—
Oliana Figure 194 R. Segre	74	3	Gravity arch	Conglomerate	Bores, 1,780 m cement, 1,220 tons
Generalisimo R. Turia	112	1·9	Gravity		
El Molinar R. Jucar	18	—	Earth	40m sands and clays on limestone. Maximum settlement of dam, 80 cm after 5 years	Bores, 6,450 m cement, 3,900 tons clay, 1,250 tons
Cenajo R. Segura	84	—	Gravity	Alluvium	2,200m of piles bores, 1,550 m cement, 1,370 tons
Burguillo R. Alberche	90	—	Gravity	Gneiss	
San Juan R. Alberche	80	3	Gravity	Gneiss with granite intrusions	Bores, 2,160 m cement, 125 tons
Las Picades R. Alberche	54	3	Gravity	Fractured gneiss	1 ton per m
Tranco de Beas R. Guad Al-Quiver	84	—	Gravity arch	Jurassic limestone Trias marls and limestones	—
En el Rio Cubillas	46	—	Earth	Clay	Bores, 5,610 m cement, 540 tons clay, 1,760 m^3

Figure 194. Oliana dam. A large gravity dam nearing completion in 1956

Reliance is placed on cement and clay grouting, and the straight or curved gravity dams are the most favoured, but there are two large buttress dams. There are also three high earthen dams on very unreliable strata treated with clay grout.

Figure 194 shows the Oliana dam on the River Segre under construction in 1956.

BIBLIOGRAPHY

BECERRIL, E. Heightening of existing dams. R. 13. *6th Congress of Large Dams*, 1 (1958) 387. Presas Españolas Procedimientos Rodio, Madrid. (1955).

SMITH, NORMAN A. F. The Heritage of Spanish Dams. Madrid (1970).

26

SWEDISH DAMS

There are over a dozen large dams in Sweden between 60 and 400 ft. in height, more than half are of earth or rockfill owing to the large amount of glacial morainic material in the country.

Several of these dams are of special geological interest; for example, the multiple arch dam of Suorva; the earth and concrete dam of Bergeforsen; and the long gravel dam of Jarkvissle.

Suorva Dam

The Suorva multiple arch dam (1919–23) with 34 arches each of 40 ft. span impounding 80 ft. of water has developed cracks and in a few years will be replaced by a rockfill dam with a moraine clay core.

Bergeforsen Dam

The Bergeforsen earthen dam with a concrete spillway and power-house section is 100 ft. in height; it was constructed in 1955 just downstream of a pre-glacial gorge 170 ft. in depth filled with silt and sand which was expected to give trouble; as it happened to be covered with silty and clayey moraine material no problems arose (*Figure 195*). This was not the case, however, with the hard rock (granite gneiss) of the foundation for when this was opened out the 'Alno' dykes were found to contain limestone and free carbonic acid which softened to a weak clay when in contact with air and water. The dykes were 2–3 m in thickness and settlements of 20–30 cm were envisaged as well as leakage under the dam.

Two cement grout screens to 70 ft. in depth were made between which saturated lime water is pumped to neutralize the CO_2. In 1960 some 8,000 U.S. gal. were pumped per 24 hours and it was expected that this would have to be continued for some years. In addition, as part of the dam on the right bank had been completed, a tunnel had to be driven in order to construct the grout screen, and to enable this to be done asphalt was injected by means of vertical bore holes in the roof of the tunnel connecting to the concrete wall under the dam. In 1958, 20 U.S. gal. of asphalt were still being used for this purpose.

Järkvissle Dam

The Järkvissle dam (1958) is on permeable material. Its maximum cross-section is 795 ft. in length and 35 ft. in height, upon which a rockfill dam was placed 30 ft. in height impounding about 65 ft. of water (*Figure 196*).

It is apparent that the trend of dam building in Sweden is towards the earth and rockfill types owing to the amount of glacial material available.

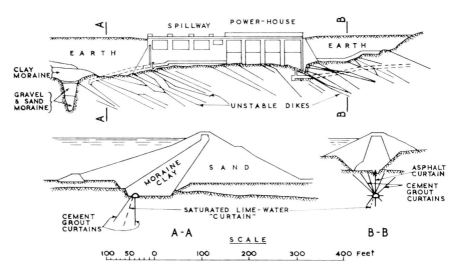

Figure 195. Bergeforsen dam. Although this dam was put on a buried glacial gorge filled with sand and gravel, there was a natural covering and no leakage or trouble from this occurred even though a considerable depth of water was impounded (A–A). The relatively hard dykes, however, were found to contain limestone and free carbonic acid which softened in water. To neutralize the acid, lime water, 8,000 U.S. gal., are pumped in daily. The grout screen on the right had to be constructed from a tunnel protected by an asphalt curtain (B-B)

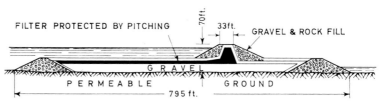

Figure 196. Järkvissle dam. This extraordinary structure, some 70 ft. in height is built on permeable ground. The dam (of clay, shown black with a sand filter supported by gravel and rockfill) is supported on gravel deposited under water between two rubble embankments. The effective storage of the reservoir and delayed action leakage is nicely balanced

BIBLIOGRAPHY

Aastrup, Å., and Sällström, S. *Bergeforsen I.C.O.L.D.* II 473 (1961).
Nilsson, T. *Engng News Rec.* **161** 60 (1958).
Pira, G., and Bernell, L. Järkvissle dam, an earthfill dam founded under water. *I.C.O.L.D.*, **IV,** 213 (1967).

27

SWISS DAMS

The geology of Switzerland produces interesting dams in the mountains and in the moraines; these give rise to the construction of dams under complex tectonic conditions and sometimes on alluvial soil of adverse physical condition.

The case histories of two early dams, Schräh (1924) and Spitallamm (1932) are of interest and brief descriptions are given below.

Schräh Dam

This is a straight gravity dam near Lachan (Canton Schwyz) and constructed between April 1922 and October 1924.

Maximum height	112 m
Length of crest	156 m
Top water level	900 m O.D.

The dam is on the Cretaceous in a ravine caused by a fault; corresponding beds show different levels on the right and left side of the ravine. The strata dip upstream. A deep channel to a depth of 45 m had been eroded below the existing bottom of the valley.

Figure 197 gives, in outline, the profiles of the up-and down-stream faces and the system of galleries in the material filling the eroded channel to ascertain the whereabouts of the valley sides buried in the filling.

The behaviour

After eight years in 1932 the downstream slope was protected with a layer of granite masonry blocks placed in concrete of a rich mix to prevent further deterioration of the original concrete by frost. The method of concrete shooting produced a porous concrete which deteriorated to a depth of 1·5 m.

Between 1950 and 1956 further tests on concrete strength, concrete temperature, pore water-pressure, injections tests, and quantity and quality of seepage water were made. Thus the concrete strength was found to vary between 300 kg/m^3 and 189 kg/m^3, the upper figure applying to the crest and the lower to the mass of the dam. The latter would not have been satisfactory had it not been covered with granite blocks at least on the downstream face. The temperature tests, after several years, showed that under severe winter conditions, the frost line is 2·5 m in the Schrän dam. The pore pressure, as far as it went, was ascertained at 25 per cent instead of the presumed 80 per cent of the full reservoir pressure in the Cretaceous. Cement injection with chemicals, such as sodium silicate, was made in the body of the dam with the object of increasing the concrete to withstand corrosion; such injection was not considered useful in increasing the resistance of the concrete but it did fill up any large voids. Finally exhaustive tests were made on the water in the reservoir, which was found generally to be

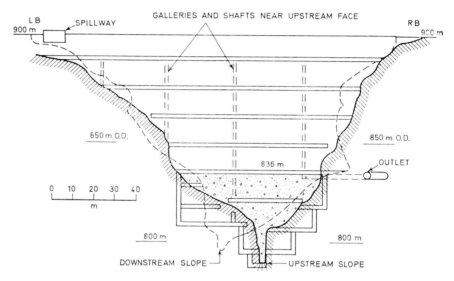

Figure 197. Schräh dam. This is the earliest high dam constructed in Switzerland, between 1922 and 1924. 112 m in height, on the Cretaceous, with a large system of galleries in the material, filling the eroded valley for a depth of 45 m. Some 259 tons of cement grout were used for grouting in about 4728 m of boreholes using 54·8 kg/m. The dam has been, and is still, examined for 'behaviour', so far, for 47 years after it was begun to be built in 1922

inocuous except at the bottom of the dam where there was a high concentration of CO^2. By comparing the water in the reservoir with that of leakage water in the lowest gallery it was inferred that the loss of substance in the concrete was estimated at 400/500 milligrams/litre of leakage water which, at 1 litre/min gave a loss of 600 grams/day for the whole of the dam. The seepage water through the dam when full was, 1927 70 litre/min.; 1930 22 litre/min.; 1951 5 litre/min. through clogging in the concrete.

In 1960 the dam was examined by dynamic auscultation; these vibration tests (longitudinal and transversal) will be repeated in 1970, to detect any alteration in the condition of the dam.

Spitallamm Dam

This is a composite arch gravity (or hick arch) dam near Interkirchen (Canton Berne), and constructed between June 1926 and June 1932:

Maximum height	114 m
Length of crest	258 m
Chord/Height ratio	about 1·5
Top water level	1909 m O.D.
Geology	Very good quality Granite

In 1922 a straight gravity dam was suggested owing to the serious difficulty of building an arch dam with a high ratio of curvature at the base of the dam and the difficulty of analysing the stresses in a thick arch, stress disturbances through heat of hydration and shrinkage of cement. (Doubts perhaps not entirely at rest today). However it was only after 35 years that a longitudinal fissure became apparent in the crest.

Nevertheless the thick arch dam had been adopted because maximum stresses would be less than those for the straight gravity dam and there was a substantial economy in the amount of concrete. It was however stipulated that the dam should be constructed in separate vertical blocks to allow for setting heat and rise in temperature, to be joined up afterwards. A thick arch dam with a straight sloping face resulted, but today on *very good quality granite* and a chord/height ratio of under 2, a cupola dam might have been adopted.

Deformation

Triangulation and measurements were undertaken for detecting movement and it was at Spitallamm that plumb-lines, now called pendulums, were used for the first time. (Cf. Amos **VII**, 7). Such records have been taken for 30 years so far and their interpretation has caused much interest. The following is an example of one of these records.

As the water in the reservoir rose and fell, the deviation of a pendulum registering the movement between 1,804 O.D. and 1,869 O.D., covering the lower two-thirds of the dam, hardly varied for 18 years. The deviation was 5·5 mm when the reservoir was full and 5 mm when empty. (*Figure 195*.)

In 1950, quite suddenly, the movement of the dam increased to 6 mm and in subsequent years up to 1962 oscillated between 6 mm and 7 mm. This change in deviation is attributed to the filling of the new Rätherichsboden reservoir distant 1,000 to 1,900 m from Spitallamm and is also on granite. The long record also showed that the granite/concrete is elastic, although there is a tendency for the amplitude of the deviations to diminish. The records of these deformations over so long a period have shown up other points of interest not only at Spitallamm but also at other reservoirs, e.g., Oberaar, Albigna. The report on the Spitallamm dam concludes:

'If the apparatus for measuring the deformations at Spitallamm had to be installed today, a somewhat different choice would be made, and in particular there would be more pendula linked together by the combined measurements along the 1901 metre arc. Some supplementary pendula were installed in 1962. Nevertheless the results obtained have been very satisfactory, since their accuracy has enabled the phenomena observed to be interpreted quite unequivocally.'

Raising the dam by 16 m is under consideration.

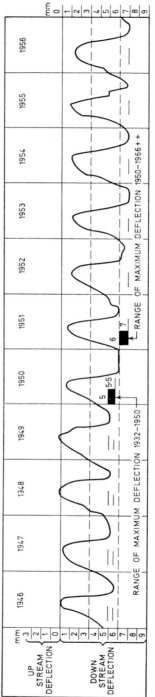

Figure 198. Spitallamm Dam. Pendulum measurements of the deformation between the levels 1804 m O.D. and 1869 m O.D.; 6.5 m, have been calculated since the year 1932. The deformation downstream, when the reservoir is full, never varied between 5 and 5.5 mm but since the year 1950, it increased to 6 and 7 mm. The cause of the sudden increase was due to the completion of the Räterichsboden dam also on Granite about the same size as Spitallamm, distant 1000 to 1900 m. The curve, for 30 years, is very instructive show that the foundations in the Granite are elastic but there is a tendancy for the amplitude to decrease; this amplitude varied with the level of the water in the reservoir, which was full generally in the last half of the year and low in the spring. Some effect also was seen at Spitallamm when the water-level in Räterichsboden Reservoir varied

MODERN SWISS DAMS

The following brief descriptions of the geological conditions in modern Swiss dams are based on the publication '*The Behaviour of Large Swiss Dams*'.

Name of dam and year completed	Type	Max. ht. m	Crest m	T.W.L. O.D. m	Main formation at dam site	Town (Canton)
Rossens (1947)	A	83	320	677	Molasse Miocene	Bulle (Fribourg)
Rätherichsboden (1950)	G	94	456	1,767	Aar-granite	Innertkirchen (Bern)
Châtelot (1953)	A	74	150	716	Jurassic Limestone	La Chaux de Fonds (Neuchâtel)
Oberaar (1953)	G	100	526	2,303	Gneiss/schists	Innertkirchen (Bern)
Castiletto (1954)	E	60	400	1,680	Greenstone/serpentine	Bivio (Grisons)
Sambuco (1956)	A	130	363	1,461	Gneiss	Fusio (Ticino)
Zeuzier (1957)	A	156	256	1,777	Jurassic Limestone	Sion (Valais)
Zervreila (1957)	A	151	504	1,862	Gneiss/clays	Vals (Grisons)
Mauvoisin (1958)	A	237	520	1,961·5	Calcareous Schists	Fionnay (Valais)
Moiry (1958)	A	148	610	2,249	Metamorphic Rock	Ayer (Valais)
Malvaglia (1959)	A	92	290	990		Biasca (Ticino)
Albigna (1959)	G	115	810	2,162·6	Granite	Vicosoprano (Grisons)
Göscheneralp (1960)	E	155	540	1,792	Aar-granite alluvium	Göschenen (Uri)
Isola (1960)	AG	45	290	1,604	Gneiss and clay	Mesocco (Grisons)
Grande dixence (1962)	G	285	695	2,364	Schists/Slates	Hérémence (Valais)
Sufers (1962)	A	58	125	1,401	Gneiss	Splügen (Grisons)
Contra (1965)	A	221	380	470	Gneiss	Locarno (Ticino)
Mattmark (1967)	E	115	780	2,197	Moraine Gneiss Granite	Visp (Valais)
Ova Spin (1968)	A	74	130	1,630	Dolomite	Zernez (Grisons)
Punt dal Gall (1969)	A	128	540	1,805	Dolomite	Zernez (Grisons)

Rätherichsboden Dam (1948–50)

This gravity dam rests on the very compact granite of the Aar massive which is of excellent quality. The foundation of the dam, however, is interesting owing to a vertical joint in the valley which causes an abrupt change of height along the axis of the dam causing 200 m of the 456 m crest to be only 10 m to 20 m in height, whereas the maximum height is 94 m. The deformation of the Spitallamm dam, about a mile distant was increased by about 2 mm when this dam was filled.

Castiletto Dam (1950–54)

This earth dam is founded on greenstone and serpentine which is generally compact, but the left bank is a landslide of serpentine detritus 70 m on which the dam sits. The core was from a moraine and was consolidated with twelve passes of a 20 ton sheepsfoot roller (10^{-7} to 10^{-9} cm./sec.) with filter (dried) from talus and embanking material in layers of 20 cm, stone over 12 cm in diameter being removed. Settlement after 2 years 150 mm; 4 years 200 mm; 6 years, 210 mm; 8 years, 220 mm.

Sambuco Dam (1952–56)

This arch gravity dam is founded on undisturbed gneiss (of augen-gneiss and schistose gneiss with biotite). As the dip is 60 degrees from the right bank to the left the right bank end of the curved dam is almost parallel to the dip and the left end is parallel to the strike. The unsymmetrical arrangement of the Gneiss relative to the curvature of the dam, accounted mainly for unsymmetrical deformations recorded in the pendulum measurements where the largest, in the centre of the dam with a full reservoir was 27 mm, on the left bank it was only 3 mm.

Zervreila Dam (1953–57)

The Zervreila arch dam (*Figure 199*) is on the Valser Rhine, a tributary of the Vorder Rhine, in the east of Switzerland. It is founded on Gneiss. The main dimensions of the dam are as follows:

Height	151 m
Crest length	504 m
Crest width at top	7 m
Crest width at bottom	35 m
Radius (Angle 110°)	251 m
Volume of concrete	620,000 m³
Storage of reservoir (useful)	100×10^6 m³
Top water level	1862 m O.D.

Necessity for investigation of the Gneiss

The formation consists of so-called para-gneiss and ortho-gneiss i.e. allied to and ordinary Gneiss. The formation and the height/crest ratio, 151/504, about 1 in 3, suggested an arch dam, but an extensive investigation had to undertaken as:

(a) There was a thin covering of debris over the Gneiss.

(b) As the general dip of the strata of 20° was into the hillside on the left bank, the strike being parallel with the valley, and although this gave rise to compact layers of rock on the left bank, on the right bank the strata were very loose and tumbled.

(c) At about 100 m from the right-hand end of the proposed dam there was a small valley filled with soft material, 'which created a certain problem'.

As no arch dam could be founded, even on Gneiss, if its stability was in question, three tunnels at three different levels of 40 m were drilled into the hill-side the lowest at 1,742 m O.D., about 40 m above the bottom of the valley; at 1,783 m O.D., about half way up the proposed dam; and at

1,822 m O.D. about 40 m below the crest. The length of these exploratory tunnels, cut into the Gneiss were respectively 208 m, 165 m, and 330 m, or about 700 m in all. The tunnels were cut with great care to enable a thorough examination of the rock. After a depth of 10 m or under had been reached sound rock could be relied upon and the tunnels were unlined. The dam foundation was therefore carried down to this level on the right bank as on the left bank.

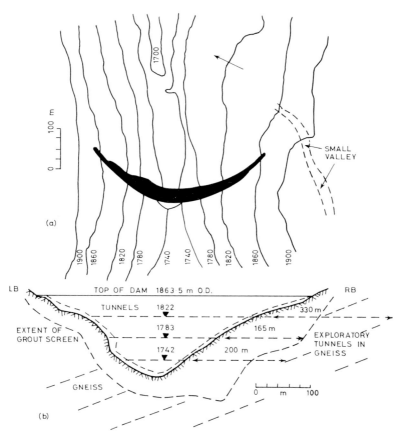

Figure 199. Zervreila arch dam. (a) Plan of dam showing small valley, dotted, on right bank which was under suspect for it may have caused weakness for the arch thrust. It is the kind of depression which could not be ignored for an arch dam 151 m in height, for a small gravity dam it might have been considered with some suspicion, perhaps. (b) Section of dam showing extent of investigation by 700 m of exploration tunnels on the right bank. The large area of grout screen and the large quantity of grout consumed on the right bank, 10 times that on the left bank, suggests some relative weakness due to this small valley

Grout consumption

In view of the foregoing resulting from the dip of the strata even Gneiss, across a valley, the consumption of grout on the right bank compared with the left, both for the curtain and contact grouting, is interesting.

Curtain. Cement consumption

	Left bank	Right bank
	92,700 kg	2,967,200 kg
	39 kg/m	569 kg/m

Total length of holes 7570 m   Consumption 3,000 tons.

Contact grouting. Cement consumption

	Left bank	Right bank
Area (horizontal projection)	5,300 m²	5,400 m²
Absorption	71,600 kg/m²	694,000 kg/m²

Total consumption of cement, 750 tons.

Leakage

The leakage is of the order of 0·2 litre/sec. at top water level.

Deformation

The deformations were measured by plumb-lines, geodetic surveys, and geoseismic tests which confirmed each other. There was greater deformability on the right than on the left where the strata dip into the hillside.

Mauvoisin Dam (1951–1957)

This arch dam is on lias lustered Calcareous schists of the Combin zone between the Dent Blanche fold upstream and the Great Saint-Bernard fold downstream. So great a depth of water impounded (the dam is 237 m in height) on such a formation necessitated an area of grout curtain around the dam foundation of 247,000 m² which consumed 32·6 kg/m² in 51 km of boreholes taking 158 kg/m (i.e. 8,000 tons in all). The permeability was estimated to be reduced from 14 to 1 lugeon. (1 litre/min. in a borehole under a pressure of 10 atmospheres or column of water of 100 m).

Deformation measurements have been recorded between 1958 and 1963 For example (*Figure 200*) a pendulum in the centre of the dam extending between 1957 O.D. (near the top of the dam) and 1685 (40 m in the rock, below the foundation), with the water level in the reservoir ranging from 1953 O.D. to 1825 O.D. gave a deformation of the order of 70 mm in 1958 reducing to 50 mm in the year 1962. Thus the central portion of the dam did not appear to return to its original position when first filled. This is 'attributed to a light permanent rotation of the entire foundation following the first filling as practically no residual displacement has been observed at the junction of the concrete and bedrock' by the 1724 m O.D.—1,685 m O.D. pendulum.

Albigna Dam (1956–1959)

The site of this gravity dam is on Granite, scoured clean by the Albigna glacier, and built on a rise in level of the granite across the valley. The old lake basin, upstream of the dam, is filled up with sand to a depth of 120 m, like some of those in Cumberland. The dam has a chord/height ratio of

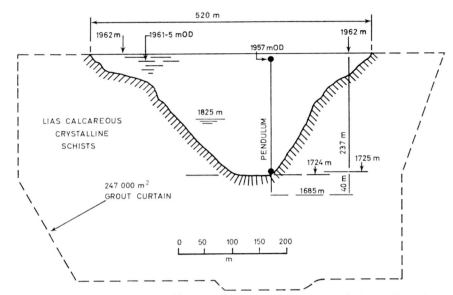

Figure 200. Mauvoisin arch dam. This is the highest (237 m) concrete arch dam in Switzerland. As the dam is founded on Calcareous Schist a very large grout screen is provided covering 247,000 m² in area in which 8,000 tons of cement reduced the leakage from 14 to 1 litre per minute. Apart from size and the permeable nature of the foundation, pendulum data are interesting. Thus a central pendulum between 1957 m O.D., near the top of the dam, and 1685 m O.D. (40 m in the Calcareous Schist below the foundation of the dam) with the water levels varying between the levels 1953 m O.D. and 1825 m O.D. showed deformation of 70 mm in 1958 and 50 mm in 1962. The 1724 m O.D.—1685 m O.D. pendulum ('15 Radial') in the rock has not shown any movement; perhaps the difference between the 70 and 50 mm, i.e. 20 mm may be a measure of the inelasticity of the Calcareous Schist and temperature effects on the dam, the latter generally being difficult to disintangle

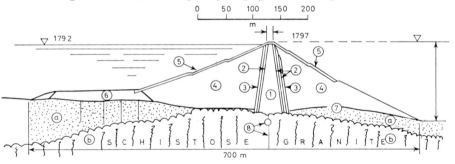

Figure 201. Göscheneralp dam. This dam, constructed in 1956–60, is founded on overburden and schistose Granite.

1. *Central core. Formed of clayey material, 0–100 mm to which was added 220 to 250 kg/m³ of Opalite clay from Holderbank (at the foot of the Jura mountains) with 22% < 0·1 mm. Water content 6%. Permeability 10^{-7} cm/sec. Weight 2·72 tons/m³. Density 2·2 tons/m³. $\phi = 36°$.*
2. *Filter grain. 0–100 mm, 8% < 0·1 mm.*
3. *Filter grain. 0–200 mm 13% < 1·0 mm.*
4. *Natural moraine material from valley sides, 0–800 mm, 12% < 1·0 mm. $\phi = 41·5°$. Cohesion 0·5 kg/cm².*
5. *Rip-rap.*
6. *Upstream berm and vertical drains in moraine.*
7. *Horizontal drainage blanket downstream.*
8. *Grout gallery and curtain in (a) Moraine. (b) Granite.*

The core was compacted by rubber-tyred rollers and the bank by dumpers. The maximum settlement became approximately stable after 3 years at 80 cm

810/115 m. Compare Vyrnwy silurian Limeston (page 100 and Haweswater, Borrowdale volcanic series, (page 26).

Göscheneralp Dam (1956–1960)

This high earth dam is founded for the most part on stable overburden e.g. alluvium, on granite, but at the upstream toe there is silt and peat which called for vertical sand drains and an upstream berm 200 m in length (*Figure 201*). The central core is taken through the overburden where it is shallow and founded on the schistose Aar Granite. As the granite is broken a reliable grout curtain (cement consumption, 250 kg/m of borehole) was necessary. Several designs of cores and filters were considered and a central clay core with a system of filters was adopted. General details of the core are shown in *Figure 201* and may be compared with the new Mattmark dam shown in *Figure 203a* constructed ten years later.

Isola Dam (Grisons) (1960–1967)

The Isola dam is a gravity-arch dam on the River Mesolcina a mile or so downstream of the village of San Bernardino in southeast Switzerland near the Italian border (*Figure 202a*).

The main dimensions of the dam are

Height	45 m
Crest, length	290 m
Height/crest ratio	1/6
Maximum thickness of arch	22 m
Volume of concrete	71,000 m³
Volume of excavation	23,000 m³

The reservoir is on Gneiss and Crystalline Schists (d'Adula Series). The dam is founded on Crystalline Gneiss, with mica and muscovite, the strike approximately parallel with the thalweg and the dip approximately 30 degrees to the east or left bank. The beds are only slightly fissured, but between them there are layers of fine clay about 10 cm in thickness which contain thin clay seams a millimetre in thickness. These clays are practically impermeable but this desirable quality was found not to prevent large uplift pressures under the barrage when the reservoir was filled.

The grout screen

As the Gneiss was so compact and the clay was impermeable, the grout selected was a weak mixture of four parts of water to one of cement to which was added 1 per cent of bentonite, no other chemical products were used. The consumption was as follows:

Distance between bores	Bores direction	Quantity of cement per metre of bore
36 m	Vertical	61·7 kg/m
12 m	Vertical	38 kg/m
20 m	Oblique	41 kg/m

The quantities of cement consumed are low and test water pressures showed a loss of only 2 l./min., (2 lugeons) under a pressure of 10 atmospheres, (c. 150 lb./in.²).

Contact grouting

Contact grouting under the concrete of the dam was carried out to a depth of 10 m, the bores being spaced 4 m apart across the valley. Pressures were kept low so as not to lift the strata as it has been recorded that even with low pressures, lifting of strata has occurred on small dams. However the rock was so dense that there was practically no visible affect, the absorption of cement being less than 8·3 kg/m of borehole.

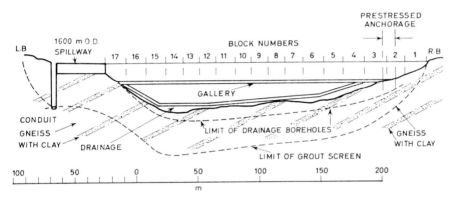

Figure 202a. Isola dam. This gravity dam on Gneiss, with intervening clay seams, was subject to high uplift pressures, soon after its construction in 1961. The grout curtain was deepened covering the area approximately shown but little grout was taken and the upward pressures remained rather high. The right end was strengthened by prestressed anchorage, the dip of the gneiss being into the valley at this end

After the grouting operations the dam was completed and the following observations were made.

Uplift measurements (Figure 202b)

In 1960, when the water in the reservoir was 4 m below top water level, the following percentages of the height of the water in the reservoir were recorded by manometer 5 m below the base of the dam:

In the gallery, downstream of the grout curtain, 35 per cent to 75 per cent; mean, 55 per cent.
At the downstream toe, 30 per cent.
At 20 m downstream of toe, at various places in the rock, 20 per cent.

In 1961, when the reservoir was full, the pressures along the axis of the dam were much the same, 35 per cent to 75 per cent, a mean of 55 per cent. It is pointed out that the individual blocks could not move independently, but it was the pressures downstream of the dam, which could detach rock, that caused some concern.

Reduction of uplift

(a) Additional grouting was therefore decided upon, particularly as a comparatively small amount of grout was consumed as indicated above. Thus boreholes at every 2 m in the drainage gallery, injected at pressures of 6 to 20 atm. (90 to 300 lb./in.²) of fine cement grout, consumed a total of only 23 kg/m of borehole which was less than the initial grout consumption. Other grouts were tried without success, or if proceeded with, the cost would have been out of all proportion.

(b) As the leakage had been proved to be so small and could not be economically reduced by more grouting and as loss of water was not of paramount importance it was decided to increase the leakage by adding

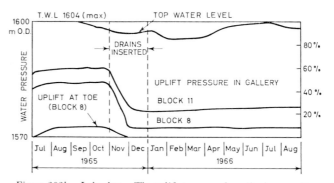

Figure 202b. Isola dam. The uplift pressures from the lower gallery were 35% to 75%; at the toe, 30%; and at 20 m downstream below the toe the pressure was 20%, enough to dislodge the rock. As additional grouting was unsuccessful in reducing the uplift, additional drainage by boreholes over the area, shown in Figure 202a, completed in November 1965, reduced the pressures along the gallery by varying amounts. Typical reductions are shown for blocks 8 and 11 in the centre of the dam. Pressures at the toe and beyond the toe were eliminated entirely. The leakage was increased from 35 litres/m into 70 litres/min which was immaterial

additional drains under the dam. The value of the loss of water would amount to only a few pence for a hydro-electric scheme such as this and it seemed most desirable to reduce the uplift particularly in the bare rock downstream of the toe. The uplift did not appear to have diminished for some years, so the following work was undertaken in 1966.

Effect of additional drains (1) Reduction of uplift

From the tunnel, about 2 m on the downstream side of the grout curtain, boreholes, 10 cm in diameter spaced at 3 m were drilled for about 15 m in length at a slope of 10 degrees downstream. In blocks Nos. 1, 2, and half No. 3, on the right bank, it was not possible to insert drains from the gallery. These blocks are anchored to the gneiss by prestressed cables which would compensate for any uplift by transferring the thrust of the abutment more directly on to the Gneiss.

In November 1966, in blocks Nos. 3 to 14, in the middle part of the valley, the drains reduced the uplift pressure substantially. Thus, in block No. 5, the 60 per cent uplift was reduced and the uplift at the toe and downstream of the dam, was completely eliminated and, in block No. 8, the reduction in uplift was nearly as good. In block No. 11 the result was not quite so good but as the blocks could not act as separate units it is fair to take the mean uplift at 30 per cent at the gallery and zero at the toe and beyond the dam.

In blocks Nos. 15 to 17, the influence of the drains is not so apparent, but in any case the original pressures were not so high as on the right bank, both in the gallery and, zero, at the toe.

There was no apparent uplift downstream of the dam.

Effect of additional drains. (2) Increase in leakage

With the reservoir full at top water level, the leakage was doubled from 35 l./min. to 70 l./min.

The leakage water is clear but should it become cloudy, coming out into the open, it would be possible to add a filter.

Grande Dixence Dam (1951–1952)

This gravity dam is built on rock which consists of Prasinites, Albites and Phylites which constitute hard rock although these names suggest a schistose-slatey nature. As the height of the dam was to be 285 m the elasticity module of the rock was tested and was ascertained to be between 100,000 and 250,000 kg/cm^2. The rock is somewhat fissured and weak to a depth of 2·5 m, occasionally to 6 m below the surface. On the whole, the foundation was considered satisfactory, although the crest/height ratio is 695/285 the gravity section was adopted, for downstream of the dam, the topography is abrupt and for the exceptional height of over 900 ft., uplift pressure had to be kept down as much as possible. For this purpose grouting was taken down to 200 m in the bottom of the valley and to a depth of 100 m below the foundations on the valley banks. In 240 m, 95 kg of cement was consumed per metre of borehole.

The aggregate for concrete for this great dam came from a moraine and consisted of prasinite (rich in chlorite and feldspar) and Gneiss (rich in quartz and cerite). The dam submerges the old buttress Dixence dam. As a result of the bombing of the Möhne gravity dam (page 54) three or four other buttress dams in Switzerland had alternate gaps between pairs of buttresses filled in with concrete.

The Mattmark Dam

The Mattmark dams, *Figures 203a, b* and *c*, was constructed between 1962 and 1967. The dam is founded on thick moraine material overlying crystalline rocks of Gneiss, Granite and Prasinite (Hornblende-Epidote schist).

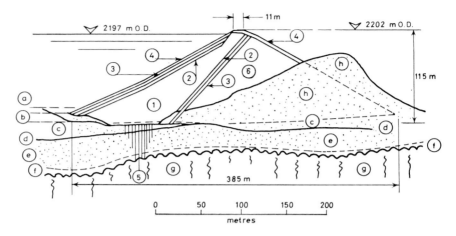

Figure 203a. The Mattmark dam. This dam was constructed in the period 1962–67. Maximum height 115 m. Founded on thick moraine material overlying to 100 m crystalline rocks of gneiss, granite and prasinite (green hornblende-epidote schist).

1. Sloping core of material from natural moraine: 0–100 mm; –13% < 0·1 mm. Density 2·50. Permeability $0·5 \cdot 10^{-5}$ cm/sec. Moisture content 3·4%.
2. Filters.
3. Drains.
4. Protecting layers, rip-rap, etc.
5. Grout curtain passing through (a) recent moraine; (b) lake sediments; (c) fluvo-glacial alluvia; (d) disturbed moraine; (e) bottom Wurmian moraine; (f) pre-Wurmian sand; (g) crystalline rocks; (h) recent Allalin glacier incorporated in downstream embankment.
6. Downstream embankment; 20% < 0·1. Density 2·41. Permeability $4 \cdot 10^{-2}$ cm/sec. Moisture content 2·6 %.

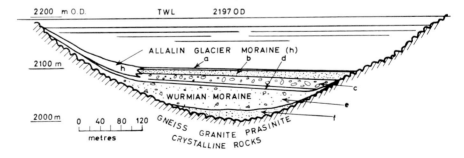

Figure 203b. The Mattmark dam. Moraine deposits in valley at dam site. These deposits overlie crystalline rocks, gneiss, granite and prasinite (hornblende-epidote schist)

(a) Recent moraine (fine and coarse sand).
(b) Lake deposits, silts and sediments.
(c) Fluvio-glacial alluvia, grits, sands and gravels.
(d) Disturbed moraine, muddy sands, sandy gravels.
(e) Bottom Wurmian moraine, blocks and cobbles of gabbro, serpentine, gneiss and granite.
(f) Pre-Wurmian sand, sedimentary lake deposits.

Table for Figure 203b. Mattmark dam

	(a)	(b)	(c)	(d)	(e)	(f)
Max. thickness, m	—	16·5	24·5	7·0	50·5	22·5
Percolation: cm/sec. natural	—	10^{-4}	10^{-3}	10^{-3}	10^{-2}	10^{-2}
with grout	—	—	$6 \cdot 1^{-5}$	$3 \cdot 1^{-5}$	$1 \cdot 1^{-5}$	$6 \cdot 1^{-6}$

Total thickness 121 m

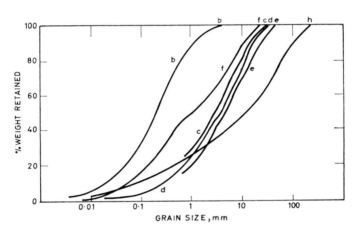

Figure 203c. The Mattmark dam. Sieve analysis of the moraines shown in Figure 200b. The chart shows broadly that the recent moraines b tend to be fine in texture; the fluvo-lacustrine strata c and d medium and the Würmian moraine e coarse

BIBLIOGRAPHY

Göscheneralp earth dam
GILG, B. Essais sur place et mesures de controle des dignes de Göscheneralp et de Mattmark. *I.C.O.L.D.*, R.8 **III**, 699 (1964).
SCHNITTER, G. and ZELLER, J. Geotechnical investigation of mixtures of bitumen, clay or bentonite with sandy gravel. *I.C.O.L.D.*, R.38 **IV**, 197 (1961).

Isola gravity-arch dam
GILG, B. Mesures prises pour l'amélioration de la stabilité du barrage d'Isola. *I.C.O.L.D.*, R.56 **I**, 923 (1967).

Mattmark earth dam
GERBER, F. P. les injections dans les alluvions et dans le rocher de la fondation pour la digue de Mattmark. *I.C.O.L.D.*, R.68 **II**, 459 (1961).

Mauvoisin arch dam
GILG, B. and DUBOIS, F. De Mauvosin (1958–63). *I.C.O.L.D.*, R.22 **II**, 393 (1964).
PARÉJAS, E. and ROMBERT. Reconnaissance geologie et traitement de le fondation du barrage de Mauvosin. *I.C.O.L.D.*, **IV**, 1179 (1955).

Rassens arch dam
GICOT, H. The deformations of Rossens arch dam during 14 years service. *I.C.O.L.D.*, R.24 **II**, 419 (1964).

Sufers arch dam
   SCHNITTER, N. Properties and behaviour of the foundation rock at Sufers arch dam. *I.C.O.L.D.*, R.39 **I,** 717 (1964).

Zervreila arch dam
   FREY-BAER, O. Subsoil exploration for Zervreila arch dam. *I.C.O.L.D.*, R.34 **II,** 261 (1961).

General
   GRUNER, E. Mechanism of dam failures. *I.C.O.L.D.*, R.12 **III,** 197 (1967).
   — Dam disaster. *I.C.E.*, **24,** 47 (1963) and **27** 343 (1964).
   SWISS NATIONAL COMMITTEE on LARGE DAMS. The Behaviour of Large Swiss Dams (1964). Berne, Switzerland.

Since the first edition, the Swiss National Committee published in 1964 'The Behaviour of Large Swiss Dams' and in compiling the above notes, the author has drawn upon this publication for a great deal of geological information affecting design and behaviour. The author is also very grateful to Mr. Edward Gruner, the Swiss correspondent of the Institution of Water Engineers, for reading his notes and making many suggestions in their presentation.

28

TURKISH DAMS

RECENT TURKISH DAMS OVER 50m IN HEIGHT

Ref. on map	Name of dam	Year completed	Type	Height m	Crest m	Purpose	Geology of foundation
E. of Ankara	Cubuk I	1936	G (Curved)	58	250	PIF	Andesite under decomposed andesite 19 to 33 m
7.	Porsuk I	1948	G	43·5	179	IF	Serpentine under 13 m of detritus
7.	Porsuk II Raised	1967	G	64·7	258·8	IF	
S. of Istanbul	Elmali II	1955	BG	49·0	238	P	Silurian arkose, quartzite and igneous dykes under 5–12 m alluvium
22.	Seyhan	1956	E	50·7	1955	HIF	Miocene siltstone under alluvial deposits
(140 km W. of Ankara)	Sariyar	1958	G	108	257	HF	Quartzite, schist, under 8–10 m alluvium
12,	Kemer	1958	G	113·5	310	HIF	Oldest Turkish metamorphic system, mica-schists. Phyllites marble schistose limestones underlying Tertiary marls clays siltstones
10.	Demirköprü	1960	E	77	543	HIF	Gneiss of variable composition. Basalt with cavities under half of dam which is spongy and permeable
32.	Almus	1966	ER	93·5	370·5	HIF	Andesite, active faults near dam, 15 m alluvium
17.	Kesikköprü	1967	ER	52·6	265	H	Igneous complex; spilites, syenite, gabbros, diorites etc. under alluvium 2–3 m
54 km N.N.E. of Ankara	Cubuk II	1965	E	69	230	P	Andesites and basaltic lavas under alluvium 10 m
30.	Keban	U.C.	R CG	167	1095	HI	Calcareous schists and crystalline limestones jointed and cracked. Alluvium 40–50 m under river
6.	Gökcekaya	U.C.	A	158	466·4	HI	Grauwacke, slate, schist, serpentine and dolomite limestone, alluvium 36 m under river
50 km N. of Ankara	Kurtbogazi	1967	E	52·6	332	IP	Pleistocene clay on Andente
4.	Hasanlar	U.C.	R	72·8	310	IF	Eocene flysch on Cretaceous limestone
8.	Caygören	1970	E	53·5	658	IF	Neogene andesite under 8 m (max.) alluvial sands and gravels
13.	Büldan	1968	R	64	295	IF	Palaeozoic gneiss with Lias drift
18,	Hirfanli	1958	ER	83	364	IFH	Gabbro. grano-diorite dyke
N. of 22.	Kesiksuyu	U.C.	E	57	520·2	I	Sand. Silt. Clay.
27.	Kartalkaya	U.C.	ER	57	205	IF	Limestone underlain by flysch
28.	Sürgö	1968	R	57	736	I	Gneiss.

```
P denotes Water-supply          I denotes Irrigation
H    ,,   Hydro-electric        F    ,,   Flood-control
```

There are two known *Roman* dams in Turkey: (1) Cavdarhisar (250 km west of Ankara) 7 m high and 80 m long, and (2) Örükaya, 200 km NE of Ankara, 16 m high and 40 m long. There are also several masonry dams for the water-supply of Istanbul still in operation dated 1651, 1723, 1756, 1796, 1837 and 1839, ranging in height up to 15·45 m and crests 95 m in length.

Other interesting dams are the Elmali (I) dam built in 1893 for the water-supply of Instanbul which is 22·0 in height with a straight masonry crest, 118·6 m, joined to a curved earth crest of 179·8 m in length. The Cubuk I dam, mentioned in the table, for the water-supply of Ankara built in 1936.

In the table recent dams over 50 m in height are shown. The information in this table has been adapted from 'Large dams in Turkey' which gives excellent geological summaries of their foundations. *Figure 204* shows the

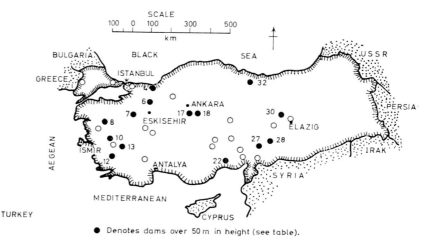

Figure 204. Map of Turkey showing the sites of some large, modern dams. Nos. 4, 7, 10, 13, 27, 28 and 32 would appear to be in more vulnerable areas than 6, 8, 12, 17, 18, 22 and 30. Nos. 4, 8, 13, 27 and 32 have been designed for earthquake resistance by variation in current practice for earth/rockfill dams. (After O.M. Üral, I.C.O.L.D. **IV**, 315 (1967))

location in black, and numbered in accordance with a paper by O. M. Üral on the design and construction of earthquake resistant dams in Turkey. These dams are all in areas subject to earthquakes in more or less degree— Nos. 4, 7, 10, 13, 27, 28 and 32 would appear to be the more vulnerable areas and Nos. 6, 8, 12, 17, 18, 22 and 30 less vulnerable.

Earthquake Provisions

In his paper O. M. Üral has given typical earthquake provisions used in the design and construction of Nos. 4, 8, 13, 27 and 32 (*Figure 204*). The following extract will give some idea of the problems of designing dams for earthquake construction, particularly for sloping core dams:

(1) Special investigation of the geology relative to the dam/axis and type of dam. Special investigation of slides of the abutments.

(2) Self-healing structures, i.e. earthen embankments most favoured.

(3) Provide positive type of cut-off with a large contact area and large clay core. Fractured or decomposed rock in the cut-off trench area, supported by concrete blocks, a metre square, in sections. Double row of grouting not considered useful.
Drainage galleries in the abutments, downstream of the grout curtain, useful. Where the cut-off cannot be guaranteed to stop leakage, a long clay blanket, say 100 m in front of the dam, may be an alternative to reduce leakage.

(4) Higher freeboard, e.g. 3 m in case of wave in reservoir due to earthquake.

(5) Wider crest as larger amplitudes are believed to take place at the crest. (This seemed to be the case at Loughborough England (page 53) where most damage was at the coping of the Blackbrook concrete dam.)

(6) High camber is provided against slumping.

(7) Filter slopes may, with advantage, be flatter.

(8) A large core gives more contact with foundations and gives a longer leakage path. Also if a fault crosses the dam a large horizontal flow can be tolerated.

(9) A more plastic clay can be specified so as to be self-healing.

(10) Filter between difficult materials, normally unnecessary, may be desirable as it avoids sudden changes of permeability.

(11) Spillways, gates and structure to be subject to a horizontal acceleration of 0.10 g and vertical 0.05 g

(12) Larger outlet capacities for lowering water quickly—generally impossible.

(13) Any suspicion of an active fault, e.g. hot springs, should rule out the site.

No dam in Turkey has been recorded as having to be abandoned on account of earthquake damage.

The only record of a partly abandoned dam which was due to leakage through excessive deposits (in area and depth) of gravel sand alluvial in the Cretaceous limestones, is that of the May Dam (see Bibliography).

This is a rockfill dam, 27·8 m in height, with a crest 420 m in length founded on Cretaceous and karstic Neogene limestones and alluvium (silt, sand, gravel, clay) 15–20 m in thickness.

Briefly, the *annual* losses recorded are:

$$\begin{array}{ll} 1962 & 22 \cdot 82 \times 10^6 m^3 \\ 1963 & 48 \cdot 92 \times 10^6 m^3 \\ 1964 & 13 \cdot 24 \times 10^6 m^3 \\ 1965 & 26 \cdot 24 \times 10^6 m^3 \end{array}$$

Nevertheless, the reservoir can be utilised for flood protection, early spring irrigation and regulation of ground water.

The Demerköprü dam (No. 10 on *Figure 204*) not far from Izmin; constructed in 1960, developed leakage in the foundations in 1969, and additional grouting is proceeding (1970).

At the Keban dam (No. 30, *Figure 204*) near Elazig in the East of Turkey, grouting has not been entirely satisfactory and a concrete cut off diaphragm wall is to be constructed.

I should like to thank Mr. Saim Kale, sometime my geological assistant, for his generous help in preparing these notes on his country's dams.

BIBLIOGRAPHY

Alpsü, Irfan. Investigation of water losses at May reservoir. *I.C.O.L.D.*, **III,** 477 (1967).
Ural, O. M. and Ungan, U. Large Dams in Turkey, General Directorate of State Hydraulic Works. (*Sponsored by* the Turkish Committee on Large Dams (1967) and (1964).
— Design and construction of earthquake resistant dams in Turkey. *I.C.O.L.D.*, **IV,** 311 (1967).

29

ALGERIAN DAMS

Algerian dams (*Figure 205*) of geological interest, fully detailed in the Proceedings of the Nineteenth International Geological Congress (1952), may be briefly referred to as follows.

Bakhadda Dam

Maximum height	45 m
Type of dam	Rockfill
Formation	Miocene on Jurassic
Geologist	A. Clair

The soft alternating Miocene (Cartenian) limestones, grits and clays at the dam site, over 20 m in thickness, lie on Callovian (Kellaway) and Oxfordian clays of the Jurassic down to which the grout curtain was taken.

Some 450 holes aggregating 20,000 m absorbed 3,500 tons of cement, and it is estimated that 125 kg per m^2 of screen was consumed. The unconformable junction of the Miocene with the Jurassic was considerably disturbed, the latter being exposed, no doubt, to weathering for many millions of years. This weathered zone in excavating the foundations caused a 'disagreeable surprise'.

D'Iril Emda Dam

Maximum height	70 m
Type of dam	Rockfill of small stones
Formation	Cretaceous (Upper Chalk Senonian horizon, after Sens, France)
Geologist	A. Lambert

A rigid structure was rejected after full scale experiments which suggested that the ground would not carry 10–20 kg per cm^2.

D'Erraguéne Dam

The D'Erraguéne dam near the D'Iril Emda dam whose position is shown in *Figure 205* is a dam built in 1956–58 on siliceous schists of the Lower Chalk, which is 86 m in height with a crest 530 m long. It is a multiple arch dam like Beni Bahdel with buttresses supported on jacks and the structure prestressed which gives a very light appearance.

Foum el Gherza Dam

Maximum height	65 m
Type of dam	Cupola
Formation	Massive Cretaceous limestone (Maestrichten from the substage of the Upper Chalk at Maestricht, Holland)
Geologist	N. Gouskov

Much experimental work for preventing leakage and a very complete grout screen was necessary in this Cretaceous limestone, a problem which might be comparable to the building of an impounding reservoir on the Chalk!

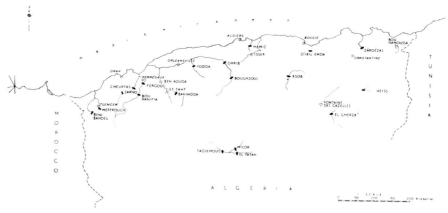

Figure 205

Foum el Gueiss Dam

Maximum height	23 m
Type of dam	Rockfill
Formation	Faulted grits and shales Burgundian horizon of the Oligocene
Geologist	L. Lessard

Certain clay schists were unstable in air and water and had feeble bearing capacity (3 kg per cm^2) so a rockfill dam was adopted.

Hamiz Dam

Maximum height	50 m
Type of dam	Gravity
Formation	Palaeozoic schists, Trias, Eocene
Geologist	G. Durozoy

This was a very difficult site on which to establish a dam as the strata were nearly vertical, the strike being nearly parallel with the dam across the valley.

The left end of the dam was on ancient schists and Permo-Triassic grits, shales and conglomerates and was sound enough. The right side of the valley, however, consisted of an unconformable junction of soft broken material between the Eocene limestones and the Palaeozoic schists, which had to be extensively treated before the dam could be put on it.

Ksob Dam

Maximum height	32 m
Type of dam	Multiple arch
Formation	Eocene
Geologist	J. Emberger

The Thanetian (with phosphatic nodules, as in England) and Yypresian horizon of the Eocene dipping 30 degrees downstream, proved to be sound enough for a multiple arch dam (32 m), although a gravity dam (of 50 m) was originally intended. The smaller dam was adopted for economy.

Zardézas Dam

Maximum height	40 m
Type of dam	Curved concrete
Formation	Miocene
Geologist	G. Durozoy

The Miocene was much broken and faulted and in 1923 the site was considered unsuitable for a dam on the grounds of stability and watertightness. In 1925, however, a rockfill dam was suggested.

Further exploratory work proved the Poudingues, conglomerate of the Miocene to be more compact below the surface than anticipated, and in 1930 a concrete dam was suggested instead of the rockfill type.

A slip nearly rendered the whole scheme abortive; a movement of 30 m horizontally and 35 m vertically between July 1933 and December 1935, the movement reaching 20 cm a day.

The left end of the bank was ultimately secured by a block of nearly 5,000 m^3 of concrete.

In addition to the foregoing, foundation difficulties experienced at the following Algerian dams are described in more detail in the following pages: Les Cheurfas, Ghrib, Fodda, Fergoug, Beni Bahdel, Bou Hanifia, Sarno.

Les Cheurfas Dam

The Les Cheurfas dam, 50 miles south-east of Oran, owes its present shape to geology. Its case history is as follows.

1882–85. Dam built with limited geological knowledge about 90 ft. high with a top water level of 225 m, impounding 72 ft. of water.

Feb. 8, 1885. Right bank washed away in a storm submerging St. Denis-du-Sig (20 km downstream) (*Figure 206*).

1892. Right bank rebuilt with geological knowledge (crest 229 m). Considerable leakage, on left bank and under dam (*Figure 207*).

1927 After the Fergoug dam disaster due to floods, a crack between the old and new masonry (on a concrete base) was noticed and the cross-section was considered to be too thin; the crest was therefore reduced from 229 to 227 m O.D.

1935. Cement injection of foundations. Strengthening the dam by pre-stressed cables. Limited flood level to 229·25 m O.D.

1956. Still standing (*Figure 208*).

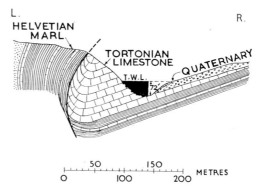

Figure 206. Les Cheurfas dam. February 8th, 1885. The right side of the dam was swept away because it was founded on unstable Quaternary beds which were mistaken for Tortonian limestone

The mistaken foundation

The foundation consists of 70–100 ft. of Alluvium and Quaternary, clays and stones, on 260 ft. of Miocene Tortonian limestone on 260 ft. of Helvetian shales and marls.

The left end of the dam was founded well enough on the Tortonian limestone, but the right end was built on the heterogeneous Quaternary (*Figure 206*) owing to deficient geological knowledge. The report after the disaster in 1885 reads: 'The Quaternary is sometimes cemented as conglomerate and one such facies, occurring on the right bank, was confused with the breccia limestone of the Tortonian'.

The examination revealed that the Quaternary conglomerate contained black eratics and pebbles (galets), whereas the Tortonian breccia conglomerate was not formed of pebbles of the same material or age as those in the Quaternary.

In 1892 the right bank end of the dam was rebuilt, but not on the same line because the Tortonian limestone was 70–100 ft. below the Quaternary rubbish which would have necessitated very heavy excavation. Hence, the dam was reconstructed with a mitre, and the right end brought upstream to meet the Tortonian limestone.

Figure 207 shows the outcrop by the limestone on plan and *Figure 208* the section of the dam, with the right-hand end founded on the limestone after reconstruction. *Figure 209a* shows a general view of the mitre pointing downstream, and Tortonian limestone outcrop as a cliff on the left bank.

Leakage

There was a considerable leakage from the Tortonian limestone on the left bank. Colour tests showed this to vary with the depth of water in the reservoir, which was not surprising when one sees the extraordinary geological section shown in *Figure 208*.

L'Entreprise Soletanche undertook the grouting work between 1931 and 1935 (*a*) in the foundations, and (*b*) in the body of the dam.

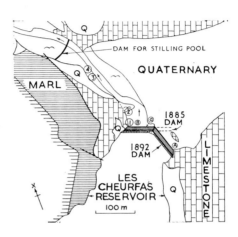

Figure 207. Les Cheurfas dam. In 1892 the right side of the dam was reconstructed to join with the Tortonian limestone. To save excessive excavation through the Quaternary the reconstructed half of the dam was turned through 45 degrees forming a mitre pointing downstream

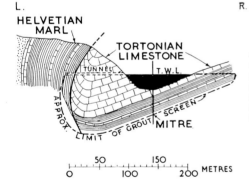

Figure 208. Les Cheurfas dam. Following the rebuilding of the dam in 1892 it has been substantially satisfactory ever since. Considerable leakage in the limestone had to be dealt with; the limit of the grout screen is shown. The mitre joint had also to be grouted and the dam strengthened by pre-stressed cables in 1935

Preliminary trials showed the necessity of taking the grout screen down to the underlying Helvetian marl. After a single line of bore holes had been grouted in the limestone several defects were found on the left bank in the laminated and fractured syncline of the Tortonian limestone; this necessitated a second or, in part, a third row of bore holes.

The regeneration of the body of the dam was done from the crest with a thin solution of cement. The same bore holes were used for the grout curtain in the limestone below, and some of them also served as pilot bore holes for the 'Coyne' pre-stressed cables.

Some of these bore holes were drilled to a depth of 180 m, but most of them were 30–40 m, on an average, under the foundations of the dam. They were usually spaced at 4 m apart.

Right bank screen

For a length of 50 m and a depth of 70 m, the bore holes were drilled in two lines, staggered, and spaced at varying distances apart between 2–8 m, the closest spacing being nearest the dam.

(a)

(b)

Figure 209. Les Cheurfas dam. (a) The left end of the masonry dam is founded on mass concrete below ground level on the upturned end of a syncline of Tortonian limestone which is visible as an arcuate outcrop above the end of the dam. After the disaster in 1885 the right-hand end of the dam was rebuilt with a mitre in order to be founded on the Tortonian limestone, only attainable at a reasonable depth upstream. The row of knobs on the top of the dam are the tops of the pre-stressed cables added in 1935 to strengthen the dam. (b) This view of the right end shows the strong Tortonian limestone which formed the right abutment of the dam with the Quaternary weak material in the foreground

Left bank screen

It was found that the left bank screen had to be constructed in two operations owing to the enormous thickness of limestone caused by the syncline.

The first operation, carried out from the top of the dam up to a construction tunnel, was to drill three rows of staggered bore holes, as the rock here was very much of a 'breccia' or 'conglomerate' for 20 m or so in depth.

The approximate distance between the bore holes is 2·5 m and the depth of the screen, for a length of 50 m, reached 90 m, that is, down to about 150 m O.D. (*Figure 208*).

The second operation was to construct a screen from a tunnel, driven in the rock of the left bank 80 m in length (*Figure 208*).

The injection of grout was made from a series of bore holes alternating with a second series all in one line. The bore holes were spaced at 2·5 m apart and the depth of the curtain was 100 m, that is, down to about 120 m O.D.

At the end of the tunnel the screen was completed by an 'aureole' of inclined bore holes into the Helvetian marl.

Method of drilling

The bore holes through the masonry of the dam and into the rock were rotary drilled, through the masonry of the dam by drilling crowns of hard prisms, and through the rock by diamonds. The diameters varied between 35–65 mm. For soft rock or conglomerate drilling was done by percussion.

Materials injected

For the masonry of the dam the cement to water ratio varied according to its sponginess, sometimes a gell of silicate was used. For rock, the ratio of cement to water varied according to the rate of absorption. For large fissures inert products such as sand or sawdust were added to the cement.

Method of injection

All the injections in the rock were executed in descending slices, each piece of rock treated being dependent upon the covering of the overlying section; the injection was made from each bore hole which was washed out and the overlying grouted part was rebored. This process was repeated until the Helvetian marl was reached, the injections were stopped when the specified pressure was reached. The pressure in the dam and its immediate neighbourhood was limited to a maximum of 10 kg per cm^2, but for the screens in the rock the pressure was limited to 50–60 kg per cm^2.

Grout consumed

Table 1 gives a summary of the quantities of dry products injected into the body of the dam, and into each part of the grout screen and, in addition, the grout consumed between the dam and the rock in the course of grouting the normal screen.

The following facts were observed.

(1) Absorption on the right bank is normal.

(2) The quantity of cement injected is expressed in tons per lineal metre of bore hole, but this includes both the first and last rows. Actually, the injection of the first row filled large voids; those of the following rows, on the other hand, filled small fissures which did not absorb anything like those of the first injections.

(3) The injection for the regeneration of the body of the dam is much less in the new part of the work than in the old.

(4) Again, this is the same for the injections between the dam and rock, where the quantities absorbed under the old part of the dam were much greater than those under the reconstructed right-hand side.

TABLE 1
GROUT CONSUMED

Grout screen	Length of bore hole (m)	Area of screen (m^2)	Amount of cement absorbed		
			Tons	Tons per m of bore hole	Tons per m^2 of screen
Centre	2,128	6,650	313·1	0·148	0·047
Right	966	3,500	50·0	0·050	0·014
Left	5,825	12,950	2631·0	0·450	0·203
Total mean	8,919	23,100	2994·1	—	—
Mean	—	—	—	0·334	0·129
In dam	1,545	3,850	83·5	0·054	0·022
Under dam	725	1,890	198·8	0·275	0·105
Total	11,189	28,840	3276·4	—	—
Mean	—	—	—	0·293	0·114

Reduction in leakage

The gauging of the springs in connection with the water in the reservoir showed that the grouting was very successful, especially that in the difficult left bank screen. The following leakages before and after the grouting works were recorded.

Table 2 shows that the losses at points (2) (A) and (B) have been completely eliminated, and those at (1) and (a) considerably reduced. Losses (3) and (4) have been reduced only by 35 per cent but this is normal for these sources are fed chiefly by rain (they are also over 100 m downstream of the reservoir (see *Figure 208*).

TABLE 2
REDUCTION OF LEAKAGE

Spring (see Figure 208 for locality)	Before grouting (1931–32) Litres/min	After grouting (1932–35) Litres/min	Reduction due to grouting %
1	110	15	87
2	402	0	100
3	250	165	35
4	500	300	40
a	15	0·3	98
A	0·8	0	100
B	0·75	0	100

After the dam had been treated it was reported that the damp patches on the wall on the downstream face, particularly those on the old part and at the junction between the old and new part, had disappeared.

Pre-stressed strengthening

In 1935 the dam was strengthened by 37 pre-stressed cables, the first example of its kind of this modern technique, the cables being anchored in the grits of the Helvetian.

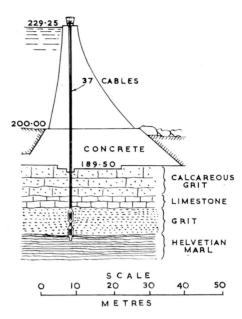

Figure 210. Les Cheurfas dam. A typical section through the dam showing the pre-stressed cables, anchored into the grit above the Helvetian marl, which were added in 1935

Figure 211. Les Cheurfas dam. In 1935 a small concrete dam was added at some 300 m from the dam (where the photograph is taken) to form a natural stilling pool for flood water passing over the dam. The very rubbishy nature of the Quaternary material on the right bank in the foreground can be discerned

Each cable (the top of which are shown in *Figure 209* and a section in *Figure 210*) was inserted in a bore hole 25 cm (about 10 in.) drilled from the top of the dam through 20 m of Tortonian limestone and 15 m into the Helvetian marl below. Each cable consisted of 630 steel wires 0·5 mm in diameter encased in plastic material and surrounded with a strong cloth sheath.

To secure the cables in the grit above the top of the Helvetian marl, a length of 3 m was reamed out to a larger diameter and the annular space between the cable and bore hole was filled with cement.

The cables were stressed to 1,000 tons.

For a few years the average loss in tension in the cables was measured thus: 1936–38—14·5 tons per year; 1939–42—1·2 tons per year; 1942–45—0·4 tons per year.

Stilling pool

To protect the toe of the dam a small arch dam was built in 1935 about 300 m downstream to make a stilling pool below the dam (*Figure 211*).

Ghrib Dam

The site of the Ghrib rockfill dam (constructed 1933–38, 60 miles south-west of Algiers) is geographically good, but geologically bad.

It is geographically good because it is in a valley with a gathering ground of 23,000 km² (9,000 sq. miles), and for a maximum height of 65 m and a crest length of only 300 m, it gives the very large capacity of 280×10^6 m³ = 61,500 million gal.

It is geologically bad because it is built upon heterogeneous deposits of part of the Miocene formation, as follows.

Upper Grit (with 'Jacob's marl' in 3 beds (1·5 m))	80 m
Upper Marl	16 m
Helvetian stage, Middle Grit	24 m
Helvetian stage, Middle Marl	+

The above strata are similar, in many ways, to those under the Bou Hanifia dam.

Figure 212 is a typical geological section on the axis of the dam; it shows several interesting features, as follows.

(1) How the Upper and Middle Grits of the Miocene (80 and 24 m in thickness respectively) carry the dam. The rockfill type was adopted because the relative bearing pressures between the grits and underlying marls are very variable: typical bearing pressure in the marls is 7 kg per cm², (100 lb. per in.²) and in the grits 180 kg per cm², (2,580 lb. per in.²).

(2) How the valley sides have formed an anticlinal bulge, the soft Upper Marl being dug out above the underlying Middle Grit.

(3) How the grouting underneath the foundation in the bottom of the valley was effected from the excavation and from tunnels, the screen being taken down to the Middle Marl.

(4) How the Upper Grit on the right and left banks was cement grouted from the surface, namely, by two lines of bore holes (2·5 m, apart in the most fissured parts); in the first line the holes inclined away from the thalweg of the stream, and in the second line were vertical (for these sodium silicate was used to help the penetration of the cement).

Figure 213 shows a typical cross-section of the dam which includes some 700,000 m³ of rock; the upstream face was covered with a bituminous concrete apron on a layer of hand-packed rubble.

The rock for the rockfill was brought from quarries, $4\frac{1}{2}$ miles away, in the larger calcareous lenses of the Cenomanian horizon of the Cretaceous.

The grout curtain

For the grout curtain under the dam the total length of bore hole drilled was 25,500 m (originally estimated at 3,700 m). The amount of cement used was 1,680 tons and, in addition, 1,250 tons of sodium silicate and hydrochloric acid were used.

The total amount of cement and chemical products consumed per metre of bore hole was 115 kg. As the area covered by the grout curtain was about 32,000 m², the amount of cement and chemical products per m² of curtain was 90 kg.

Some 1,000 m of tunnel and 3,200 m of drainage bore holes were made. There were also 5,500 m of observation bore holes. The leakage is 17·5 l. per second.

The overflow channel

The overflow works are of especial interest. The area of the gathering ground is 9,000 sq. miles to the dam site, the River Cheliff being one of the largest in Algeria, stretching from the Saharian Atlas to the Mediterranean.

This gathering ground consists largely of Cretaceous rocks and, nearer the dam, Miocene. It was estimated that in one week, in January 1951, 500×10^6 m³ flowed, or an average at about 850 m³ per second.

The overflow provided is capable of taking 4,000 m³ per second. It consists of a channel about $\frac{1}{2}$ mile in length with an effective width of 170 m (558 ft.). *Figure 214* shows its position in relation to the dam, as well as the contours of the ground and the outcrops of the Upper and Middle Grits and Marls.

The lip of the channel at (C) (*Figure 214*) is constructed over the junction of the Upper Grit and Upper Marl, a grout screen through the Grit was necessary for the protection of the lip. This necessitated some 2,900 m of bore hole into which 900 tons of cement and 120 tons of chemical products were injected.

The overflow channel was dug in the Upper Grit, which formed an old river course, and a large slip of several hundred thousand m³ occurred near (A). This slip was due to a lenticular seam of marl, although only 1·5 m in thickness, called 'Jacob's marl' after M. Ch. Jacob, the geologist investigating the matter. This necessitated very heavy walls of the channel to maintain stability in the region of (A) (*Figures 215b and 216b*).

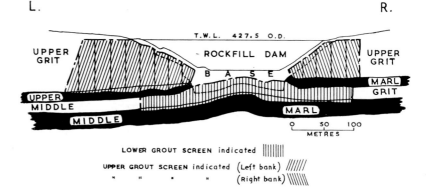

Figure 212. Ghrib dam foundations. Owing to differential pressure between the Miocene grits and marls a rockfill dam was adopted. There are two grout screens; the one in the Upper Grit down to the Upper Marl constructed from the surface, and the other constructed from the base and tunnels at the top of the Middle Grit down to the Middle Marl

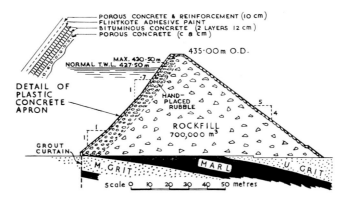

Figure 213. Ghrib dam. Diagrammatic section of rockfill dam built on Miocene grits and marls. The material for the rockfill was obtained from the calcareous limestones in the Cretaceous $4\frac{1}{2}$ miles from the dam. The grout screens, follow the upstream curved toe (on plan) of the dam, to which the concrete waterproof apron is keyed. Height, 65 m; length, 340 m; volume, 700,000 m³. As the dip is downstream, the rockfill was placed on concrete slabs keyed with concrete piles into the grit

As the strata between the reservoir and the channel are permeable, extensive drainage works were also necessary between (A) and (B) (*Figure 214*) to protect the channel, needing 450 m of galleries and 4,800 m of drilling.

Again, in the hill of permeable Upper Grit between (C) and (D) and the overflow channel, several bore holes were put down for cementation and observation of underground water levels.

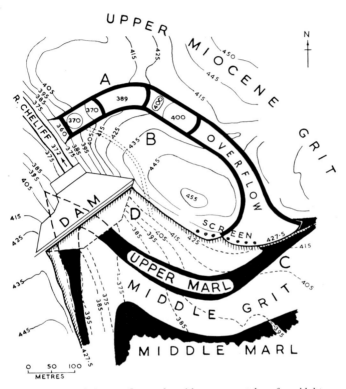

Figure 214. Ghrib dam overflow works. Advantage was taken of an old depression in the Miocene Upper Grit to construct the overflow works. A slip (at A) occurred involving several hundred thousand m³ due to an unsuspected lense of marl (1·5 m in thickness) in the Upper Grit. Extensive drainage works were necessary to protect the channel in the permeable Upper Grit from water pressure from the reservoir towards (B). The cill (C) of concrete was partly on the Upper Grit overlying Upper Marl, so a grout screen had to be constructed. Between (C-D) the reservoir and the overflow channel there were many bore holes, some grouted, others used for observation for flow through the Upper Grit

Figure 215a shows a general view of the dam, power-house, and overflow channel; *Figure 215b* shows a part of the overflow channel, while *Figure 215c* shows a model of the overflow built on site. *Figure 216a* shows the Upper Grits (on the left) which were grouted, that the crest of the overflow weir (beyond) should not be undermined, and *Figure 216b* is taken at the lip looking down the overflow channel.

(a)

(b)

(c)

Figure 215. Ghrib dam. (a) Rockfill dam, power-house at foot, and overflow channel above power-house on left. The slipped area is to the left of the channel which necessitated heavy walling. On the right, under the dark patch of trees between the channel and dam, a threat of excessive water pressure necessitated an extensive system of drainage. (b) Part of the overflow channel and above the wall the permeable Upper Grits which necessitated extensive draining. (c) Model of the overflow works. In the channel were three stilling pools of large dimensions as shown in Figure 214 at the River Cheliff

(a)

(b)

Figure 216. Ghrib dam. (a) *Lip of overflow channel* (left) *and site of grout curtain in foreground—reservoir in Miocene country to the right.* (b) *Part of lip of the overflow channel, 558 ft. in length, left bottom corner, constructed partly on Upper Grit and Upper Marl of Miocene age. Special grouting had to be done in the Grit to prevent leakage under such a large overflow channel*

Fodda Dam and Reservoir

The Fodda dam (gravity, concrete) was built between 1928–33 and commands a gathering ground of 340 sq. miles. The reservoir capacity is nearly 50,000 million gal. and the dam is 89 m (292 ft.) high above the stream. Crest, 170 m; width of base, 65 m.

Figure 217. Fodda dam. (a) Well-bedded Upper limestones of left bank to 150 m in thickness. (b) Evenly bedded Middle limestones of right bank to 100 m in thickness. First point, on left, Oxfordian (Couches Rouges) marls

(a)

(b)

The dam is founded on Jurassic limestones, probably of Upper Liassic age, which dip steeply upstream obliquely to the thalweg so that the Middle Lias (?) appears downstream and strata of Oxfordian and Neocomian age upstream (Figure 217a and b).

The general sequence is as follows.

Neocomian	(7)	Marls (schistose) with beds of quartzite and limestone	100 m
Neocomian	(6)	Limestones	200 m
Oxfordian } Callovian }	(5)	Couches Rouges, marls with red and green horizons	50 m
Upper Lias	(4)	Upper limestones, well-bedded . . .	150 m
Upper Lias	(3)	Middle limestones, evenly bedded . . .	100 m
Middle Lias	(2) (a)	Lower limestone, siliceous, massive	80 m
Middle Lias	(1) (b)	Lower limestone, siliceous, massive	100 m

Note: The terms Upper and Middle Lias must be treated with reserve.

Dam site

The valley narrows downstream of the existing dam giving rise to what appears to be a better and more economic site for a dam. The Triassic limestone, however, is full of grottos and springs, and is dolomitized, being readily attacked by water which disintegrates the rock leaving sand. When the sand is washed out the rock becomes karstic. There are also deposits of soft tufa, calcium carbonate deposited from the limestone springs. Hence, a larger dam has been constructed upstream on the Jurassic to avoid these difficulties. Although the crest-height ratio is about 2, the limestones would be too weak in themselves to take the thrust of an arch dam.

Earthquake threat (1954)

The Fodda dam was thought not to be affected by the earthquake which occurred in the Orléansville district on September 9, 1954, the epicentre of which was 30 km away. Nevertheless oscillations of about a 1 m were observed in the reservoir and the reservoir keeper's house was damaged although solidly built on rock.

Grouting

The tilted fissured Jurassic limestones necessitated much grouting (for example, it is said, 10 times as much as for the Chambon dam of the same size founded on Gneiss).

Four galleries were driven. Large fissures were injected with cement and small fissures with silicate of soda and aluminium sulphate at a pressure of not less than 50 kg per cm^2.

Grout consumed	7,700 tons
Length of bore holes.	9,500 m
Hence the Aml (amount metre lineal)	705 kg per m

The reservoir topography upstream of the dam is affected by the dip of the strata, being oblique to the thalweg and the alternating hard and soft strata of Tertiary age; for the reservoir water area widens out first on the left side and then on the right with a series of staggered cliffs or promontaries due to hard material (*Figure 218*).

Silting

Owing to this soft material, dolomitic and other chemical action causes silting which is rapid and troublesome and it is estimated that the reservoir has been reduced in capacity by 50 per cent between 1940 and 1960.

Figure 218. *Fodda reservoir. Alternating hard and soft strata of marls and limestones of Neocomian and Upper Jurassic age; the limestones forming promontaries. Oxfordian (Couches Rouges) promontaries in foreground. Some will see in this an excellent natural example of 'landscaping'*

Fergoug or Perrégaux Dam

Perrégaux dam, 40 miles west of Oran, is below the Bou Hanifia dam. The gathering ground is very large—3,280 sq. miles—from which floods may be enormous.

Although there are alternating pudding-stones, marls, grits, clays of the Lower Miocene (Cartennian), with a general dip downstream, the dam 1,040 ft. in length, is on a good foundation of rock. Upstream the widening out of the valley is due to the softer Tortonian marl.

The case history of this unfortunate dam is as follows.

1871. Dam about 110 ft. in height.
1872. March 10, 165 ft. of the overflow was washed away by a storm of 25,000 cusecs. Cause attributed (wrongly) to insufficient foundation.
1873. Reconstructed.
1881. Dec. 15, 410 ft. of dam washed away by a storm of 30,000 cusecs. The cause attributed (wrongly) to insufficient foundation.
1885. Dam repaired.
1927. Most of dam washed away (510 ft. of it to 13 m below top water level) by a flood attaining perhaps 141,000 cusecs. The cause was attributed to original

inadequate design of 1865, deterioration of mortar between 1865 and 1927, inadequate estimate of catastrophic storms, and weight of mud.

1934. Rebuilt, but with a lowered crest by 45 ft. with pre-stressed concrete strengthening.

1939. Raised by 6 ft.

Figure 219. Fergoug or Perrégaux dam. In the foreground can be seen the original dam. The gap left by the part which was washed away is shown in the centre

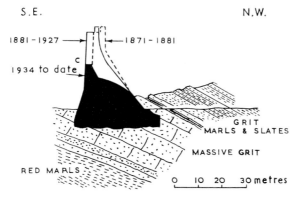

Figure 220. Fergoug or Perrégaux dam. This dam failed several times between 1871 and 1934 owing to floods causing excessively high depths of water going over the crest and to poor construction. The geology and design were wrongly accused from time to time

The geology after all was quite good, the dam being largely built on a very substantial bed of gritstone; it was the quantity of storm water that was underestimated.

Figure 219 shows the 1873 reconstructed overflow weir (125 m) in foreground, and the breached main wall beyond.

In 1956, French engineers considered using siphons to reduce the head of storm water and the raising of the crest to its original level.

Figure 220 shows a typical cross-section of the dam and Lower Miocene on which it is founded with the profiles of the original and later variations of the dam.

Beni-Bahdel Reservoir

There are three dams near the Moroccan border of Algeria which enclose the Beni-Bahdel reservoir for the water supply of Oran, for irrigation, and for hydro-electric power.

Figure 221. Beni-Bahdel reservoir. Looking from Beni-Bahdel dam (Upper Jurassic), towards Col Nord dam (Middle Jurassic), Digue de la Route dam in distance (Lower Jurassic)

The data relating to the reservoir are as follows.

Area of gathering ground	1,016 m²
Capacity of reservoir	63×10^6 m³
Surface area	365 hectares
Top water level before 1938	645 m O.D.
Top water level after raising	654·25 m O.D.
Average annual rainfall (1924–46)	279 mm
Average annual run off	75×10^6 m³
Estimated maximum flood (1,000 years)	1,200 m³ per second
Highest flood experienced since 1925	170 m³ per second
Overflow provided	
(a) 'Duckbills'	1,000 m³ per second ⎫
(b) Penstock	200 m³ per second ⎬ 1,200 m³ per second
Minimum flow of river	22×10^6 m³

All three dams, of similar design, are in the Jurassic (*Figure 221*), namely: Beni-Bahdel dam, 180 ft. in height, on Upper Jurassic, Lusitanian grits, alternating with marls; Col Nord dam, 50 ft. in height, on Middle Jurassic,

Callovo-Oxfordian marls cleaved with thin grit beds; and Digue de la Route dam, 50 ft. in height, on Lower Jurassic, Bajocian gritty clays and Aalenian limestone (top of Upper Lias).

Beni-Bahdel Dam

At the site of the high Beni-Bahdel dam (*Figure 222*), the Upper Jurassic grits and marls have variable bearing pressures and permeability; they are unconformable on each other, faulted, variable in thickness, texture, strike and dip.

Figure 222. Beni-Bahdel dam. The dip is not only downstream as indicated by the country above the dam, but it is also oblique, giving rise to potential side-thrust, hence the struts between the counterforts of the dam. The top 7·25 m of the dam was raised in 1938. The vertical part of the buttresses were pre-stressed with 'saddles' at the top

Hence many problems of great interest arise, some of which are as follows.

(1) A superficially better site for the dam downstream of the present one was not suitable owing to 'weak' tufa, on the right bank, having been deposited by springs coming from the massive limestone of the Upper Jurassic (*Figure 223*).

(2) The fact that the dip of the Lusitanian beds is 45 degrees downstream, oblique to the thalweg (the residual dip downstream being 25 degrees) sliding of the dam was feared (*Figure 224*).

(3) The Lusitanian grits and marls, with their oblique dip, also varied in bearing pressure, and a gravity concrete dam was rejected.

(4) The marls disintegrated in air.

(5) Uplift was feared owing to so many permeable beds under the dam.

A multiple arch dam was therefore adopted with large counterforts taken down to the several grits, thus getting equality of pressure, circumventing the obliquity of dip (by joining the counterforts with beams), avoiding dependence upon the weak marls, and, being open, effectively dealing with the drainage and uplift problems. Thus, on the right bank (*Figure 224*) the foundations were built largely on solid gritstone with subsidiary layers of marl.

On the left bank, however, there was a fault, and when the ground was opened out the marls were found to be much worse than expected, they

Figure 223. Beni-Bahdel dam. 'Weak' tufa on right bank below the dam. The power-house below the dam is on the right. This site was a much better one geographically but had to be abandoned owing to deposits of masses of soft calcareous tufa (from spring water)

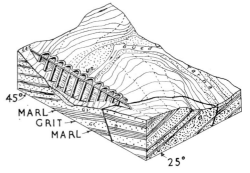

Figure 224. Block diagram showing Lusitanian grits and marls of the Upper Jurassic dipping obliquely to the thalweg at 45 degrees, the general residual dip downstream being 25 degrees

broke up in air. The three marl layers are shown on the right-hand front corner of *Figure 224*.

Foundation tests

It was felt that extensive tests on the site of this left side area were essential, and these were fully described by Drouhin in 1936. (The foundations must remain especially rigid for a multiple arch dam.) Results were as follows.

Loading up to 20 kg per cm^2 depressed the ground by some 2·5 mm, and when the load was removed the permanent deflection was about 1·6 mm.

In about half an hour the following deformations occurred: 6·5 kg per cm², 0·1 mm; 12 kg per cm², 0·15 mm; 20 kg per cm², 0·2 mm. In 7 hours at 20 kg per cm² the deformation was 0·6 mm.

These experiments seemed to indicate that the foundations of the counterforts at the left end of the dam should be broadened (from what was originally contemplated) with a limitation of pressures to 6 kg per cm² on the foundation; actually it was 5 kg per cm² for the dam having a crest of 647 m O.D. (before raising in 1938).

Figure 225. Beni-Bahdel dam. The crest of the dam showing the tops of the curved multiple arches, covered to give a walkway

For the prevention of sliding, it was considered a favourable factor that (*a*) the arches of the dam were inclined upstream, (*b*) there were the grits to which pressure would be transmitted by the marl, (*c*) much of the marl could be excavated, (*d*) the foundation could be given a definite rise towards the downstream, and (*e*) the foundations of the superstructure could be well keyed into the cut-off wall.

These preliminary studies took place prior to construction in 1934 for a dam having a crest of 647 m O.D.

The dam was raised very ingeniously in 1938 to a crest of 654·25 m, that is, by 7·25 m (*Figure 225*). The method was fully described by G. Ribes and P. Boutin in 1958. Briefly, the arches were raised maintaining generally the same dimensions. The counterforts were raised by pre-stressed bars and their feet were strengthened by heavy blocks, downstream, between which were inserted powerful jacks to test sliding movement.

For the first three years, after construction, the toes of the counterforts remained against the jacks to measure any movement of the grits. As no settlement was recorded, the jacks were replaced by concrete.

Dimensions

The data relating to the Beni-Bahdel dam (1958) after the raising operations are as follows (*Figure 222*).

Eleven concrete arches with two gravity concrete ends.

Maximum height above old river bed	55 m
Length of crest (at 647 m O.D.)	320 m
Length of crest (at 654·25 m O.D.)	350 m
Length of crest arches	220 m
Length of crest left gravity abutment	80 m
Length of crest right gravity abutment	50 m
Thickness of arches (top)	0·7 m
Thickness of arches (base)	1·3 m
Thickness of counterforts (top)	3 m
Thickness of counterforts (base)	5 m
Spacing of counterforts	20 m
Batter of counterforts (upstream)	1 to 1
Batter of counterforts (downstream)	0·3 to 1
Depth of trench (concrete)	6 to 20 m

There are 'primary' and 'secondary' grout screens below the concrete cut-off trench.

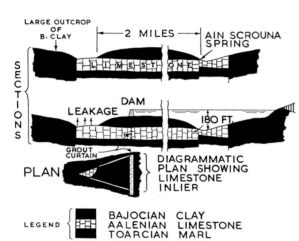

Figure 226. Beni-Bahdel reservoir. A spring (see top diagram) when submerged was converted into a swallow. The top diagram shows how the Ain Scrouna spring originated from a large outcrop of Aalenian Limestone and drainage from Bajocian Clay two miles to the left. The middle diagram shows what would happen when it was submerged by 180 ft. of water (the leakage for which was estimated at 10–20 million gal. per day). The plan shows how the grout curtain, shown on the middle diagram, was effectual in substantially reducing the leakage, owing to two faults bringing up the Toarcian marl

Spring becomes a swallow

When impounding water to a depth of 180 ft. by the three dams, it was not surprising that a spring was converted to a swallow (*Figure 226*).

The spring, Ain Skrouna, was thrown out from the Bajocian junction with the Aalenian Limestone, 2 miles away from the dam. Thus, when the reservoir was filled to a depth of 120 ft. there was a considerable leakage of 80 l. per second (1·5 million gal. per day), with cloudy water appearing under the Digue de la Route dam. This would be equivalent to a leakage of 500–1,000 l. per second (10–20 million gal. per day) when the reservoir was full.

An extensive grouting screen (15 per cent cement, 35 per cent sand and 50 per cent clay) was put down under the Digue de la Route dam to the Toarcian Marl, fortunately the Aalenian Limestone had a limited outcrop being surrounded by faults (*Figure 226*).

Overflow: 'duck bills'

An interesting overflow weir has been built at the Digue de la Route dam with a very large stilling pool.

The weir consists of a series of rectangular sloping lips over which the water flows into tunnels through the dam and thence into the stilling pool below the downstream face (*Figure 227*).

These channels are known as duck bills and being at right-angles to the dam a considerable amount of space is saved, as in the aggregate the 20 bills

Figure 227. Beni-Bahdel reservoir. Duck bills, Digue de la Route dam. There are 20 of these enormous overflow weirs aggregating 30 m (98 ft.) each or equivalent to a weir 4,000 ft. in length. They are placed at right-angles to the dam, thereby saving considerable space. The water is rather low, but when overflowing it passes over the two parallel lips of each duck bill, falls over the black shadows (concrete walls) and down on to the central floor sloping towards us, and into tunnels passing under the dam

are each some 30 m (98 ft.) in length, the total length of weir being about 1,250 m, or over 4,000 ft. They permit a flow of 1,000 m³ per second at a depth of 0·5 m (1·65 ft.).

Bou Hanifia Dam

The Bou Hanifia dam, 62 miles south-east of Oran, had to be founded on very unreliable strata in order to impound the water of a large gathering ground covering 7,850 km² (3,000 sq. miles).

The geology of the site of the dam consisted of alternating grits and marls and conglomerates of the Upper Miocene on impermeable Miocene Marl. The strata are similar to those under the Ghrib dam but are much more disturbed and there are many faults (*Figures 228 and 229*).

As the strata at the dam site are so variable and as the maximum height of the dam had to be 53 m to impound 72×10^6 m³ (22,000 million gal.), great care had to be taken to determine the most suitable type of structure as well as certain details of construction.

Figure 228. Bou Hanifia dam. Left bank showing rockfill dam and cliff of grits and marls of the Upper Miocene

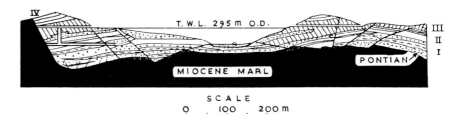

IV   GRITS & MARLY GRITS.
III   GRITTY MARLS WITH CONGLOMERATE, UNEVEN DISCONTINUOUS BASE.
II   GRITTY MARLS WITH MICACEOUS MARLS.
I   SANDY CLAY & SAND & GRITTY MARLS.

Figure 229. Bou Hanifia dam. The strata on which the dam is built, are a mixture of grits, marls, sands and clays resting on Miocene impermeable marl. On the left end (downstream) there is a large fault which very conveniently brings up the marl as shown (the section is developed and follows the curve of the dam and the grout screens). On the right, the marl is nearer the surface (upstream) and reliance is placed for watertightness on the length of the screen

Owing to the availability of grits and marls, a rockfill dam was chosen, but these materials necessitated (*a*) a carefully designed bituminous upstream facing with a protective layer of reinforced concrete, and (*b*) a well-drained broad foundation with filter upon which the dam would stand, thus:

	Rockfill course (average 1 m in thickness) on	Thickness (m)	Grain size (mm)
Drain	Rubble masonry	0·5	over 60
	Broken stone	0·5	25–60
	Gravel	0·5	6–25
Filter	Sand (coarse)	0·25	3·5–6
	Sand (fine)	0·25	0·5–3·5
	Sand (very fine)	0·50	0·1–0·5
	on Maiden ground	2·50	

Figure 230 gives, in outline, a profile of the dam.

The broad foundation was taken down into the faulted heterogeneous beds of the Miocene grit, marl, sand and clay (*Figure 229*); the variable bearing capabilities of these strata confirmed the decision made in 1930 to adopt a rockfill dam.

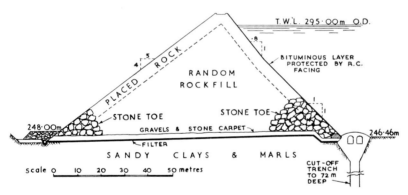

Figure 230. Bou Hanifia dam. Profile of rockfill dam constructed in 1936. The random rockfill is put upon a filter of fine material, 1 m in thickness, surmounted by 12 m of course gravel and stone to act as a drain; there is no clay suitable for puddling in the vicinity

During construction the maximum settlement recorded at the centre of the dam at a height of 174 ft. was 20 cm (8 in.) and after 10 years a further 8 cm (3 in.) was observed.

The cut-off trench (4 m in width) was taken through a considerable thickness, from 19 to 72 m in depth, in this heterogeneous strata and was constructed at the toe of the upstream embankment.

The grout curtains

The grout screens are instructive and their correct alignment is the fruit of careful experimental work, and it will be noticed on the plan (*Figure 231*) on the left side of the valley, that there are two screens the directions of which point, generally, downstream. This enabled the screens to end in the Miocene impervious marl far sooner than if they had been taken in any other direction; for the marl is raised by a fault of 300–400 ft. as will be apparent from *Figure 229* (the developed section).

The second screen on the same side, called an 'ecran de securité', was constructed up to a distance of 100 m from the primary screen because it was thought safer not to rely on a single screen in such strata.

On the right bank, owing to the rise of the surface the drilling of bore holes and grouting had for economy to be carried out from an adit. The primary screen (150 m in length) pointed upstream (see *Figures 229 and 231*); and thence for another 150 m pointed downstream and, in addition, a security screen pointing downstream was also constructed.

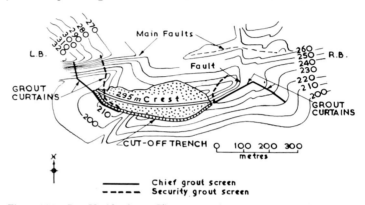

Figure 231. Bou Hanifia dam. The contours show the top of the Miocene marl below the surface relative to O.D. in m. Top water level is 295 m O.D. On the left bank the impermeable Miocene marl is brought up by a fault quite near the surface to 300 m O.D. (that is, above top water level 295 m), so the grout screen was taken downstream. The second screen (dotted) was added for security. On the right side there was no such fault and after following a course of 150 m pointing downstream, the screen was bent pointing in a direction upstream for another 150 m where the marls were nearer the surface. There was also a short secondary screen (dotted) pointing downstream. Some indication of the surface levels is given by the section in Figure 229

As neither of these curtains could be designed to meet any fault bringing up the Middle Miocene clay to above top water level, such as on the left bank, the best that could be done was to zig-zag the screen along the line where the surface was lowest and the marl highest. This procedure was what Dr. Herbert Lapworth called 'wire-drawing', that is, that any material leakage would be dissipated in an enforced circuitous route necessitated by the grout screen.

The grouting data were as follows.

Length of bore holes	62,371 m
Cement	14,412 tons
Chemical products (sodium silicate)	5,033 tons
Special clay gel	13,633 tons
	(29,000 m³)
Area of screens	97,000 m²

Construction of grout screen

The boreholes were drilled to a diameter of 45–85 mm, and were lined through the sandy material which occurs especially on the left bank. The injection was done in stages of 3–4 m necessitating numerous reperforations. It was found that a type of tube (patented) with clack-valves was satisfactory in expediting the work.

For grouting any coarse conglomerate or firm sandstones classical methods were used, but for fine sands or grit silica-gel consisting of sodium silicate plus a coagulant made by a patented process (Rodio) resulted in a product only a little more viscous than water.

During World War II, however, this patented process had to be replaced by a solution of clay made as follows. After the dry clay was ground it was put into suspension in water by mixers with horizontal axes, and the whole mass passed through a screen which retained the largest particles and the liquid then passed on to closed circuit digestors. The complete dispersion of the clay particles was effected by the addition of a special enzyme product plus a very vigorous mixing produced by high speed pumps in the digestors.

The colloidal suspension obtained passed into decanting vats which eliminated any further solid particles, and thence into a final mixer through which sodium silicate and a flocculating agent flowed.

The resulting product was a grey fluid which made a good gel. It was quite stable for mechanical handling owing to the gel, and for resisting chemical actions owing to the clay.

The overflow channel

As the Fergoug or Perregaux dam, only about 30 km upstream of the Bou Hanifia dam, was damaged several times by floods the provision for safely by-passing large quantities of water was a major operation.

The area of the gathering ground to Bou Hanifia is 7,850 km² (3,050 sq. miles), for which the provision for flood water is 6,000 m³ per second, 5,500 m³ per second over the main spillway and 550 m³ per second at the dam.

About a mile north of the dam there is a tributary valley which was utilized for the main overflow to cope with a flood of 5,500 m³ per second. The tributary was converted into a channel cutting through the Quaternary and Middle Miocene marl as far as the main valley, which was a mile below the dam (*Figure 232*).

To carry the 5,500 m³ per second (192,500 cusecs) a semi-circular concrete channel of 22 m radius with a sill of 80 m and a length of 1,425 m (over a mile) was constructed.

At the entrance to this channel there are 16 movable hydraulically operated gates (each 5 m in length, in all 80 m) which, when shut, sustain a water level of 295 m O.D. (top water level in the reservoir) and flood water over them may be any quantity up to 2,000 m³ per second.

For larger floods the gates are lowered permitting the water to flow at 289 m O.D., and with a flood level of 300 m O.D., the full quantity of 5,500 m³ per second can flow over the weir. (*Figure 233* shows part of these gates).

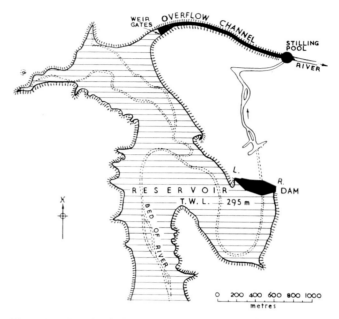

Figure 232. Bou Hanifia dam. Showing relative position of the dam and overflow channel. The overflow channel and stilling pool presented great difficulties in construction owing to their magnitude and the soft Quaternary rocks which had to be stabilized. Chemical action of the water also caused difficulties

Figure 233. Bou Hanifia dam. Flood gates. The overflow, 80 m in length, will permit a flood of 2,000 cusecs, and with the 16 gates lowered (by hydraulic power) the channel will carry 5,500 cusecs (see also Figure 234)

The outlet end of the canal where it joins the river bed is at a level of about 250 m O.D., or 45 m below top water level. There were trials and tribulations during the construction of this overflow. First, owing to the sulphates in the water in the Quaternary sands being liable to attack ordinary cement, ciment fondu was used. Secondly, owing to Quaternary water in permeable material overlying impermeable Miocene clay, very heavy concrete and drainage works were necessitated. Thirdly, in order that the 5,500 m³ flow could be safely discharged into the river without causing serious damage, a stilling pool was formed in the soft Quaternary materials designed in accordance with a model to a scale of 1 in 50. This pool is no less than 100 m in diameter (*Figure 234*) and receives the maximum flow

Figure 234. Bou Hanifia dam. Stilling pool. For the safe discharge of 5,500 m³ per second, at a velocity of 25 m³ per second, the stilling pool (100 m in diameter) has been dug on soft Quaternary materials. The discharge channel has a radius of 72½ ft.

coming into it at a speed of 25 m per second with a quantity of 5,500 m³ per second; up to 1955 the maximum flow recorded is 1,200 m³ per second.

Sarno Dam

The Sarno dam, completed in 1952, is about 50 miles south of Oran; the water in the reservoir (holding 4,900 million gal.) is used for irrigation. The dam is rather high (92 ft.) for an earthen embankment on soft material.

On the left bank (*Figure 235*) there is a thickness of 100 ft. of soft Pliocene conglomerate (Poudingues) on about 160 ft. of Pliocene sandy marl which is not impermeable; and on the right bank there is a thickness of 130 ft. of Pliocene conglomerate resting on inclined Oligocene grits and clays. The inclined Oligocene strata below the surface may be attributed to part of a buried valley side of pre-Pliocene age.

Any form of dam, other than an earthen one, would hardly have been possible on such a foundation. For a rockfill alternative, stone would have had to be fetched for a distance of 10 miles, and as earth material was at hand the earthen embankment was adopted.

These nearby soft materials contained little clay, as may be seen from the typical mechanical analysis.

	Per cent	mm
Pebbles	100–50	150–20
Gravel	50–25	20–2
Sand	25–15	2–0·05
Silt	15–12	0·05–5 μ
Clay	12–5	5–1 μ

The compaction was effected by eliminating all stones over 150 mm. Consolidation was found to be best with a 6-ton sheepsfoot roller giving a

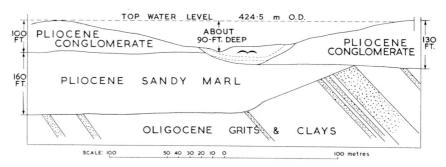

Figure 235. Sarno dam. Section along the toe wall. This extraordinary formation of the Oligocene underlying the Pliocene is partly permeable, particularly the grits (shown dotted). The overlying conglomerate (Poudingues) and sandy marl are permeable

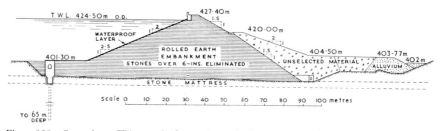

Figure 236. Sarno dam. This was the first earthen embankment constructed in Algeria. A puddled clay core could not be constructed with local materials, and reliance on water retention was placed on a bituminous facing, with an elaborate system of drains. The embankment material generally had 48 per cent materials above 20 mm, but nevertheless proved satisfactory with 8 passes of a sheepsfoot roller

pressure on the soil of 24 kg per cm² (a 25-ton vibrating pneumatic tyred roller giving a pressure of 5·4 kg per cm² was not satisfactory). The sheepsfoot roller passed 8 times with wetting and harrowing on 25–30 cm layers. A section of the embankment is shown in *Figure 236*.

The cut-off trench

In order to prevent water with a pressure of over 90 ft. forcing its way through the permeable Poudingues, and underlying Pliocene sandy marl, under and around the ends of the dam, an extensive cut-off trench and grout curtains down to the permeable Oligocene marls (250 ft. below top water level) were constructed. Any scouring in such a soft formation with such pressure would be highly dangerous for the earthen embankment above (*Figure 237*).

The inclined Oligocene alternating grits and marls under the Poudingues on the right bank certainly did not help the grout screen, for the included grits would be submerged upstream in the reservoir; they also appear downstream and they are permeable.

Very briefly, at the toe of the embankment there is a concrete wall, up to 20 m in depth right across the valley in the Poudingues. Below this wall were drilled many bore holes for grouting the sandy marls down to the Oligocene beds.

On the right bank, bore holes were drilled for a distance of no less than 1,700 ft. from the stream for grouting down to the top of the sloping Oligocene marls and grits (*Figure 237*).

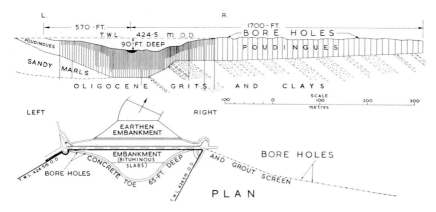

Figure 237. Sarno dam. The grout screen, owing to the conglomerate being thinner, was taken for a distance of 570 ft. on the left bank, whereas for the thicker conglomerate and sandy Oligocene measures underlying on the right bank, it was taken for a distance of no less than 1,700 ft.

On the left bank, where the Poudingues are thinner and rest comfortably on the Pliocene sandy marl, grouting was carried out for a distance of 570 ft. from the thalweg.

In all, no less than nearly 6 miles of bore hole were drilled; 1,600 tons of cement and 1,300 tons of clay being used in the Poudingues, or 320 kg per m of cement and clay.

The area of the grout curtains covered by the 30,000 ft. of boring is about 34,600 m^2, giving a value of 85 kg of cement and chemical products per m^2 of curtain.

The leakage in a well designed system of wells and adits under the dam amounts to about 200,000 gal. per day, chiefly from the left bank, including leakage through the 20 cm thick bituminous concrete apron on the sloping face of the dam.

Figure 238. Sarno dam. The downstream embankment with a building on the Poudingues outcrop at the end of the embankment

Figure 239. Sarno dam. It is seen, even from the photograph, that the Poudingues are soft and permeable

Figure 240. Sarno dam. The 'Marguerite' overflow, the 8 petals 17 m in length, take a flood of 500 m³ per second. Another overflow takes the remaining 500 m³ per second required. The outline of the soft Pliocene rocks in the distance is noticeable

It would seem that in working a job of this nature to a successful conclusion, French engineers must understand as much about geology as geologists about engineering.

Figure 238 shows a general view of the right bank side of the dam with a building on a Poudingues cliff, while *Figure 239* shows the nature of this conglomerate.

Reservoir data include the following.

Area of gathering ground	276 km²
Capacity of reservoir	22 × 10⁶ m³
Surface area of water	260 ha.
Capacity of 'Marguerite' overflow (at 1·5 m depth)	500 m³ per second
Capacity of other weirs	500 m³ per second
Total capacity of overflow	1,000 m³ per second
Largest storm known	350 m³ per second

Figure 240 shows the interesting form of overflow known as a 'Marguerite'; the 8 petals being 17 m in length.

BIBLIOGRAPHY AND REFERENCES

Les Cheurfas dam

l'Entreprise Soletanche. Le Barrage des Cheurfas. *XIXe Congrès Géologique International* (1952).
GIGNOUX, M., and BARBIER, R. *Géologie des Barrages* (1955) Masson et Cie.

Ghrib dam

CHEYLAN, G. Le barrage du Ghrib. *XIXe Congrès Géologique International* (1952).
GIGNOUX, M., and BARBIER, R. *Géologie des Barrages* (1955) Masson et Cie.
MARTIN, M. La lutte contre les erosions souterraines au barrage du Ghrib. *2nd Congress on Large Dams*. 4 (1936) 50.
THÉVENIN, J. The Ghrib dam. *Travaux* 141 (1958). (Special Number, 286, Aug.)

Fodda dam and reservoir

GIGNOUX, M., and BARBIER, R. *Géologie des Barrages* (1955) Masson et Cie.
GOURINARD, Y., and TRÉVENIN, J. Le barrage de l'Oued Fodda. *XIXe Congrès Géologique International* (1952).
THÉVENIN, J. les effets du seisme de Septembre 1954 sur deux barrages de la regione d'Orléansville (Algerie). *I.C.O.L.D.* **II,** 214 (1964).

Fergoug or Perrégaux dam

CORNET, A. Barrages de l' Oued Hammam, Perrégaux. *XIXe Congrès Géologique International* (1952).
GIGNOUX, M., and BARBIER, R. *Géologie des Barrages* (1955) Masson et Cie.

Beni-Bahdel reservoir

DROUHIN, M. Essais Géotechniques des terrains de fondation. 2e *Congrès des Grands Barrages* (1936).
GAUTIER, M. XIXe *Congrès Géologique International* (1952).
GIGNOUX, M., and BARBIER, R. *Géologie des Barrages* (1955) Masson et Cie.
RIBES, G., and BOUTIN, P. Surélévation du Barrage des Beni-Bahdel. 6e *Congrès des Grands Barrages*. C.16. IV (1958) 245.

Bou Hanifia dam

CHAGNAUD, L., and S.E.C. (RODIO). Le Barrage de Bou Hanifia. *Service de la Colonisation et de l'Hydraulique* (1951).

CORNET, A. Les barrages de l'Oued El. Hammam. XIXe *Congrès Géologique International* (1952).

DROUHIN, M. La lutte contre les erosions souterraines au Barrage de Bou Hanifia en relation avec les essais de permeabilité et l'étude des ecoulements. 2e *Congrès des Grands Barrages.* IV (1936) 29.

GIGNOUX, M., and BARBIER, R. *Géologie des Barrages* (1955) Masson et Cie.

Sarno dam

GEVIN, P. Le barrage de l'Oued Sarno. XIXe *Congrès Géologique International* (1952).

GIGNOUX, M., and BARBIER, R. *Géologie des Barrages* (1955) Masson et Cie.

SALVA, J. Méthodes de compactage du Sarno, et teneur en eau. *I.C.O.L.D.* **III,** 337 (1958).

— Methodes pour la mesure des tassements, *I.C.O.L.D.* **III,** 367 (1955).

d'Erranguéne dam

DURAND, MAX. Le Barrage d'Erranguéne, *I.C.O.L.D.* **III,** 613 (1961).

30

IRAK

Dokan Dam*

The Dokan dam is an arch dam in the north-east of Irak.

Crest level	516 m O.D.
Crest length	350 m
Maximum height	111 m
Crest-height ratio (approximately)	3
Radius of upstream face	120 m

The dam base is founded on Dolomite approximately horizontally bedded, shale and limestone higher up the valley sides.

The left bank is a kind of peninsula against which the left abutment thrusted.

The shale was rather soft and the ground disturbed, there was a collapsed cavern, a fault or system of joints, and it was decided to make the area into a solid mass (comparable to the right bank which was treated in a similar manner for the Bort dam).

Some 575,000 m³ of rock were injected with 1,707 tons of cement and 471 tons of sand (4 kg per m³) effected by 28,000 m of bore holes drilled from various tunnels, from the surface, and from the large cavern. The injection pressure was 10 kg per cm² applied by stages so as not to loose cement when approaching the cliff face. This operation was probably the most essential and the most intangible foundation problem to decide upon in the whole scheme.

Another major problem was that of securing watertightness in the reservoir.

As shown in *Figure 241*, the general sequence of strata is as follows.

Alluvium on

			Metres
	Marl		1,000
	Limestone		123
Cretaceous	Shale ⎫ contact zone		1
	Limestone ⎭		3
	Dolomite		130+

*I am very grateful for the suggestions of Mr. G. M. Binnie and Mr. P. F. F. Lancaster Jones concerning these notes on the Dokan dam and again to Mr. Binnie for his notes on the additional successful grouting since 1960.

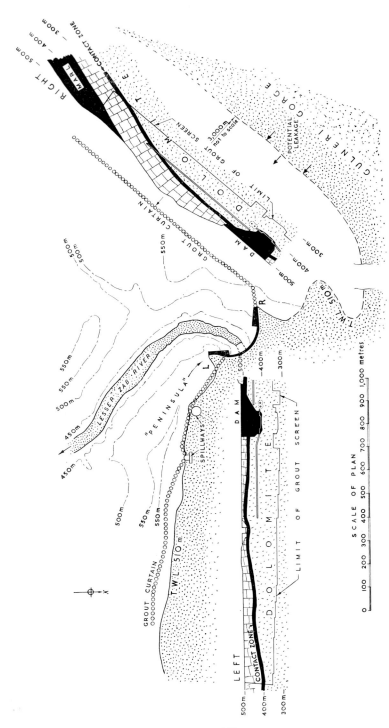

Figure 241. Dokan dam. The river turns almost at a right-angle downstream of the left abutment and thereby forms a natural 'peninsula' against which the left abutment thrusted. As the limestones were much disturbed, a considerable amount of strengthening had to be done here. As the reservoir submerges permeable limestone, long grout curtains were necessary; that on the right had to be carried downstream to avoid leakage from the Gulneri Gorge. The grouting programme cost £2,400,000

As the valley on the left bank curved downstream to the left, a long grout screen (1,348 m) was necessary to retain the water in the reservoir; no sign of leakage has so far developed round this curtain.

The curtain on the right bank was designed to stop leakage through the Dolomite, not only around the end of the dam, but also from a gorge known as the Gulneri Gorge which runs parallel with the river at a distance of 2 km. For this it was necessary for the grout screen (1,033 m) to run parallel with the river after making a right-angle at an appropriate distance from the end of the dam to save loss of cement into the valley when grouting downstream.

This is probably the largest grouting operation ever carried out for a reservoir. The data recorded in 1958–59 are as follows.

Area of curtain	4,528,000 ft.^{2}
Length of bore holes	601,000 ft.
Total cement injected	45,600 tons
Total sand injected	32,200 tons
Total water injected (approximate)	100,000 tons
Cost	£2,400,000

It was not until 1960 that a serious leakage of 6 m^3/sec. occurred beyond the left bank grout screen and emerged in the lesser Zab river about a kilometre away in a series of springs. The screen had to be extended from 1,348 m by 336 m over a depth of 150 to 160 m covering a dolomite-limestone unconformity. Primary holes were spaced at 8 m with intermediate secondaries and, where necessary, tertiary holes at 2 m spacing. The quantities involved in sealing the leakage of 6 m^3/sec. for a grout screen of 50,800 m^2 were:

Length of boreholes	22,882 m
Cement	10,718 tons
Sand	8,061 tons
Other products	149 tons

Cement, sand and products per m of borehole 827 kg cement, sand and products per m^2 of curtain 373 kg.

Mr. J. F. F. Clark reported in discussion that this grouting was carried out successfully except for a number of holes where grout escaped and appeared in the river downstream.

It was impracticable to form a gravel filter through 2 in. holes, 100 m in depth, but Mr. A. D. Humphreys of the Iraq Petroleum Company who 'postulated two faults or major fissures at which he expected the leakage would be occurring' and his company suggested a soft plug should be formed using a bentonite-oil mixture with a filler added. This special grout consisted of 50 kilo of bentonite with 30 l. of diesel oil and $\frac{1}{2}$ kilo of cotton flock.

These injections were carried out when the water level in the reservoir was half full and finally when lowered. Proof of the effectiveness of the grouting was the appearance of a spring in the area of the leakage ingress caused by rainwater seeping through the hillside being diverted to the upstream side of the curtain.

At one time injection with hot bitumen was considered (after consultation with a chief engineer of an oil company) but this was rejected on the grounds of cost.

Grouting techniques are in a world of their own and it is thought that much information and expense could be used to the mutual advantage of industries of water and oil exploration.

BIBLIOGRAPHY

BINNIE, G. M., CAMPBELL, J .G., EDGINTON, R. H., FOGDEN, C. A., and GIMSON, N.H. The Dokan project: the dam. *J.I.C.E.*, 14 (1958–59) 157.

JONES, P. F. F. LANCASTER and GILLOTT, C. A. The Dokan project: the grouted cut-off curtain. *J.I.C.E.*, 14 (1958–59) 193.

PERROTT, W. E. and LANCASTER JONES, P. E. E. and CLARK, J. F. F. (in discussion). *Case Records of Cement Grouting. Grouts and Drilling Muds in Engineering Practice*. Butterworths. Symposium arranged by the British Nat. Soc. on Soil Mechanics and Foundation Engineering at the I.C.E. Aug. 1963, pp. 80 and 112.

31

RHODESIA

Kariba Dam*

The Kariba dam, constructed between 1955–60 on the River Zambesi, is some 300 miles downstream of the Victoria Falls. It commands a gathering ground of 250,000 sq. miles which produced a flood of 570,000 cusecs in a day in March 1958. Average dry weather flows are of the order of 10,000 cusecs (1948–58).

At top water level 2,000 sq. miles of country are submerged, a distance of 175 miles up the valley from the dam.

The main sequence of strata is as follows.

Triassic	{ Upper Karroo sandstone { Escarpment grit and conglomerate
Permian	Madumabsa argillaceous mudstone
Pre-Cambrian (late) ? . . .	Quartzite
Archaen	Gneiss, amphibolites, schists

Between the Victoria Falls and Kariba, the Triassic and Permian strata lie in a broad syncline and the Madumabisa mudstones (with the impermeable gneisses which form the outer rim of the broad valley) ensure watertightness.

The dam is founded on a horst of Pre-Cambrian quartzite and Archean gneiss (*Figure 242a*) 2 miles from the head of the 16-mile Kariba gorge, the report states that the floor and sides of the valley at the dam site are gneiss but on the right bank, 300 ft. above the river bed, the gneiss is overlain unconformably by open jointed quartzite. The report goes on to say that mineralogically the gneiss consists of feldspar, quartz, and biotite mica, though it is by no means uniform in grain and composition and shows strong foliation. In the river bed the rock is very sound but at higher levels on the banks of the gorge there is considerable superficial weathering.

The type of dam ultimately adopted is a double curvature arch dam having a maximum height of 420 ft. and giving a crest chord-height ratio of about $3\frac{1}{2}$; the length of the curved crest is 2,025 ft. and has a circle of reference across the valley of a radius of 800 ft. The maximum thickness of the dam is 43 ft. at the top and 83 ft. at the base; these dimensions were arrived at by tests on 16 different plaster models—checked by trial load analysis.

*I am grateful to Mr. T. A. L. Paton who kindly read these notes on the Kariba dam.

A grout curtain was established under the dam for a depth of 80–120 ft. Cement absorption was comparatively small (8,000 tons in 30,000 ft., or 11 lb./ft^2 of curtain). Pressure relief holes 4 in. in diameter and 150 ft. in depth were drilled downstream of the dam.

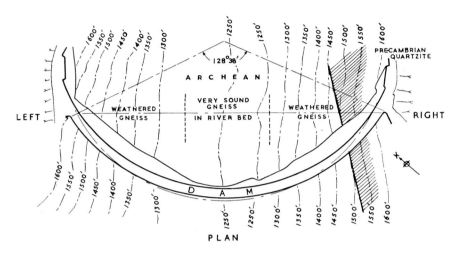

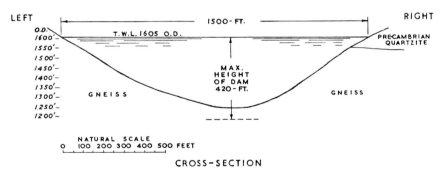

Figure 242a. Kariba dam. The dam is founded on a horst of Pre-Cambrian quartzite and Archean gneiss which is sound in the river bed and weathered on the valley sides. The reservoir area lies in a syncline and relies mainly on Permian mudstones for watertightness

Right (or south) Bank Weakness

Figure 242a shows a layer of pre-Cambrian quartzite overlying weathered gneiss on the right abutment, whereas on the left abutment, sound gneiss was overlain by 10 m of weathered gneiss the removal of which was a straightforward proposition.

On the right bank, however, the quartzites overlying the gneiss dipped steeply towards the river and, in 1956, were found to be very fractured, sheared and slickensided with bands of micaceous quartzite and gneiss, as indicated in the shaded area of *Figure 242a*.

This area on the right bank was, however, further investigated in 1960 (*Figure 242b*), and I am especially indebted to Dr. J. L. Knill for a brief description from his paper (with K. S. Jones) which shows how wise it is to make further investigations on things which appear to be doubtful; in this case the structure of the quartzite, the presence of a mica seam within it and the extent of weathering between the quartzite and underlying gneiss.

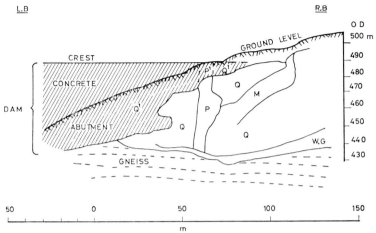

Figure 242b. Kariba dam. Strata on right bank were found to be variable, very fractured and were specially investigated in 1960; the weak quartzite, pegmatite and mica seam were removed or specially grouted. The thrust of the concrete abutment was thus transmitted satisfactorily to the sound gneiss at a lower level

   Q.   Quartzite (*Q'* unsound).
   P.   Pegmatite (*P'* unsound).
   M.   Mica seam (*grouted or excavated and replaced by concrete*).
   WG. Weathered gneiss.

A typical example of the exploratory work is described as follows:

'Owing to the fractured condition of the quartzite, surface material had been washed into the rock fissures. In addition, the quartzite is sufficiently permeable to allow ground-water to percolate along the margin with the underlying gneises. This situation has led to the development of a zone of weathered gneiss immediately below the quartzite. The gneiss has become partly converted to clayey material and this was not removed during the jetting and consequently not replaced by grout. The thickness of weathered gneiss varies from 1–2·5 m below the dam to as much as 30 m downstream of the dam; the surface of the sound, weathered gneiss has, therefore, a different form to the quartzite-gneiss boundary.'

Figure 242b shows how the concrete abutment of the dam transmits the arch-thrust to sound gneiss lying under the quartzite.

REFERENCES

ANDERSON, D., PATON, T. A. L., and BLACKBURN, C. L. Zambesi hydro-electric development at Kariba. *J. Instn civ. Engrs*, 17 39 (1960).
KNILL, J. L. and JONES, K. S. The record and interpretation of geological conditions in the foundations of the Kariba dam. *Geotechnique*, **15** 94 (1965).

32

PAKISTAN

The Mangla Project

Only the barest outline of the geological problems can be given for this immense scheme, costing in the order of £165,000,000. (At the rate of exchange current 1962–67).

Geological

The area of the reservoir is in the Tertiary formation, the Siwalic series originally some 10,000 ft. thick which runs parallel to the Himalayas. The Siwalic clays, silt stones, sands and subsidiary gravels were uplifted in early Pleistocene times and we find them as anticlines such as the Changar anticline which runs west to east, 1 to 3 km south and downstream of the Mangla dam. As some 4,000 to 6,000 ft. of formation has been eroded away we find them overconsolidated.

The bedrock, i.e. the Siwalic clays and siltstones dip 10–15 degrees to the north-east at Mangla, 45 degrees at Jari and south of the Changar anticline, vertically.

Reservoir

The reservoir area with a conservation water level of 1,202 ft. O.D. (to be raised later to 1,252 ft. S.P.D.) is enclosed by the Mangla dam on the west, an intake embankment from Mangla dam to Sukian dam and thence following the natural level of the land for several miles to the Jari dam on the east. The rest of the reservoir area is not shown on the diagram (*Figure 243*), neither are the dams which are shown correct to scale for, as they are earth dams, they conform with the natural level of the land giving them irregular shape both on plan and in section.

Dams

The total volume of fill in the three dams is 109×10^6 m^3, excavation 92×10^6 m^3 and their main dimensions and composition are:

	Mangla	Sukian	Jari
Maximum height, m	138	44	83
Crest length, m	3,130	5,130	3,740
Crest level, ft S.P.D.	1,234	1,234	1,234
Crest level, ft. S.P.D., future	1,274	1,274	1,274

S.P.D. denotes survey of Pakistan datum.

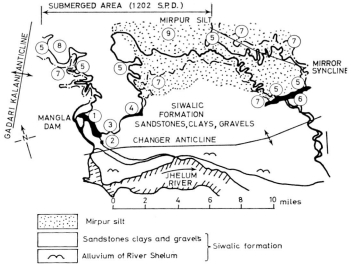

Figure 243a. The Mangla project. This project for power and irrigation is 120 miles north of Lahore, Pakistan, and the Mangla reservoir is enclosed by three dams—Mangla, Sukian and Jari. There is an intake embankment between Mangla and Sukian dams and the perimeter of the reservoir then follows the 1202 O.D. contour eastwards to the Jari reservoir. The dams are all built on the Siwalic formation which consists mainly of sandstones with clays which have a low shearing strength for which special precautions had to be taken

1. Mangla dam (*conservation level 1202 S.P.D.*)
2. Power house
3. Crest road
4. Sukian dam
5. Top water level } (*Conservation level 1202 ft. S.P.D.*
6. Jari dam } *may be raised 1252 ft. S.P.D.*)
7. 1250 ft. contour
8. Jhelum river (*submerged*)
9. Khad river (*submerged*)

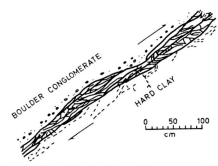

Figure 243b. Mangla dam. Shear zones. During construction it was found that shear zones extended for distances of the order of 500 m. The section shows a length of 4 m and a 'parting seam' or shear zone between clay and conglomerate in the Siwalic formation. Within this zone of 20 to 30 cm major and minor shear lenses were formed. Zones varied in thickness from a parting seam of a few millimetres to over a metre

Mangla. Constructed with clay and sandstone bedrock from excavations such as spillways and gravel imported from river and terraces.

Sukian. Foundations with bentonite and shear zones with dip on cross-sections normal to the dam varying from 10 degrees upstream to 5 degrees downstream. Excavation carried to 50 ft. below surface, a sandstone core and flatter slopes than those first thought to be necessary.

Jari. Silt core and gravel shoulders. Clay-bed on the bedrock joined the clay of the core on the axis of the dam and thus formed an impermeable cut off.

The River Jhelum. The area of the gathering ground down to Mangla is 13,000 miles^2 (33,700 km^2); minimum flow, 6,000 cusec (170m^3/sec); maximum flood (29/8/29), $1\cdot 1 + 10^6$ gall (31,000 m^3/sec).

Shear Fissures in Bedrock

The feasibility of the scheme and the exploratory work having been undertaken for the dams and embankments of total length in terms of several miles, revealed discontinuities in the bed-rock of clays and sandstones.

Mr. Geoffrey Binnie in a brief description of the project has separated these discontinuities into three types of fissures namely:

(1) *Random fissures.* These may be up to a metre in length, polished and/or slickensided, in surrounding clay of little strength.

(2) *Thrust shear joints.* These also have little strength although they are at right-angles to the bedding planes.

(3) *Shear zones.* These are in clay strata where the strains produced by folding have caused relative movements with, and along, the bedding plane which can be likened to a fault, sometimes several metres in length.

Zones may be a few millimetres to 1 m in thickness and where thin, they may be found in one plane or as a number of lenses or in shear planes parallel with the shear zone to the bed-rock for several metres (*Figure 243b*). The diagram shows a typical shear zone in the Siwalic formation at the Jari dam site with shear lenses and shear planes drawn by Prof. A. W. Skempton and P. G. Fookes.

In such a long length of enclosure dams and embankments on uneven slopes of bedrock, some steep, and as the natural water level underground in this arid country is at the level of the river, when the reservoir is filled to a level of 1,202 ft. S.P.D. (later 1,252 ft. S.P.D.) varying dangerous pressure effects or pore pressures will be introduced corresponding or due to the water levels in the reservoir, and hence the strength of the bedrock will be diminished to half its strength as suggested by the results of laboratory experiments.

Pore pressure reduction

To reduce the pore pressures and to arrest the tendency of any excessive shear effects which may be produced, the following precautions were adopted in those parts of the project where they are applicable.

(1) Additional lining in those tunnels in the bed-rock subject to reservoir head 9·5 m in diameter (which were necessary for diverting the river) with tubular steel to make them water-tight, to a reduced diameter of 7·9 m, sufficient for permanent purposes.

(2) Placing compacted sandstone ever the fissured sandstone bedrock below to reduce its permeability, a gain estimated to be 90 per cent in impermeability.

(3) Constructing additional adits and wells particularly on the intake embankments.

(4) Removing certain clay beds.

(5) Building upstream and downstream toes to the dams to resist potential deep slips.

Alterations due to geology during construction

Irrespective of size, several instances of change during construction necessitated by geological conditions are noticed on pages 19, 62, 169, 197 and 390.

BIBLIOGRAPHY

BINNIE, G. M. Mangla Dam Project. *I.C.E.* **36** 213 (1967).

BINNIE, G. M., CLARK, J. F. F. and SKEMPTON, A. W. The effect of discontinuities in clay bedrock on the design of dams in the Mangla Project. *I.C.O.L.D.* **I,** 165 (1967).

ELDRIDGE, J. G. and LITTLE, A. L. The seismic design of earth dams of the Mangla Project, *I.C.O.L.D.* **IV,** 39 (1967).

THOMAS, A. R. and GWYTHER, J. R. Diversion of the River Jhelum during construction of Mangla Dam. *I.C.O.L.D.* **II,** 121 (1967).

33

JAPANESE DAMS

There is great activity in dam construction in Japan as indicated in Table 1.

TABLE 1

Type		Height—top to foundation (m)		Number constructed			
		Maximum	Minimum	Before 1939	After 1939	Under construction (1958)	Total
Concrete gravity (over 60 m in height)		157	60	3	42	15	60
Hollow gravity	⎫	119	47	0	1	5	6
Buttress	over	33	20	8	0	0	8
Arch	15 m	194	18	0	7	9	16
Rockfill	in height	131	18	0	3	4	7
Earth	⎭	41	30	4	13	7	24

TABLE 2
GRAVITY CONCRETE DAMS

Name	Maximum height (m)	Crest (m)	Crest-height ratio	Geology of site	Observations
Sakuma	155·5	294	2	Granite	25 m of gravel entailed deep excavation.
Ogochi	149	345	2·3	Greywacke	
Arimine	140	505	3·6	Conglomerate Greywacke	Some slate intrusion and a liparite dyke. Jointed rock necessitated extensive consolidation grouting, tested by geophysical exploration.
Ikari	112	267	3·3	Granite, tuff breccia Layers of greywacke and shale	Right bank cracked and brecciated necessitating excavation up to 50 m. Curtain grout to 50 m. Bores, 11,700 m; 1,950 tons of cement.
Maruyama	98·2	260	2·7	Greywacke Clay-slate	

Mr. Tai Kobayashi, Development Section, Ministry of Construction, Tokyo, has very kindly read these notes on Japanese dams.

TABLE 2: GRAVITY CONCRETE DAMS—continued

Name	Maximum height (m)	Crest (m)	Crest—height ratio	Geology of site	Observations
Akiba	89	270	3	At river bed up to 50 m of gravel on granite	(1) On left bank 20 m of talus began to slide necessitating 456 m long drain, 23,200 m bores, 14,700 tons of cement grout. Removal of $1 \cdot 2 \times 10^6 m^3$ talus. (2) Consolidation grouting. Bores, 57,000 m, 4,200 tons of cement. (3) Curtain grouting. Bores, 19,600 m, 1,100 tons of cement. The difficult geological conditions necessitated in all nearly 100 km of bores and 20,000 tons of cement grout.
Tase	81·5	320	4	Serpentine	
Nukabira	78	290	3·7	Andesite tuff	
Odomari	74	163	2·2	Quartz porphyry	Built in 1935, raised 11 m in 1959, necessitated: (1) curtain grouting from 2,000 m of bore holes with cement, 15–20 kg per cm²; (2) defective concrete in dam repaired by 1,000 m of bore holes with cement injection at 10 kg per cm² (3) joining old existing concrete of upstream face with new concrete a rich reinforced concrete layer 70 cm in thickness.
			HOLLOW GRAVITY DAMS		
Ikawa	104	240	2·3	Slate greywacke chert	These alternating strata are hard but fractured, variable in dip and occasionally faulted. The dam was chosen in case of earthquakes and the rock was covered entirely with concrete to increase stability against shear, with vertical drains in the hollow section: 19,500 m grout bores were necessary for curtain and 33,000 m were used for consolidation. Including all grouting 5,200 tons of cement were used.
Omorigawa	72	197	2·7	Quartz-graphite and schists	Small folds but no faults

TABLE 3
ARCH DAMS

Name	Maximum height (m)	Crest (m)	Crest-height ratio	Geology of site	Observations
Kurobegawa (No. 4)	194	437	2·3	Granite	Cupola or dome type, under construction 1959.
Kamishiiba	110	340	3	Greywacke clay-slate with intrusions	Arch (constant angle). Graphite-filled fault 1·8 m in width covered by thrust block to prevent direct arch thrust on left bank.
Futase	95	290	3	Breccia Phyllite	Thick arch chosen for economy and safety, for although right bank is solid, left bank is weathered and cracked.
Narugo	94·5	215	2·3	Granodiorite Granite with joints and partly weathered	Nearly vertical face. Thrust blocks were necessary at both abutments to transmit load to solid rock.
Sazanamigawa	67·4	125	1·9	Quartz porphyry	No foundation weaknesses enabling cupola type to be adopted.
Tonoyama	64·5	118	1·9	Hard conglomerate greywacke	No foundation weaknesses Thick arch type three-centred.

TABLE 4
ROCKFILL DAMS

Name	Maximum height (m)	Crest (m)	Crest-height ratio	Geology of site	Observations
Miboro	131	426	3·3	Altered Quartz porphyry Granite porphyry	A crushed fault zone 40 m in width on the right bank and the difficulty of transporting cement determined the type of dam.
Makio	106	264	2·5	Greywacke	Complex foundation on left bank and faults and fissures suggested a rockfill dam as being the most suitable; CO_2 and H_2S gases issued from the foundation. An extensive investigation of the permeability of the strata was necessary.

TABLE 4: ROCKFILL DAMS—continued

Name	Maximum height (m)	Crest (m)	Crest-height ratio	Geology of site	Observations
Ishibuchi	53	345	6·5	Liparite	The reasons for a rockfill dam were unfavourable geological conditions and difficulty of procuring cement. Suitable dacite rock was obtainable a mile upstream.
Nozori	44	160	3·7	Anderite propylite (an altered volcanic rock)	Suitable rock was on site and the severe climatic conditions (top water level 1,514 m O.D.) necessitated a rockfill dam.

TABLE 5

EARTH DAMS

Name	Maximum height (m)	Crest (m)	Crest-height ratio	Geology of site	Observations
Ainono	41	120	3	Shale	As shale and clay were the only materials available for the embankment, vertical sand-drains were made and a good deal of flow-net investigation carried out.
Hatori	37	170	4·6	Tuff	As embankment material available was mainly clay some crushed stone was added. Centre concrete core to increase stability and resistance to shear and reduce pore water pressure.

The figures in Table 1 are interesting as they show the unpopularity of the low (relatively speaking) buttress dam.

The notes (Tables 2–5) from the Japanese National Committee's publication on large dams (1958) give some indication of the geological reasons leading up to the choice of dam.

A paper by T. Mizukoshi, H. Tanaka, Y. Inouye with geological descriptions by T. Taniguchi (see Bibliography) records slides and potential slides due to a special investigation as a result of the Vaiont slide.

The Kamishiiba arch dam (Table 3) on grauwacke clay-slate with intrusions. A slide of 70 m wide occurred in very weathered Mesozoic rocks of grauwacke and clay-slate near a contact with volcanic welded tuff.

The Futase arch dam (Table 3) on breccia, phyllite. The movement occurred in this Mesozoic, phyllite, sericite schist and graphite schist on a gentle slope of the bank extending over a width of more than a kilometre.

The Narugo arch dam (Table 3) on grano-diorite and granite. The Miocene tuffaceous rock and intruded andesite covered by thick talus in the valley produced several slides, aggregating 1,120 m in width, in the talus area.

Two other dams (a) *Kanogawa and* (b) *Hitotsuse*. In the reservoirs impounded, slides were produced in alternating shale and sandstone respectively, (a) of Palaeozoic age fractured by many faults, producing a slide of 300 m in width and (b) of Mesozoic age with thick talus producing a slide of 170 m in width.

The Ishibuchi dam (Table 4). Although this rockfill dam is founded on liparite, the slip of 400 m in width in the reservoir area occurred in Neogene sandy tuff and tuffaceous sandstone underlying Quaternary pyroclastic material, sandwiched between loam, sandy clay and pumiceous clay.

Elsewhere other similar troubles, in the reservoir area, are set out in Section 14 (pages 63 to 72).

BIBLIOGRAPHY

Dams in Japan. Japanese National Committee on Large Dams (1958) Tokyo.

Kondo, M., and Kakitani, M. Odomari Dam. *6th Congress on Large Dams.* R.23, 1 (1958) 469.

Mura, Y., Nakamura, K., Ohkubo, T., and Iida, R. Design of arch dams in the multi-purpose projects of Japan. *I.C.O.L.D.* **IV,** 533 (1961). (Geological effects on models are dealt with in this paper.)

Tashiro, N., and Kimishima, H. Kamishiiba Arch Dam. *6th Congress on Large Dams.* R.24, 2 (1958) 277.

Mizukoshi, T., Tanaka, H. and Inouye, Y. (with geological description by Taniguchi, T.). A geological investigation on the stability of reservoir banks. *I.C.O.L.D.* **I,** 47 (1967).

34

U.S.A. DAMS

In the U.S.A., 2,635 dams varying in height between 50 ft. and 770 ft. are listed in the Register of Dams up to 1963. Between 1963 and 1966, 298 dams were completed and 238 were under construction. There are more than twice as many earth and rockfill dams as concrete dams of all types and these are now more popular than concrete dams owing chiefly to the improvement and availability of earth moving, quarry, and compacting machinery during the past few years.

Earth/Rockfill Dams

It has been asked on several occasions how high can one build an earth/rockfill dam. After considering the various materials which can be used for construction and the foundation and the fact that by compaction one can convert a true rockfill material into an earthen embankment, it is best to turn to the views of P. F. Baumann (U.S.A.), who has discussed this subject in his paper 'Limit height criteria for loose-dumped rockfill dams'. His considered opinion is that the maximum height could be 500 ft. *provided* that:

(1) The quarried rock is sound and uniform.
(2) The angle of friction is not less than 45 degrees.
(3) The upstream slope is not steeper than 1·25 to 1 and the downstream slope not steeper than 1·50 to 1.
(4) The foundation rock is equal in strength to the quarried rock and free from pore pressure.

However, it would appear that Mr. Baumann would prefer a 250 ft. limitation rather than 500 ft.

Typical Foundation Problems

A few typical foundation problems are set out later to illustrate conditions which may reasonably be encountered.

Two non-typical, unreasonable and dishonest geological conditions, however, are cited by Mr. C. V. Davis as follows:

(a) Recently during the construction of the Box Canyon dam on the River Pend Oreille, Washington, a buried gorge filled with sand, 100–150 ft. in width and 150–200 ft. in depth, was discovered. As shown in *Figure 244* the gorge was spanned by a concrete arch on which the dam was built.

(b) In 1928, under the site of the Rodriguez dam, Mexico, a fault-producing 100 ft. width of soft material was exposed. As the dam was to be 267 ft. in height at this point, the bad material was spanned by a concrete arch on which four buttresses were built; the arch transmitted the load on to the rock which was, fortunately, sufficiently strong to carry the load and thus relieve the pressure on the intervening fault zone.

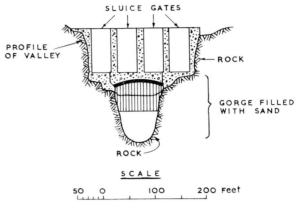

Figure 244. Box Canyon dam, Washington. Owing to the gorge being filled with a great depth of sand, the dam was built on an arch abutting concrete and rock; watertightness was secured by sheet-piling

Figure 245. Cheoah dam. This gravity arch dam was constructed in 1919 and is on Cambrian quartzite with sandstone conglomerates, calcareous shales and slates. The height is 225 ft., and the length of crest is 750 ft.

Cheoah Dam, North Carolina

The Cheoah dam is an interesting curved concrete gravity dam which was completed in 1919. It is in a gorge of Lower Cambrian quartzite, sandstone,

Figure 246. Cheoah dam. The power-house is on a ledge of rock on the left bank. It was enlarged in 1949 owing to the redundancy of the overflow weir, with a further addition in 1958

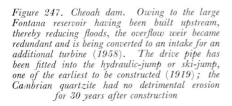

Figure 247. Cheoah dam. Owing to the large Fontana reservoir having been built upstream, thereby reducing floods, the overflow weir became redundant and is being converted to an intake for an additional turbine (1958). The drive pipe has been fitted into the hydraulic-jump or ski-jump, one of the earliest to be constructed (1919); the Cambrian quartzite had no detrimental erosion for 30 years after construction

conglomerate, calcareous shales and slates dipping upstream, in which the grouting curtain was taken to a depth of 40 ft. with bore holes at 10 ft. centres. The leakage is estimated at some 75,000 U.S. gal. per day. The maximum height is 225 ft. and the crest 750 ft., the chord-height ratio is therefore about 3 (*Figure 245*).

On the left bank there was a convenient ledge of rock for the powerhouse, which was enlarged in 1949, and again in 1958 (*Figure 246*).

Owing to the construction of the Fontana reservoir, 5 miles upstream, which has a capacity of $1 \cdot 444 \times 10^6$ acre-feet and commands a gathering ground of 1,571 sq. miles, the overflow hydraulic jump weir at Cheoah became unnecessary and the space was converted in 1949 into an intake for an additional turbine which increases the generating capacity by 37·5 per cent (*Figure 247*).

Although the hydraulic jump has been dismantled, it is interesting to note that it was one of the earliest to be constructed; it was based on model experiments, and proved satisfactory on the Cambrian quartzite for 30 years.

Other pioneering works of the Aluminum Company of America (Alcoa) were as follows.

(1) The Santeetlah arch dam, supported by two gravity buttresses, where the concrete was vibrated and controlled by the water-cement ratio (1928).

(2) The Nantahala rockfill dam with a sloping core (1942).

Norris Dam, Tennessee

The Norris dam is on the Clinch tributary of the River Tennessee about four miles east of Lake City and 20 miles north-west of Knoxville. The dam, constructed in 1933–1936, is a straight gravity structure 81 m in height and 567 m long. It is on the Cambrian formation and at the base of the Knox Dolomite, both abutments being in the Middle Copper Ridge which consists of cherty magnesian limestone and dolomite in beds varying in thickness from a few inches to 20 ft. or averaging say 5 ft. Generally the area is free from faults; but joints and bedding planes are filled with clay. The dip is 4 degrees S.E. downstream, almost horizontal, at the dam site.

The grout curtain (*Figure 248*) is 21 m to 61 m deep at the maximum section at the end of the abutments with holes 3 m apart and pressure 10·5 kg/cm². Consolidation grout, 6 m deep at 3 m centres under the whole of the foundation. Under the stilling pool and blocks Nos. 25 and 26 on the right abutment there are additional holes 12 m deep. From the gallery there are fourteen test holes in the foundation and 21 drainholes for foundation investigation and observation but are spaced too far apart for providing any uplift relief.

Uplift cells consist of gravel-filled boxes with perforated pipes about 1 m deep in rock from which pressure-pipes lead to gauges in the gallery.

The drain relief holes drilled in 1936 at the toe of the dam and under the apron or stilling pool were found to be choked and the pressure at the toe and under the apron threatened excessive uplift and damage especially over the large area of floor.

In 1960 twenty-eight new holes were drilled, five on the sloping part of the apron near the toe. These holes reduced the pressure considerably as shown by the pressure line C.

The increase in uplift beneath the spillway apron which occurred between 1936 and 1960, which was three times the calculated allowable pressure for the unwatered condition, is attributed to: (a) the grout curtain was not

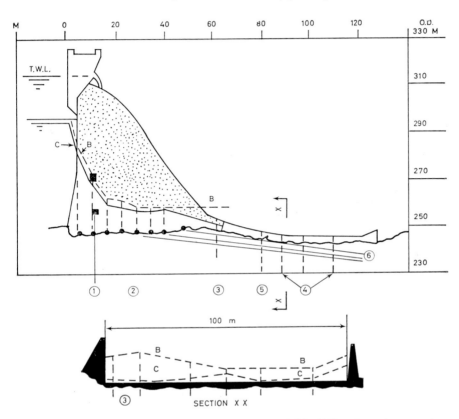

Figure 248. Norris dam. The uplift pressure (B) was recorded in 1960 and was known to have increased since the dam was built in 1936. This increase was considered excessive, especially at the toe of the dam and under the stilling pool as shown in section XX. The uplift pressure reduction is shown at C after the additional drains were drilled in 1960, in the dam and stilling pool.

1. Grout curtain, from gallery (1936). Pressure gauges from cells in gallery. 14 test and 21 relief holes from gallery.
2. Uplift cells (1936).
3. Test holes (1936).
4. Relief holes (1936).
5. Twenty-eight relief holes, 9 m into rock (1960), drilled into (6) three water-bearing seams.

effective in stopping the flow, through the three water-bearing seams under the dam and apron; (b) the consolidation under the apron up to 12 m in depth caused the pressure build up; (c) the original drainage system had not enough relief holes drilled deeply enough (especially into the three water-bearing seams) to relieve the pressure.

Hiwassee Dam, North Carolina*

The Hiwassee dam is a straight concrete gravity structure, built between 1936–40, on the River Hiwassee 60 miles south of Knoxville.

The reservoir is part of the T.V.A. hydro-electric system and one of the two turbines can be used in reverse to pump $1\cdot75 \times 10^6$ U.S. gal. per minute into the reservoir to produce 8,500 kW during the hours of maximum demand (*Figure 249*).

The dam, which is in a valley, is 307 ft. (94 m) in height and has a chord-height ratio of about 4. Owing to its thickness, low heat cement was

Figure 249. Hiwassee dam. The dam is a massive gravity concrete structure 307 ft. (94 m) in height with a crest of 1,376 ft (419 m) in length. The view shows the top of the powerhouse where the turbine can be used in reverse for pumping water into the reservoir for use in the hours of maximum demand—'pumped storage'

used, and it was constructed in blocks with vertical joints both at right-angles and parallel with the crest.

The site of the dam happened to be in close proximity to a good aggregate for concrete making; because at a ¼ mile downstream from the dam on the right embankment, there is a medium to coarse grained micaceous quartzite. This rock has a small percentage of mica and is composed of 70–80 per cent of silica and 10–15 per cent of alumina. Associated with it are schists (SiO_2 = 50–60 per cent; Al_2O_3 = 15–27 per cent).

Geology of site

The axis of the dam, as will be seen from *Figure 250*, crosses nearly vertical micaceous quartzite and mica schists of pre-Cambrian age in which over 25 small faults were encountered.

These strata are known as 'low rank', metamorphic pre-Cambrian or the Great Smoky greywacke from the German term 'grauwacke' which now

*Mr. George Palo, Chief Engineer, T.V.A., has very kindly made some important corrections to these notes on the Hiwassee dam.

generally denotes a grit of hardened bits of feldspar and felsite, although there is no felsite in the Great Smoky greywacke. (The term greywacke was formerly used by De La Beche to denote the Devonian, the grey slates and earthy rocks of the English west country.)

As seen in *Figure 250*, the dam section is on the side of an anticline whose axis is on the right bank. The dip at the crest is almost horizontal and becomes vertical in the bottom of the valley.

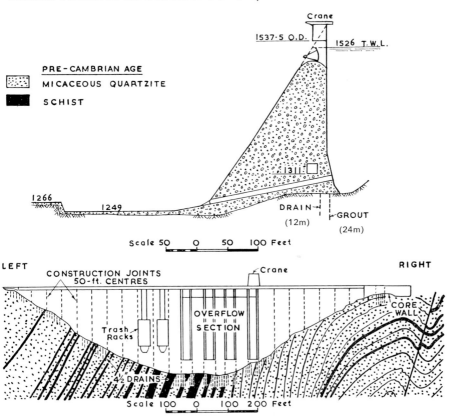

Figure 250. Hiwassee dam. A gravity dam was adopted on account of the unstable quartzite on the right bank of the valley being on the flank of an anticline. The dam is over 300 ft. in height and there was differential bearing pressure between the quartzite and schist

There was an average depth of 10–12 ft. of overburden on the sides of the valley, although in some local pockets and in the bottom of the valley talus amounted to as much as 30 ft. The overburden problem, therefore, was of little consequence, although some of the rocks *in situ* had weathered to a greater depth.

Although the compressive strength of the quartzite is 10–15 tons per in.^{2}, the dip of the strata on the right bank (*Figure 251a*) was not favourable and was considered too weak to sustain the thrust of a concrete arch. A gravity

(a)   (b)

Figure 251. Hiwassee dam. (a) The quartzite on the right abutment at the crest is nearly horizontal and at the base nearly vertical, the variation giving rise to open joints and scree. It was considered to be unsuitable for an abutment of an arch dam. (b) The alternating rocks near the base on the left abutment dip into the hill; although in themselves weaker rocks than the massive quartzite which dips into the valley on the right bank, they present a more stable face

Figure 252a. Hiwassee dam. A great deal of grouting and draining provision was necessary in the pre-Cambrian metamorphic quartzites and schists, and this large gallery was parallel with and near the upstream face and provided access to enable further work to be done should the need arise

type was therefore adopted as buttress and rockfill types of dam were considered to be more expensive than the gravity type.

At the bottom of the valley the schists with a crushing strength of 1–4 tons per in.² were decayed badly for a depth of 50 ft., and this necessitated extensive grouting and drainage.

The river section and half-way up the left (*Figure 251b*) bank the beds dip nearly vertically and consist of alternating layers of mica schist, micaceous quartzite, quartz-mica-schist, and conglomerate. From half-way up the side of the valley to the top the solid micaceous quartzite is present.

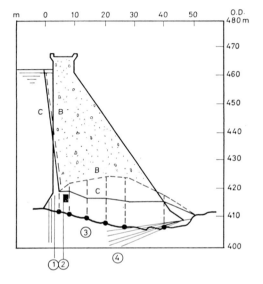

Figure 252b. Hiawassee dam. B denotes the uplift pressure which had increased by 10% between 1947–9. C denotes uplift pressure after drilling inclined boreholes in 1949 at toe at (4)

1. Grout curtain, three lines (*1940*).
2. Drain reliefs (*1940*).
3. Uplift cells (*1940*).
4. New inclined drains (*1949*).

Four types of joints were noticed, namely: strike joints, parallel or nearly parallel to the strike; dip joints, parallel or nearly parallel to the dip; oblique joints, parallel with neither dip nor strike; and sheet joints, parallel or nearly parallel with the surface.

Although from laboratory analyses the rocks were found to be able to withstand the pressure of a 300 ft. high gravity dam, and there was no threat of chemical deterioration or softening by the action of water, a considerable amount of grout was necessary for filling the joints and fissures to prevent subsidence and erosion through leakage.

In all, some 30,000 ft of bore hole was drilled, half of which was for the grout curtain and half for consolidation, and nearly 5,000 ft.³ of grout was used for each (*Figure 252a*).

Uplift

Seven years after the dam was finished in 1940, increase in pressure in the downstream half of block 7 began to develop, and during 1947-8-9, there was increase of 10 per cent. Boreholes in a fan-shaped pattern were drilled in 1949 and no material increase of pressure has occurred since (*Figure 252b*). The reason for the build-up has not been ascertained and the general statigraphy (*Figure 250*) is not easily interpreted.

Douglas Dam, Tennessee

The Douglas dam is on the French Broad tributary of the River Tennessee, about 20 m east of Knoxville. The dam constructed 1942–1943, is a straight gravity structure, 62 m in height and 520 m in length. The dam is founded on the upper part of the Knox Dolomite in the Ordovician formation which consists of crystalline and cherty dolomite limestone of a cavernous nature.

Case history of uplift (see *Figure 253*)

1944. Calcium deposits began to form in foundation drain holes.
1950. Deposits reduced flow enough to justify cleaning holes but removing soft deposits with a water jet did not noticeably increase the drainage flow or reduce the pressure.

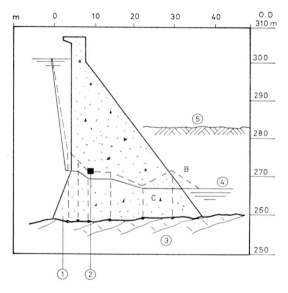

Figure 253. Douglas dam. B denotes uplift pressure from cells before drilling holes in 1957, 14 years after the dam was finished. C denotes uplift pressure after drilling diamond coreing holes in 1957
1. *Grout curtain.*
2. *Drain holes.*
3. *Uplift pressure cells.*
4. *Tail water level.*
5. *Earth-fill level (at block 11).*

1953. Uplift pressures increasing, drain flow reduced and drain holes found to be 44 per cent filled with deposits.
1957. Drain holes redrilled with diamond coreing and all deposit removed and the holes were deepened slightly. This resulted in increase of flow and reduction of pressure as shown on *Figure 253c*. The switchyard earth fill against the dam toe is believed to have increased the uplift and decreased the flow under the construction and at block No. 11 by preventing effective relief at the toe.

Fontana Dam, North Carolina

The Fontana dam, like the Hiwassee dam, is a straight concrete gravity structure. It was built between 1942–44 on the Little Tennessee River 50 miles south of Knoxville as part of the T.V.A. hydro-electric system. It is a short distance upstream of the old Cheoah reservoir, and the large storage at Fontana—1,444,300 acre-feet or nearly 400,000 million U.S. gal.—rendered the overflow weir at Cheoah redundant.

At Fontana the main overflow gates on the dam deal with 250,000 cusecs, and there is also a so-called emergency overflow for 100,000 cusecs some little way from the left of the dam.

The dam is 480 ft (147 m) in height and the crest 2,365 ft. (721 m) in a valley with a chord-height ratio of about 5.

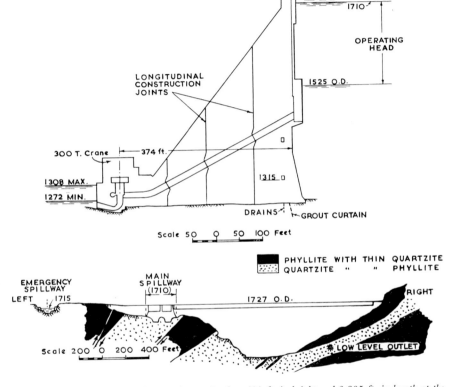

Figure 254. Fontana dam. The massive gravity dam 480 ft. in height and 2,365 ft. in length at the crest is put on alternating pre-Cambrian and Cambrian metamorphic quartzite and phyllite (slate) of variable bearing pressures. Up to 100 ft. of superimposed overburden and talus were removed

Great difficulty was experienced in finding a suitable aggregate for concrete near the site as the quartzite was so interstratified with phyllite and this was considered to be deleterious (quartzite: $SiO_2 = 77·5$ per cent; $Al_2O_3 = 8·9$ per cent; phyllite: $SiO_2 = 57·6$ per cent; $Al_2O_3 = 19·9$ per cent). A satisfactory area of quartzite was eventually found 1·2 miles downstream.

Geology of site

As will be seen from the section across the valley at the dam site (*Figure 254*) the dip varies between 15 and 75 degrees; it is part of an anticline like that of Hiwassee in the same Great Smoky formation but slightly lower rank metamorphism, namely, alternating phyllite with thin quartzite and quartzite with thin phyllite of pre-Cambrian and Cambrian age.

Bearing strengths of rocks

	lb. in.^{2}
Quartzite with thin phyllite (slate)	13,100–28,741
Average	22,182
Badly weathered	4,280–7,690
Phyllite with thin quartzite (Arkose)	9,313–17,797
Average	11,534
Maximum requirements of gravity dam	520

The phyllite at the site contains little or no pyrite or sulphides.

Overburden

Right abutment (Figure 255a)—A maximum of 52 ft. was reached, the average being 12 ft. Weathering extended to 91 ft., but the average depth at the abutment was 36 ft.

Base of valley—With a maximum of 19·6 ft. the average depth to good rock was 3 ft. although slight weathering went to 114 ft. below the river bed.

Left abutment (Figure 255b)—The overburden was thickest (70 ft.) near the crest, but the average was 18 ft. Weathering was over 100 ft., but the average depth, like the right side, was also 36 ft. The depth to good rock, therefore, was very much greater than that at Hiwassee.

Structure

Faults are numerous, particularly in the phyllite, and strike, dip, and sheet joints converted parts of the foundation into a multitude of small rhombic blocks.

Grout

Over 300,000 ft. of bore holes up to 150 ft. in depth were drilled, of which two-thirds were for consolidation and one-third for the curtain (*Figure 256*).

The grout curtain is at the upstream face and behind it, downstream, is a row of vertical drain holes leading into a large gallery like that at Hiwassee (*Figure 252a*, page 370).

On the left bank there was an interesting occurrence which called for some ingenuity, for it was found that there were beds of sound quartzite separated by badly disintegrated beds about a foot in thickness. It was decided to drill (vertically) several 36-in. holes large enough for access and to mine out the bad material for a distance of about 3 ft. around each hole, between the good seams; this space was then filled with concrete thereby making plugs of soild concrete each 9 ft. in diameter. Grouting was carried

Figure 255. Fontana dam. (*a*) *The right bank is not so steep as the left as the main dip of the beds, mainly phyllite, follows the slope of the valley. There was a good deal of overburden to be removed on this side to secure a satisfactory abutment but little overburden was met with in the bottom of the valley.* (*b*) *The left bank is steep and the general dip of the beds is into the bank. The top of the abutment is mainly on quartzite with a thick covering of overburden removed. Near the left end there is a mitre and the top few feet of the wall are turned round to meet the quartzite which rises in the hill on the right. The main overflow and visitors room is at this end*

Figure 256. Fontana dam. Lower Cambrian metamorphic beds near the left abutment. The grout has plenty of work to do in strata like these

Figure 257. Fontana dam. There were seams of weathered slates a foot in thickness between large masses of quartzite on part of the left bank and these, where exposed, were excavated out for 5 ft. and packed. In addition 36 in. access bore holes were put down and the seams excavated by hand and packed with concrete giving support pillars 9 ft. in diameter

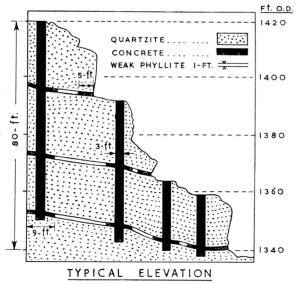

out between the plugs as an additional safeguard against any settlement of the sound quartzite seams and dam (*Figure 257*).

Uplift

The dam was finished in 1944 and after eight years, in 1952, it was found that 45 per cent of the total length of all drains were filled with deposits and uplift had increased and drainage flow had decreased. In 1953 the

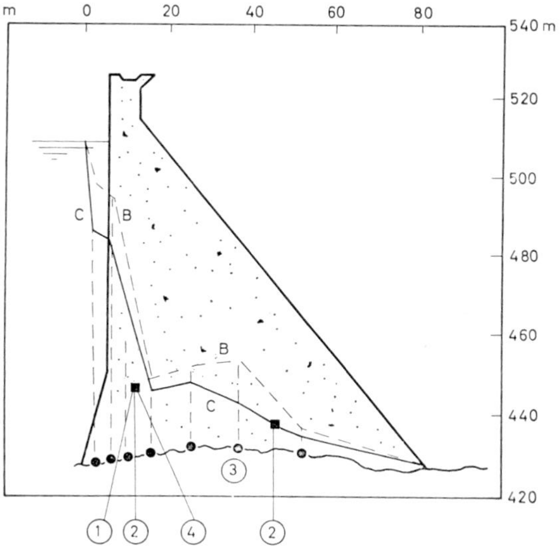

Figure 258. Fontana dam uplift. B denotes pressure uplift from cells before drilling slanting holes in 1956. C denotes uplift pressure after drilling holes, 1956.

1. Grout curtain.
2. Drain reliefs (1944).
3. Uplift cells (1944).
4. New slanting drains (1956).

drain holes were redrilled which increased the total flow but did not reduce the pressure in the area of the left bank where foundation contact grouting had been extensive.

In 1955 new holes were drilled between the existing holes without reducing the pressure. In 1956 deep holes were driven in a slanting direction downstream into the high pressure zone, which resulted in the reduction in pressure, as shown on *Figure 258*.

Chilhowee Dam, Tennessee

The Chilhowee, constructed in 1955–57, is a composite dam, the lowest on the Little Tennessee River.

Gathering ground	1,977 sq. miles
Average annual run-off	4,729 cusecs
Area of reservoir	1,747 acres
Designed flood	230,000 cusecs
Capacity (useful)	6,805 acre-feet
Type	Concrete, gravity and rockfill
Rock and earth (left)	665 ft.

Intake section	165 ft.
Tainter gate section (height 91 ft.)	266 ft.
Concrete (right)	287 ft.
Total length of crest	1,373 ft.
Height	78 ft.

Figure 259. Chilhowee dam. Downstream view of right-end abutment on Lower Cambrian quartzite. The gate overflow section and part of the power-house (faced with aluminium) with the water coming from it in the foreground are shown. There are 6 Tainter gates each 35 ft. in width and 38 ft. in height designed to pass a flood of 230,000 cusecs

Figure 260a. Chilhowee dam. Part of rockfill section over 600 ft. at the left end looking towards the power-house section and right bank. The rockfill was obtained from the tail-race area.

The dam is on Lower Cambrian partly crystalline rocks of sandstone with quartzite conglomerate shales and slate dipping slightly downstream. Grouting was taken to 24 ft. below ground with a high pressure of 140 lb./in.[2]

Figure 260b. Chilhowee dam. Rockfill part of composition dam at left end on Cambrian sandstone in quartzite. The rockfill was taken from the overflow channel excavation on the right bank

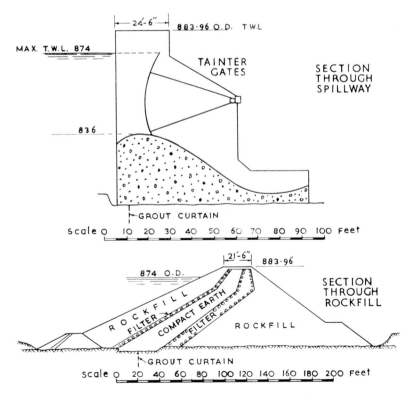

Figure 261. Chilhowee dam. Concrete section of dam above and rockfill section with a sloping core below

380 U.S.A. DAMS

The overflow section is near the right side (*Figure 259*); the 6 opening Tainter gates are each 38 ft. in height, 5 of which are visible as well as a section of the Lower Cambrian quartzite beyond. Part of the power-house section is also shown with the water from the turbines in the foreground.

The long rockfill part of the dam looking towards the right bank and power-house section of the dam is shown in *Figure 260a*, and *Figure 260b* shows the end of the rockfill at the left bank. This rockfill section has a sloping core (*Figure 261*) the rock for which was obtained from the tail-race area (*Figures 259* and *260a*).

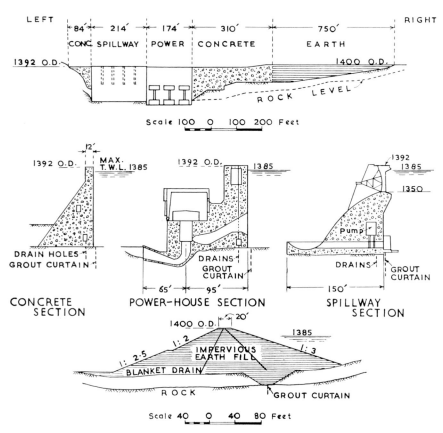

Figure 262. Boone dam. The limestone is near the surface on the left side of the valley, so a concrete gravity dam was built. Towards the right the limestone gets deeper and hence an earthen embankment was adopted. Grout curtain 35 m

Boone Dam, Tennessee

The Boone dam, constructed between 1950–52, is a composite dam below the Watauga and South Holston dams on the River Holston.

It is in an area of 'Knox dolomite' at the base of the Ordovician and top of Cambrian which has undergone metamorphosis.

The Knox dolomite (up to 3,500 ft. in thickness) is a kind of magnesian limestone light to dark blue with chert nodules. These strata produce a red and grey loam suitable on low ground for farming and on high ground for orchards. The Upper Cambrian calcareous shales, limestones, sandy shales

Figure 263. Boone dam. Showing the highly inclined Cambrian Ordovician rugged limestones at the left end of the concrete portion of the dam

Figure 264. Boone dam. This view, taken from the top of the concrete dam above the spillway, shows the relatively tame scenery of the shales and sandstones on the right bank in contrast to the limestones on the left (see Figure 263). In the foreground the dark shadow of the top of the concrete dam across the power-house shows the inclination of the concrete dam; beyond is the long 150 ft. earthen embankment with a berm half-way up, curved downstream, at the foot of this is the control building with the pylons behind

and sandstones form parallel ridges and valleys and are up to 5,000 ft. in thickness. The limestones form valleys and low hills, and decompose to deep, fertile, red clayey soils; the shales and sandstones form grey and yellowish thin generally poor soil.

On the left bank a concrete gravity section has been adopted, incorporating the spillway and power-house, where the limestone is near the surface. Although the limestone becomes deeper to 30 ft. below the surface, the gravity section is maintained until the depth of water becomes less than 100 ft. (*Figures 262 and 263*).

From this point, for the remaining 750 ft. length of the dam, until the right bank is reached a curved earthen embankment below 90 ft. in height has been adopted (*Figures 262 and 264*).

There were two grout systems, one to a depth of 40 ft. and the other from 62 to 152 ft. Holes were spaced 10 ft. apart and, with pressures of 1 lb./in.^{2} per ft. of bore hole; 7,500 tons of cement and clay were consumed.

Nantahala Dam, North Carolina

The Nantahala dam is noteworthy for being the first large rockfill dam constructed with a sloping impervious core.

High up in the Smoky Mountains the dam, with a crest level of 3,012 O.D., rests on Lower Cambrian quartzite, a metamorphic sandstone known as Arkose. The strata are massive with few cracks (*Figure 265*).

Figure 265. Nantahala dam. The Cambrian quartzite (Arkose) face of the 'quarry' which is part of the overflow channel where the rock was obtained for the dam. The rock is massive and the bedding generally is ill-defined

The dam was constructed in 1940–42 to a height of 251 ft. and a length of 1,042 ft.; it has a top width of 30 ft. and a base width of 1,120 ft. Its volume contains over $2\frac{1}{4}$ million yd.^{3} (*Figure 266*).

The greater part of the rockfill and rip-rap for the dam was quarried from the overflow channel (*Figure 267*).

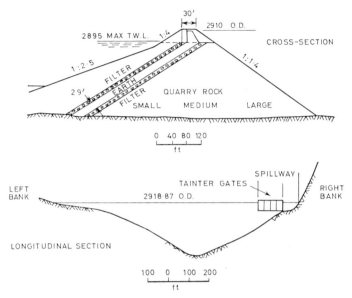

Figure 266. Nantahala dam. Section of rockfill dam constructed 1940-42, with sloping clay core believed to be the first of its kind. The sloping core of clay-earth is protected by a sand layer or so-called filter to prevent the clay flowing into the voids of the very coarse rockfill of the downstream toe. Similarly, for the upstream slope the filter protects the clay from being adversely affected by the water, such as wave action or sudden lowering in level

Figure 267. Nantahala dam. The overflow channel for 55,000 cusecs from 108 sq. miles looking downstream from which most of the rock for the dam was quarried

The sloping impervious core has a width of 30 ft., it would have been made thicker but clay was scarce.

The leakage is estimated at 2·75 ft.³ per minute; 24,000 English gallons per day in dry weather when the reservoir is full. The maximum settlement after 16 years is 2·85 ft. vertical, and the greatest lateral movement is 1·53 ft.

Grouting was adopted with holes 25 ft. in depth; 20 ft. centres at 75–150 lb./in.², and holes 125 ft. in depth, 100 ft. apart at 250 lb./in.², followed by holes 25 and 12½ ft. apart; 8,000 barrels of cement were used.

The scheme was considered economical owing to the proximity of the rock which was excavated from, and for the overflow channel, and the saving of much work by the voids in the rock providing natural drainage and eliminating uplift.

The sloping clay core is also an advantage as it transmits the water pressure directly on to the foundation and reduces sliding.

WATAUGA AND SOUTH HOLSTON DAMS, TENNESSEE*

The Watauga and South Holston rockfill dams are on the River Holston approximately 100 miles north-east of Knoxville.

The dams are almost identical in cross-section (*Figures 268 and 270*) and their main dimensions are as follows.

	Watauga	South Holston
Date of construction	1947–48	1948–50
Top water level (ft. O.D.)	1,975	1,742
Area of gathering ground (square miles)	7,200	8,750
Capacity (acre-ft.)	778,800	744,000
Surface of water (acres)	6,430	7,580
Perimeter (miles)	106	188
Length of lake (miles)	17	24·3
Maximum height (ft.)	318	285
Length of crest (ft.)	900	600
Volume of dam (million yd.³)	3·5	5·9

The situations of the dams, which are only 10 miles apart, both topographically and geologically are entirely different. Watauga is in a gorge in wild back-of-beyond semi-mountainous country (*Figure 269*) and South Holston is in a wide valley in comparatively tame country (*Figure 273*).

Although both sites were favourable for concrete gravity arch dams, the geological conditions as indicated in *Figures 268, 270 and 272* are very different, as may be imagined from an inspection of *Figure 272* alone. In both cases rockfill dams were adopted.

Watauga Dam

The dam is located on Cambrian quartzites tilting 55 degrees downstream; although the geology of the area is very complicated, for example, there is a fault with a throw of 4,500 ft. about 100 yd. upstream of the dam and a horizontal displacement of strata of several miles, the following are typical strata at the dam.

*I am much indebted to Mr. George Palo, chief engineer of the T.V.A. for having read these notes on Watauga and South Holston dams.

Cambrian: part of the Unicoi formation (see *Figure 268*)

	Unit No.	Varying quartzites above to about 1,400 ft.	
Above crest level	(6)	Medium to coarse grained, thick-bedded light greenish-grey quartzite	60 ft.
	(5)	Medium bedded medium grained, dark grey quartzite (with some shale)	55 ft.
Below crest level	(4)	Medium to coarse grained, thick-bedded light greenish-grey quartzite (pinkish when oxidized giving a beautiful effect)	35-50 ft.
	(3)	Fine grained, dark greenish-grey quartzite, medium to coarse grained, light grey quartzite and Arkose; and medium to fine grained, medium bedded, dark grey quartzite, much broken	170 ft.
	(2)	Hard, fine grained, dark greenish-grey sandy shale (broken)	12 ft.
	(1)	Medium to coarse grained, light to greenish-grey quartzite	235 ft.

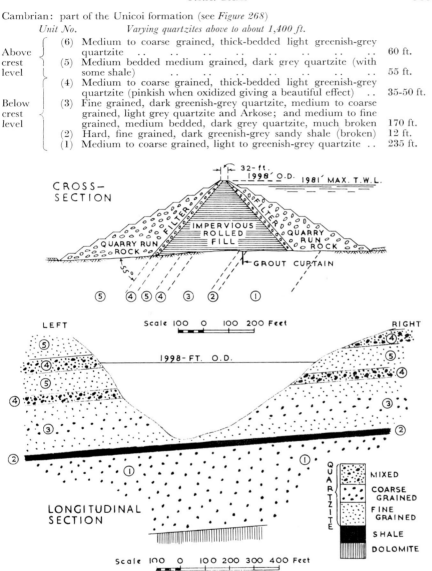

Figure 268. Watauga dam. The Cambrian strata dipped steeply downstream (see section of dam above). As the strata at the dam site were normal the shale bed '2' became a natural cut-off screen and little grout was used in the grout curtain. The normality of the strata across the valley is surprising, particularly as there is a fault of 4,500 ft. 100 yd. upstream of the dam

Pre-Cambrian below (cut off by thrust fault)

The quartzite bearing pressure is of the order of 20,000 lb./in.^{2} and the shale 9,000 lb./in.^{2}. The quartzite is resistant chemically but parts of beds 3 and 2, the shales containing sulphides, tend to break down on exposure.

Overburden and weathering of the rock did not present any difficulty although the talus at one place on the right bank extended to a depth of 50 ft., otherwise it was only 4–5 ft.

The rockfill type was chosen as it needed less excavation than that required for a concrete dam.

The beds 5 and 4 in *Figure 268* are very well seen on the right bank end pitching downstream (*Figure 271a*); similarly, the same beds on the left end are seen to outcrop at the same level as the strike is across the valley.

The rock for constructing the dam was obtained (*a*) a little way downstream on the left bank from a bed of the Unicoi quartzite, medium to coarse

Figure 269. Watauga dam. Quarry on the left bank a little below the dam in mountainous country of the Cambrian quartzite from which the rockfill of the dam was obtained

in grain, known as division No. 9, over 1,000 ft. in thickness and part of the Snowbird formation (see *Figures 269 and 271b*), (*b*) a little way upstream on the right bank, a dolomite limestone (*Figure 271a and b*).

A liberal allowance for camber for settlement of the dam was given (9·2 ft. maximum): after 8 years the settlement was 2·6 ft., or 0·8 per cent of the height of 318 ft.

Comparatively little grouting was necessary in the formation (461 bags) even with pressures of 30 lb./in.^{2} no grout was necessary for the top 123 ft. on the right and 153 ft. on the left.

South Holston Dam

The dam is on the north-west flank of a syncline in a wide valley underlain by sandstones, conglomerates and shales. As the strike is almost parallel with the valley, the section (*Figure 270*) nearly represents the maximum dip but there is great variation in detail.

The strata are Middle Ordovician in age, namely the Tellico (to 10,000 ft. in thickness?) which is a deltaic deposit with lenses of sandstone shale and conglomerate. The sandstone is hard, with 25 per cent of calcium carbonate, 60 per cent quartz sand, and has a crushing strength between 8,000 and 27,800 lb./in.^{2}. It is grey in colour. The calcium carbonate can be leached out, it is reported, leaving a weak sandstone, but the process is so slow that the useful life of the structure is not affected.

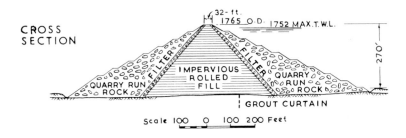

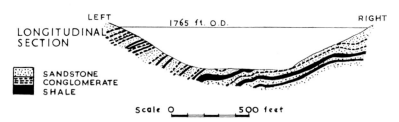

Figure 270. South Holston dam. This dam, 285 ft. in height, was constructed in 1948–50, mainly on a syncline on soft strata on which a rockfill dam was considered to be the most economical

The conglomerate contains pebbles (in beds and lenses) of limestone, quartzite, and chert in a matrix of calcareous sandstone. The conglomerate generally stands up to weathering.

The shale, in seams of 1 in. to 60 ft. in thickness, is dark blue-black in colour, soft, fissile and calcareous, and weathers into a soft, loose, yellow, splintery deposit. Like most shales it is generally a weak foundation, particularly when thick seams are sandwiched between the harder sandstones.

The rocks generally at the dam are naturally very much weathered, particularly down to a depth of 10 ft. below the surface. Slight weathering was proved down to 40–50 ft., although in one case it was 300 ft.

The rock for the fill was obtained from two quarries, one on the right bank half a mile downstream (*Figure 273*) in a lens of Tellico sandstone and conglomerate, and the other, 900 ft. upstream on the right bank, a sandstone-conglomerate-shale outcrop in the Tellico formation.

The allowance for settlement was for a camber of 10·1 ft., and after 6½

Figure 271. Watauga dam. (a) The right bank upstream showing inclined Cambrian quartzites at the end of the rockfill dam. On the right is a quarry in the Shady dolomite limestone; between this and the inclined quartzite, there is a fault of 4,500 ft. bringing up the Shady limestone to the surface. The top berms of the upstream face are seen, also the control tower, the bridge and the morning glory bellmouth overflow (128 ft. in diameter). (b) The right abutment with the bellmouth in the foreground, and the quarries on the right

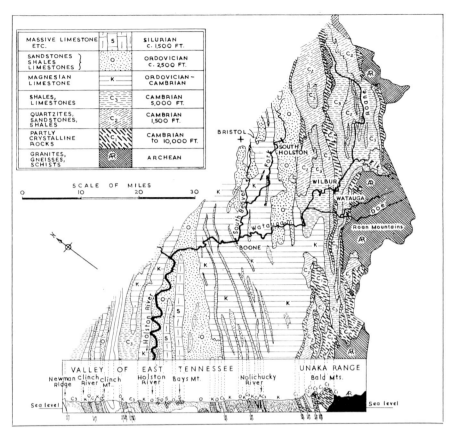

Figure 272. South Holston and Watauga dams. Geological map (based on U.S.G.S. map) showing complexity of the Cambrian-Silurian-Archean strata in which some of the dams in the Tennessee Valley are situated

Figure 273. South Holston dam. Looking downstream on the right bank towards the quarry in Ordovician (Tellico) sandstone which supplied part of the rockfill for the dam. The power-house and surge tower and South Fork Holston River are also shown

years the settlement amounted to a maximum of 2·85 ft. (1 per cent of the maximum height of 285 ft.).

A curious record is that a rise of 100 ft. in water produces a forward upstream movement of half an inch, attributed to the tilting of the dam on the soft sandstone and shale.

As would be expected a comprehensive grouting programme was necessary, and in addition to ground consolidation grouting the curtain was taken down to 120 ft. below the surface; by 1942 over 7,000 ft. of bore hole were drilled, consuming over 7,000 ft.³ of cement. No grouting was done within 18 ft. of the crest.

Tims Ford Dam, Tennessee Valley*

A practice which seems to be coming into the literature of dams is to describe a dam only in the design stage and to include it with those already built. Here is an example from George P. Palo, Chief Engineer of the Tennessee Valley Authority, who calls attention to a list of dams headed 'Some major U.S. dams put into design or construction since preparation of General Paper No. 9 (8th I.C.O.L.D. Conference 1964)'.

The dam was originally intended to be a concrete gravity dam, 175 ft. high; 1,335 ft. length; and volume, 390,000 yd³.

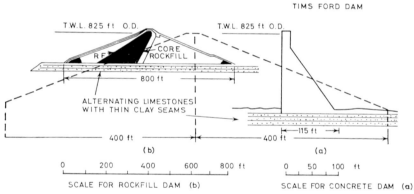

Figure 274. Tims Ford dam. Tennessee Valley. After exploratory boreholes had revealed alternating limestones and thin clay, a concrete gravity dam (height 175 ft. crest 1335 ft.) was proceeded with as shown on the right-hand side of the figure. The opening out of the site suggested danger of sliding. After ten large scale tests of 36 in. diameter limestone blocks, 4 ft. long containing natural clay seams were tested for shear it was decided that the resistance to horizontal sliding for a concrete dam with a base length of 115 ft. and 175 ft. in height would be unsatisfactory. A rockfill earth dam with a base length of 800 ft. was substituted, as shown on the left hand side of the figure

Mr. Palo, who has very kindly verified these notes, gives a diagram of the intended concrete dam (*Figure 274*) and the earth rockfill dam 'now designed and under construction'.

* Mr. G. H. Kimmons, Manager of Engineering Design and Construction, Tennessee Valley Authority, has very kindly read the notes for my friend, Mr. Palo, who has now retired. The notes briefly refer to the Tims Ford Dam and the behaviour of uplift pressure of the T.V.A. Dams, Douglas and Norris, Hiwassee and Fontana, which I visited in 1958.

Generally, the overburden and weathered sandstone as quarried, when rolled with the 50-ton rubber-tyred roller, would be as good a method as any to obtain a permeability of $2 \cdot 5 \times 10^{-4}$ cm per sec.

Large-scale tests for optimum compaction

Large-scale tests in an area of 130×30 ft. were also made to determine the number of passes of the roller necessary to secure good compaction, the amount of water necessary, and the shear strength which could be expected for the rock filling.

These experiments included 4 and 8 passes of a 50-ton rubber-tyred roller and at 100 lb./in.^{2} tyre pressure with ordinary rock: (1) as quarried, (2) with half volume of rock to water added, (3) with twice volume of rock to water added compacted in 18 in. layers.

It was found that at least 6 passes of the 50-ton rubber-tyred roller were needed and at least half the volume of the rock of water should be added—more if the materials were coarse—to obtain the maximum density and the highest shear and angle of repose of 45 degrees. The shear strength of the rockfill was determined by a most interesting full-scale experiment, as follows.

Some of the compacted rock was enclosed, *in situ*, in a box $6 \times 6 \times 3$ ft. high, in two 18 in. layers. The vertical or normal load was applied by jacks against beams.

The horizontal shear was applied by horizontal jacks on the top half of the box. With the direct load applied vertically on the area of 6 ft., 76,000 lb., sliding or failure in shear was indicated when a horizontal load was applied which produced continuing horizontal movement between the top half of the box relative to the bottom half, and this took place with approximately the same load of 76,000 lb. on the shear section. Similarly, when a load of 148,000 lb. was applied vertically in the 36 ft.^{2} section, a similar load of about 148,000 lb. applied horizontally produced continuous sliding.

As the normal stress applied vertically was equal to the shear stress applied horizontally, the angle of repose of the large-scale experiment on the compacted rockfill was 45 degrees, which confirmed the more pessimistic laboratory small scale tests of 42 degrees.

For this test 6 passes of the 50-ton rubber-tyred roller were made on the sandstone rock which was sluiced by half its volume of water.

MISSOURI BASIN EARTH DAMS

Compaction methods and moisture content for materials used in the construction of earth core and supporting fill for earth and rockfill dams was a subject discussed at the 6th Congress on Large Dams, 1958, and much information is available in Vol. III of the proceedings of the conference. For example, in a paper by Bennett on materials and compaction methods, the ground conditions for four dams are described.

Garrison Dam, North Dakota

The Garrison dam is 210 ft. in height, with a length of 11,300 ft. Tertiary thin-bedded sands and clay, including 13 per cent of lignite excavated from the spillway channel, formed the material for the dam (using large amounts of excavated material for the spillway to provide material for the dam was adopted at Nantahala dam and the Lewis Smith dam).

The compressive strength of these rocks varied from 9 to 30 tons/ft.^{2}. Dry density from 115 to 100 lb./ft.^{2}, and the moisture content 15–23 per cent. When compacted in 8-in. layers with 8 passes of tamping rollers at 640 lb./in.^{2}, the dry density is 110–95 lb./ft.^{2} and the moisture content much the same as the natural rocks (15–23 per cent).

Gavins Point Dam (S. Dakota)

The material for the Gavins Point dam consisted of Bedrock Niobrara Chalk, a grey argillaceous chalky shale; dry weight 80–115 lb./ft.^{3}, moisture content 20–35 per cent, compressive strength about 1,000 lb./in.^{2}.

The dam (height, 74 ft., length, 8,700 ft.) has a central rolled impervious core supported by chalk of which the upstream slope is 1 to 3 and downstream slope 1 to 2·5. The chalk was spread in 12-in. layers broken down by 4 passes of a spike-tooth roller, weight 50,000 lb., with teeth 10½ in. in length. Compaction was best obtained by 3 passes of a 50-ton rubber-tyred roller. The rolled chalk had a dry weight of 88 lb./ft.^{3} and a moisture content of 28 per cent. The angle of internal friction was 25–44 degrees, for which the design value was 26·5 degrees.

Tuttle Creek Dam, Manhatten Blue River, Kansas

Tuttle creek dam is 150 ft. in height and 6,700 ft. in length. Excavation of Permian shale and limestone from the overflow channel passed through 116 ft. of variable beds:

	A		B	C	D
	ft.	in.	Compressive strength (lb./in.^{2})	Dry density lb./ft.^{3}	Moisture content (per cent)
Easly Creek shale	22	2	225–870	115–133	7·4–13
Middlburg limestone	3	0	—	—	—
Hooser shale	9	3	1,802	137	7
Eiss limestone	9	5	1,610	125	7
Stearns shale	11	8	488–668	133–139	8·5–10·5
Morrill limestone	3	5	8,850	162	2
Florence shale	7	6	846–1,427	170–131	8·5– 8·7
Cottonwood limestone	6	9	2,080–4,260	134–147	7·9– 4·1
Eskridge shale	25	4	862–1,440	136–140	8·4– 5·8
Neva limestone	17	8	1,353–3,670	137–150	8·0– 4·9
	116	2			

(In the shales there is subordinate limestone and vice versa.) The maximum and minimum figures in columns B, C and D correspond.

About half the section was in limestones and half in shale of various strengths, and these were utilized as follows.

Rock and shale upstream above impervious core; although impermeability upstream does not matter, it is thought that the shale will tend to soften and create a degree of impermeability. (Upstream pore pressures at Sutton Bingham tended to confirm this.)

The strong Neva limestone is used for the rockfill topping and the Cottonwood limestone for the upstream slope rip-rap.

On weathered shale little difficulty was encountered during placing and consolidation, but the stronger shales caused difficulties when broken down and compacted by various types of machinery which, even at 580 lb./in.², rode over hard chunks, or, in the case of rubber-tyred rollers, good on a smooth surface, tended to overturn on rough surfaces.

The Hoover shales after preliminary raking were best compacted in 18-in. lifts by 3 passes of the 50-ton rubber-tyred roller to 112 lb./ft.³

The Florence shale was similarly treated and compacted to 111–116 lb./ft³. of dry density.

The other weaker shales weathered or unweathered could be broken down more easily with little or no raking or spike-tooth rolling and were consolidated as above.

Oahe Dam (S. Dakota)

Oahe dam (height, 242 ft.) is constructed on a foundation which requires flat slopes for stability, has a central rolled earth core fill of sands, clayey sands, silts and clays supported by berms of partly compacted shale for stability. As the rainfall is only 16 in. the strata are too dry (15–12 per cent deficiency) for the best stability conditions. Rather than adding water to the material when compacting, it was found best to water the borrow pits by sprinklers comparable to sprinkler irrigation. This reduced the cost of placing the rock by half.

Thus, a glacial deposit under grass had an infiltration rate of 0·3 in. of water per hour; sprinklers at 8 U.S. gal. per minute were spaced on a 40 × 60 ft. grid. For 30 ft. depth of strata it took at least 30 days for a curing period before excavation was begun. Best uniformity in wetting was given when sprinkling was done intermittently on rock which was covered by grass.

Trinity Dam, California

The Trinity dam completed in 1960 is a rockfill dam, height 537 ft., crest 2,450 ft., volume 33 × 10⁶ yd.³ founded on Copley meta-andesite or greenstone, a crystalline rock resulting from low grade metamorphism of basaltic and andesitic lavas.

Weathering on the higher part of the abutments of the meta-andesite was deep and severe along faults and joints, but much of the rock was adequate to support the dam.

Sand and gravel in the river channel and the more severely weathered rock and residual soils on the banks were excavated for the dam foundation. Under the impervious zone, material was removed to firm bedrock to ensure tight contact, under which there was a grouted area.

The dam (*Figure 279*) is built up in four main gradings of material, central clay core, filters, tailings and rock toes, upstream and downstream.

Instrumentation for pore pressures and vertical and horizontal movements in the embankment is provided. By 1964 bedrock settlement is stated to be 0·2 ft. and the maximum settlement 0·5 ft. at the crest which seems very little for a rockfill dam of 537 ft. in height.

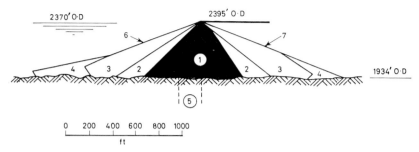

Figure 279. Trinity dam, California. Sectional elevation.

1. *Core. Selected weathered rock of meta-andesite broken down to clay, silt, sand, gravel compacted by tamping rollers to 6 in. layers.*
2. *Filters of similar rock but broken down to silt, sand, gravel and cobbles and compacted by tamping rollers to 12 in. layers.*
3. *Alternating layers of related dredger tailings, gravel, cobbles and boulders and selected dredger screenings compacted by crawler type tractors in 18 in. layers, or, alternatively, selected undredged sand and gravel compacted in 12 in. layers.*
4. *Rockfill toes, in 3 ft. layers, from sound rock of hard Copley meta-andesite, from spillway excavations.*
5. *Grout holes at 10 ft. apart for 200 ft. under core.*
6. *Rip-rap (3 ft.) on upstream slope between operating water levels.*
7. *Rock surfacing as 4.*

Oroville Dam, California

The Oroville dam is 770 ft. in height from foundation to crest or 730 ft. above the bed of the River Feather. It contains 77×10^6 yd.³ of gravel sand and clay.

The foundation bed-rock of this ambitious rock-fill dam is a metamorphic fine crystalline rock underlying marine and continental sediments and recent gravels and sands which were deposited by streams coming from the Sierra Nevada mountains. These superficial deposits now overlying the bed-rock are the 'tailings' of old gold mining operations (extending to a depth of 60 ft.) and are the result of excavations, workings, separation, redeposition, since 1880. As the gold has been extracted, the gravel was separated from the sand and the gravel and sand will be the source of the zones 3 and 2, the embanking material and the filters of the dam.

In addition, nature has provided the material for the clay core, zone 1 of the dam, by a flood-plain deposit of Pleistocene adjacent to the site, the Red Bluff formation of the Feather River. (*Figure 280* shows the disposition of these zones.)

A concrete dam was originally proposed but, after a detailed examination from 1961 to 1966 and testing of the deposits in the laboratory and on site, the concrete dam was rejected on grounds of cost. It was found that the volume of the respective zones was available; the permeability and grading of the filter sand and the permeability of the core were satisfactory and the strength, shear, compressibility, compactibility of all gravel, sand and clay, zones 3, 2 and 1, rock, filters and core, were adequate for a rockfill dam.

Modern 'shake' tests were made, for although the dam itself is not in an earthquake area, it is in California, an earthquake State. These tests suggested some modifications in shear allowances and slip-circles; taken care of in the design of slopes.

A grading specification and compaction recommendations for a contract were thus arrived at, for which contractors were open to make alternative suggestions, within limits and subject to approval that the end-product would be complied with.

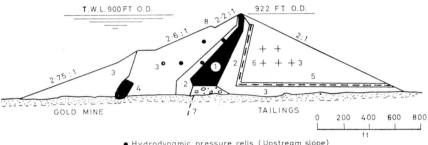

• Hydrodynamic pressure cells (Upstream slope)
+ Dynamic stress meters (Downstream slope)

Figure 280. Oroville dam. This rock-fill dam is constructed on and with the 'tailings' of old gold-mining operations. The dam is 770 ft. in height and has a volume of 80×10^6 yd^3. or $61 \cdot 2 + 10$ m^3. For an unprecedented height for a rock-fill dam there is an unprecedented number of instruments, nearly 400 in all.

1. *Impervious core from main borrow area.*
2. *Filters, well-graded silts, sands, gravels, cobbles and boulders up to 15 in. maximum size.*
3. *Sands, gravels, cobbles, boulders up to 24 in. maximum size.*
4. *Impervious core from abutment stripping.*
5 and 6. *Drainage zones, gravels, cobbles, boulders.*
7. *Concrete core block and grout curtain.*
8. *Rip-rap.*

SPECIFICATION AND RECOMMENDATIONS

ZONE 3. Rock-fill (Unrestricted use of coarse tailings).

    Grading. 25 per cent finer by weight than 4·76 mm.
                100 per cent finer by weight than 610 mm.
                Sands, gravels, cobbles, boulders up to 610 mm.

    Compaction. It is suggested that:
        (a) no advantage is gained on compaction by adding water;
        (b) vibrating rollers best, frequency of vibrations immaterial;
        (c) density of compaction not affected by lift thickness (within limits);
        (d) density of compaction not affected by the number of passes.

    Most of the dam is in this zone.

ZONE 2. Filters.

    Grading. 5 per cent finer by weight than 0·074 mm.
               20–50 per cent finer by weight than 4·76 mm.
               100 per cent finer by weight than 380 mm.

400   U.S.A. DAMS

Graded silts, sands, gravels, cobbles, boulders up to 380 mm.
Compaction in 15 in. lifts, 4 passes, 8,000 lb. by vibrating rollers.
1,100—2,400 vibrations/minute.

ZONE 1. Clay-core red bluff material to be used.

Grading. 5 to 20 per cent finer by weight than 5 micron.
10 to 40 per cent finer by weight than 0·074 mm.
40 to 70 per cent finer by weight than 4·76 mm.
100 per cent finer by weight than 76·2 mm.

Compaction. To be in 10 in. layers when compacted by 8 passes of a towed 100 ton pneumatic tyred compactor.

The embankment was substantially completed in 1968.

ST. LAWRENCE RIVER IMPROVEMENT SCHEME*

In June, 1959, Queen Elizabeth and President Eisenhower opened the long-contemplated enlargement of the canalization scheme of the St. Lawrence River.

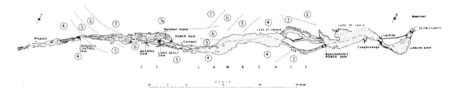

Figure 281. St. Lawrence River dams. The St. Lawrence Seaway is now controlled by five major works, the Beauharnois power dam near Montreal, the Barnhart power dam, the Long Sault dam, the Massena dam, and the Iroquois control dam, the four last-named being west of Cornwall. The Ordovician rock 'formations' are denoted by numbers of which No. 2, the Nepean, rests more or less horizontally on uneven pre-Cambrian (No. 1) metamorphic rocks, for example, crystalline limestone and dolomites, gneisses, quartzites and an assemblage of deformed igneous rock all without fossils. No. 2. The Nepean (Ordovician?), named after the town of Nepean, renowned for its sandstone quarry, is mainly a silica sandstone to 500 ft., calcareous at the top. No. 3. The March, named after the March township, comprises calcareous sandstone and dolomite which is water-bearing, fossiliferous and although only 30 ft. in thickness is persistent over a large area, including Beauharnois. No. 4. The Oxford, named after Oxford, Ontario consists of limestones and dolomites with calcite, over 200 ft. in thickness, slightly fossiliferous. The Iroquois control dam is within this area; few fossils. No. 5. The Rockcliffe, named after the park east of Ottawa is for the most part shale, believed to be 150 ft. in thickness with lenses of sandstone, slightly fossiliferous. No. 6. The St. Martin, contains shales and limestones to over 150 ft. and is fossiliferous in part. No. 7, 7a, The Ottawa, consists of shales below and limestone above forming a total thickness of 700 ft. The Power dam, Long Sault dam and Massena dam are in the synclinal area of Nos. 5, 6 and 7. The aggregate for the Barnhart, Sault, Massena and Iroquois dams was obtained from the dolomite limestone of the Beekmantown formation

By 1927 the 46 locks of the Welland Canal, to by-pass Niagara Falls, had been enlarged to take boats 800 ft. in length and 81 ft. in width with a draught of 30 ft. These locks enabled shipping to pass from the Great Lakes, Superior, Michigan, Huron and Erie (top water level, 572 O.D.) into Lake Ontario (top water level, 246 O.D.). The new scheme enables boats of this size and draught to proceed the additional 183 miles down the St. Lawrence

---

*My thanks are due to Mr. E. R. Peterson, engineering adviser, Canadian Section, International Joint Commission, for information on the geology of the St. Lawrence scheme, and to the librarian of the Institution of Civil Engineers for help in this matter.

from Lake Ontario (246 O.D.) to Montreal (20 O.D.), a fall of 226 ft., and thence to the Atlantic.

In addition to many miles of canals and dykes, several locks and bridges, five large dams have been built (*Figure 281*), namely: the Beauharnois power dam (Canada); the Barnhart power dam (Canada and U.S.A.); the Long Sault dam (U.S.A.); the Massena dam (U.S.A.); and the Iroquois control dam (Canada and U.S.A.).

The estimated cost of this fabulous scheme is $600 million.

The 80-mile stretch of the St. Lawrence River between the Beauharnois power dam and the Iroquois control dam passes over the edge of a syncline in the Ordovician rocks. The formation, mainly Ordovician, is as follows.

Locality of outcrop		Formation	Description
Ordovician	Sault Massena Barnhart Iroquois Beauharnois	7 Ottawa 6 St. Martin 5 Rockcliff 4 Oxford 3 March 2 Cambrian/Ordovician 1 Pre-Cambrian	Limestone, dolomite, sandstone, etc. Limestone and some shale Shale and some sandstone Limestone and dolomite Sandstone and dolomite Nepean sandstone Quartzite, gneiss, etc.

These strata are very free from faults and the dips are slight (under 10 degrees). The harder rocks produce 'rapids' in the river which are not navigable. There is a covering of Glacial Till (compact clays, sand, gravel) for an average depth of 20–30 ft. (the greatest depth recorded is 100 ft.).

Beauharnois Power Dam

The Beauharnois power dam (*Figure 282*) impounds some 80 ft. useful hydro-electric head between the levels of 150–70 O.D., and has a length of 2,836 ft. The mean flow at Beauharnois is 240,000 cusecs.

The dam is in the region of the Cambrian-Ordovician sandstones (March and Nepean formations) overlain with a glacial deposit of clay, underlain by the cream silica. As the sandstones are water-bearing, some difficulty was experienced with the foundations. (The Nepean formation, some 500 ft. in thickness, overlies the pre-Cambrian quartzites and altered rocks which occasionally appear, or nearly appear, at the surface in the St. Lawrence syncline which has a trend north-west–south-east.

Barnhart Island Power Dam

The Barnhart Island power dam (*Figure 283*), otherwise known as the Robert Saunders (Canada) and Robert Moses (U.S.A.) power dam (the frontier is on the middle of the dam), lies between Barnhart Island and the U.S.A. mainland.

It has a maximum height of 167 ft. (crest to foundation) and a length of 3,300 ft. The operating hydro-electric head is 83 ft., from 238 to 155 ft. O.D.

It is on the Ordovician near the outcrop of the Oxford overlain by the Rockcliffe and St. Martin formations in an area of glacial deposits. The

Figure 282. Beauharnois power dam. Showing the downstream side of the dam which has 39 power units in a length of 2,836 ft. It enables a head of 80 ft. of water to be used for power, the mean flow is 240,000 cusecs. The outcrop of Ordovician sandstones can be seen at the right end of the dam; 250 million yd.³ of rock were excavated for the foundations (15 per cent more than that required for the Panama Canal). The machinery is housed in a very high building to enable the generators and turbines to be lifted for repairs

Figure 283. Barnhart Island power dam. There are 32 generators which are housed in large pits, the covers of which can be lifted by the travelling gantry and the machinery taken up for repairs; the building comparable to Beauharnois is omitted. The right-hand gantry at the lower level is for the machinery, the left-hand gantry is for lifting the screens. The dam is on the Ordovician shales and limestones

Oxford (230 ft.) is largely a dark grey limestone with some shale, whereas the Rockcliffe (160 ft.) consists of grey shale with sandstone lenses, and the St. Martin an impure limestone.

Long Sault Dam

The Long Sault dam (*Figure 284*) is a large, curved, gravity structure $3\frac{1}{2}$ miles upstream of the Barnhart Island power dam and retains the level of the 100 square mile lake at 238 O.D. The height (crest to foundation) is

Figure 284. Long Sault dam. This dam is necessary to retain the water between the western end of Barnhart Island and the U.S.A. mainland. It is 2,250 ft. in length, 114 ft. in height, and has thirty 50 ft. control gates

114 ft. The length of the crest is 2,250 ft. There are thirty 50 ft. gates. The Rockcliffe and St. Martin shales and limestones of the Ordovician overlain by glacial deposits are in the vicinity.

Massena Dam

The Massena dam (*Figure 285a and b*) is a small, curved, concrete, gravity structure retaining the level of the lake with the Barnhart Island and Long Sault dams nearby, and preserving the River Massena for water supply. There is a small power-house below the dam.

Iroquois Dam

The Iroquois dam (*Figure 286*) controls the water level of Lake Ontario, 246 O.D., and is 28 miles above the Barnhart Island dam which works under a level of 238 O.D. Although the area of the gathering ground down to Lake Ontario is 295,000 sq. miles, the rainfall and flow does not vary much, as will be seen below.

	Rain (in./yr)	Flow (cusecs)
Maximum	43	323,000 (1870)
Minimum	25	154,000 (1936)
Average	33	241,000 (95 years)

The dam is in the region of the Oxford division of the Ordovician which consists largely of dark grey dolomite limestones with shales. The dam is

(a)

(b)

Figure 285. Massena dam. (a) A small dam necessary to control an industrial stream and maintain the level of the lake produced by the St. Lawrence and Long Sault dams. (b) When the gates are fully opened, remarkable vortices in the water are produced (upstream)

Figure 286. Iroquois control dam. Although the difference between the maximum and minimum flow (323,000 and 154,000 cusecs respectively) is not great, the dam is necessary to control a range of 8 ft. (246–238 ft. O.D.). There are 32 gates, one is shown in the up position, lifted by the travelling gantry. This dam is on Ordovician dolomites, the training bank on Glacial Till, in the distance on the right

founded on the dolomite. The approach bank at the end of the dam is on the glacial Till which was found to have a bearing pressure of 6 tons/ft.² with a settlement of 1 in.

NIAGARA FALLS CONTROL DAM
Technical Data of Niagara Falls

Average flow of river	222,000 cusecs
Minimum flow of river	176,000 cusecs
Height of Horseshoe Falls (Canadian side)	153 ft.
Length of Horseshoe Falls (Canadian side)	2,600 ft.
Length of American Falls	1,400 ft.
Denudation of Falls (maximum)	5 ft/yr
Geology of Niagara area	Glacial deposits on Silurian limestone on Ordovician shales and sandstones, generally horizontally bedded.

Figure 287. Niagara Falls scheme. The soft Ordovician limestones and shales below the hard Silurian limestone under the Falls

Typical Section of Niagara District

	ft.
Silts, sands, gravels, boulders	to 100
Silurian	
Guelph (Lockport) dolomite limestone ⎫ Gasport limestone. Decew shaley dolomite ⎭	100

Ordovician	ft.
Rochester shale	70
Limestones, shales, sandstone	30
Grimsby sandstone	40
Power Glen shale	30
Whirlpool sandstone	30

(a)

(b)

Figure 288. Niagara Falls scheme. The Horseshoe Falls have been made more uniform and the denudation reduced by the control jetty seen on the horizon on the right, in the lower picture (and in Figure 289)

Hitherto, the erosion of the rocks by the Falls has been, in places, up to 5 ft./yr along the length of 2,600 ft. of the Canadian Horseshoe Falls: the soft Ordovician shales below the 60 ft. ledge of Silurian limestone tend to disintegrate (*Figure 287*) causing the overlying limestone to collapse from place to place, making the flow over the Falls uneven (*Figure 288a*).

As the Falls are such a valuable source of revenue from tourists and hydroelectric power, it has been agreed between Canada and the U.S.A. that a minimum of 100,000 cusecs (that is, less than half the average flow) should be left to flow over the Falls during the hours of daylight for the tourist season, and 50,000 cusecs should be left at night and during the off-season period.

Hence, to retain the beauty of the Falls with the smaller quantity of water which is to flow over them in the future, and to arrest the uneven erosion, a control dam (*Figure 288b*) has been built half-way across the river on the Silurian limestone bed of the river, about a mile upstream of the Falls. This has a series of gates, each capable of being operated separately, which will 'spread out' the reduced flow of the Falls.

NIAGARA FALLS PUMPED STORAGE RESERVOIRS

In order to utilize the water which flows over Niagara Falls by night and during off-season periods there are two similar schemes: one on the Canadian

Figure 289. Niagara Falls scheme. The Niagara Falls control jetty looking from Canada towards the U.S.A. mainland is on the Silurium limestone, and extends half-way across the river at a point about a mile upstream of the Falls. As more than half the average flow over the Falls is to be diverted to generate electric power the appearance of the Falls would be ruined but for this dam. The view is downstream of the jetty, the Falls are a mile downstream on the left. Each weir is raised to a level which will ensure an even distribution of flow as seen in the photograph

side, with an estimated cost of over $350 million, and the other, on the United States side, which will cost a similar amount.

The Canadian scheme consists of the following.

(1) An intake about 2 miles above the Falls or a mile above the control dam shown in *Figure 289*.

Figure 290. Niagara Falls scheme. The reservoir (3,000 million U.S. gal.) under construction, an earthen embankment with rip-rap pitching to store the off-season and night waters of Niagara Falls

Figure 291. Niagara Falls scheme. The control dam for the pumped-storage reservoir which is under construction in the distance (top left). The water is lifted by pumps from the two 45 ft. diameter pipes which brings it $5\frac{1}{2}$ miles from the river, and when letting the water out during the day the same pumps generate power, in reverse, acting as turbines. The water then goes on to generate power at the main Adam Beck Power Station

(2) Two concrete-lined tunnels, each of 45 ft. internal and 51 ft. external diameter, 5½ miles in length, 250 ft. apart, and up to 300 ft. in depth.

(3) A canal, 2,200 ft. in length, conveying the water to the storage reservoir and thence to the drive-pipes of the turbines.

(4) A large open storage reservoir with a capacity of 3,000 million U.S. gal. (*Figure 290*).

(5) A control outlet and inlet works to the reservoir, filled by low-lift pumps which act in reverse as turbines when water is drawn out of the reservoir (*Figure 291*).

(6) A canal from the outlet works to the Adam Beck Power Station.

(7) An extension of the Adam Beck Power Station.

Figure 292. Niagara Falls scheme. The general view looking down the Niagara gorge showing the Adam Beck Power Station in the left bottom corner. The Lockport Dolomites are well seen above the talus on the right cliff on the United States side

The tunnels are driven through beds of limestone, dolomites, sandstones and shales under a thickness of 60–70 ft. of Rochester shale as follows.

Rochester Shale 60–70 ft. above

About 50 ft. {
 Irondequoit limestone . . Roof (280 ft. below surface)
 Reynales dolomite
 Negha shale
 Thorold sandstone
 Grimsby sandstone . . Floor (320 ft. below surface)
}

Power Glen Shale below

During the course of the work interesting experimental data were recorded.

It was found that the width of the tunnel (about half-way up), before the concrete lining was inserted, decreased by 1 in. in 100 days, and if the increase with time is plotted to a logarithmic scale and it is accepted that

Figure 293. Niagara Falls scheme. The water is brought from the storage reservoir (Figures 290 and 291) in a canal 2½ miles in length to the drive pipes of the Adam Beck Power Station No. 2 at Chippawa. As these Ordovician strata of the cliff are so unstable they have to be covered with a blanket of concrete (compare with figures 287 and 294)

Figure 294. Niagara Falls scheme. Preliminary foundation work in progress for the American pumped storage scheme, September, 1958. The Lockport Dolomites of the Silurian are unbared under the Drift at the top of the section (above the talus). These limestones were quarried for concrete aggregate in the distance. The talus covers the alternating sandstones and shales of the Ordovician

the rate of movement goes on, the decrease in diameter of the tunnel would be $\frac{1}{4}$ in. in 30 years. The necessary concrete lining to resist this movement would require a crushing strength of 1,500 lb./in.^{2}, and therefore the 3 ft. thickness of concrete that was provided was ample to sustain this pressure.

The vertical movement at the crown of the tunnel was infinitesimal—0·05 in.—the Irondequoit limestone is very strong in compressive stress, about 15,000 lb./in.^{2}, and free from joints and fissures. The uplift of the Grimsby sandstone (14,000 lb./in.^{2}) in the floor of the tunnel was $\frac{1}{2}$ in.

A curious feature was that similar measurements taken in the second tunnel, 250 ft. distant, were only half those in the first tunnel, and it is thought that the first tunnel restricted the movement of the strata in the second tunnel.

The cliff overhanging the new Sir Adam Beck Power Station (*Figure 292*) is cut well back and the rock-face covered with a massive concrete wall (*Figure 293*), for although the rock is horizontally bedded, the glacial material at least might cause local falls and damage. This seems a very necessary precaution to protect the cliff from weathering, particularly as on the American side a power-house was partly destroyed, with loss of life, by a cliff fall.

The new American Lewiston power-house on the right bank is opposite the Sir Adam Beck Power Station, and *Figure 294* shows the beginning of the work (September, 1958) for the power-house, cliff, treatment and storage reservoir above.

REFERENCES AND BIBLIOGRAPHY
(U.S.A. AND ST. LAWRENCE)

BAUMAN, P. F. Limit height criteria for loose-dumped rockfill dams. *I.C.O.L.D.* **III**, 781 (1964).
DAVIS, C. V. *Engng News Rec.*, 161 82 (1958).
MERMEL, T. W. *Register of dams in the United States*. McGraw-Hill (1958).
U.S. National Committee. *I.C.O.L.D.* **IV,** 255 (1964).

Cheoah dam
    Alcoa's *hydro-electric developments in the Smoky Mountains* (1958).

Douglas and Norris dams
    LACY (JR.), F. P. and VAN SCHOICK, GARY, L. T.V.A. Concrete gravity dams. Uplift observations and remedial measures. *I.C.O.L.D.* **1** 487 (1967).

Hiwassee and Fontana dams
    LACY (JR.), F. P. and VAN SCHOICK, GARY, L., T.V.A. Concrete gravity dams. Uplift observations and remedial measures. *I.C.O.L.D.* **1** 487 (1967).
    U.S., T.V.A. Technical Report No. 22, pp. 449 and 410. *Geology and foundation treatment* (1949) Washington; Government Printing Office.

Chilhowee dam
    Alcoa's *hydro-electric developments in the Smoky Mountains* (1958).
    GROWDEN, J. P. Symposium of rockfill dams. *Amer. Soc. civ. Engrs*, papers 1743 and 1744, Vol. 84, P.O. 4 (1958).

Nantahala dam
    GROWDEN, J. P. Symposium of rockfill dams. *Amer. Soc. civ. Engrs*, papers 1743 and 1744, Vol. 84, P.O. 4 (1958).

Watauga and South Holston dams
    LEONARD, G. K., and RAINE, O. H. Symposium of rockfill dams. *Amer. Soc. civ. Engrs*, Paper 1736, Vol. 84, P.O. 4 (1958).
    U.S., T.V.A. Technical Report No. 22, pp. 371 and 357. *Geology and foundation treatment* (1949) Washington; Government Printing Office.

Lewis Smith dam
  SOWERS, G. F., and GORE, C. E. JNR. *Preconstruction field tests of embanking materials Lewis Smith dam*. Alabama Power Company (1958).

Missouri Basin dams
  BENNETT, P. T. Materials and compaction methods. Missouri Basin dams. *I.C.O.L.D.* **III,** 299 (1958).

Trinity dam
  U.S. National Committe. *I.C.O.L.D.* **IV,** 253, 262 (1964).

Oroville dam
  GORDON, B. B. and WULFF, J. G. Design and methods of construction of Oroville dam. *I.C.O.L.D.* **IV,** 877 (1964).
  GOLZE, A. R., SEED, H. B. and GORDON, B. B. Earthquake resistant design of Oroville dam. *I.C.O.L.D.* **IV,** 281 (1967).
  U.S. National Committee. Instrumentation. *I.C.O.L.D.* **IV,** 762 (1967).

St. Lawrence River dams
  MACCLINTOCK, P. Glacial geology of the St. Lawrence Seaway and power projects. *N.Y. St. Mus. Sci. J., Albany*, (1958).
  RIPLEY D. M., The St. Lawrence Seaway. *Engng J., Montreal*, 39.9 1134 (1956).
  PECKOVER, F. L., and TUSTIN, T. G. Soil and foundation problems (St. Lawrence Seaway). *Engng J., Montreal*, 41.9 69 (1958).
  WILSON, ALICE E. The geology of the Ottawa St. Lawrence lowland, Ontario and Quebec. Dept. of mines and resources. *Geol. Surv.*, 241 (1946).

Niagara Falls scheme
  HOGG, A. D. Some engineering studies on rock movements in the Niagara area. In *Engineering geology case histories*, No., 3 Ed. by P. D. Trask (1959) Geological Society of America.
  NIEMEN, E. J., and DOUGLASS, C. T. Concrete aggregate for St. Lawrence and Niagara projects. *I.C.O.L.D.* **I,** 597 (1961).

Tims Ford Dam
  PALO, G. P. The Safety of Dams from the point of view of Foundations. *I.C.O.L.D.*, **IV,** 166 (1967).
  U.S. NATIONAL COMMITTEE. Major U.S. dams put into design or commission. *I.C.O.L.D.*, **IV,** 737 (1967).

35

CANADIAN DAMS (British Columbia)

Important recent developments in British Columbia have included the construction of three interesting 'gravelfill' dams, Arrow, Mica and Portage on the Peace River.

The dams range in height between 190 ft. and 800 ft., the storage of the reservoirs vary between 7.1×10^6 acre-feet and 62×10^6 acre-feet; they are all founded on sands and gravels; they are all constructed with gravel, sands and clay but their sections are entirely different.

The accompanying table gives a comparison of the three dams and their sections are shown *Figures 295 to 297*.

Dam Construction	Arrow 1965–1969	Mica under construction 1965–1973 (approx.)	Portage (Peace River) 1964–1968
Maximum height	190 ft. 58 m	800 ft. 244 m	600 ft. 183 m
Foundation	Sands and gravel	Fluvial sands and gravels up to 150 ft. on micaschists	Sandstone, sands, gravel, boulders. Dipping 10° downstream
Length of foundation	about 2,000ft (750 m) (with upstream blanket)	over 3,000 ft over 900 m	2,500 ft 770 m
Basic geology	Grano-diorite L.B. 0 to 400 ft. below surface R.B. 100 to over 500 ft. below surface	Pre-Cambrian mica-schists and gneisses inter-bedded with granite	Cretaceous Dunlevy sandstones overlying Gething shales
Crest length level	2,850 ft 1,459 ft. O.D.	2,550 ft. 2,500 ft. O.D.	6,700 ft. 2,230 ft. O.D.
T.W.L.	1,444 ft. O.D. ⎱ 76ft	2,475 ft. O.D.	2,220 ft. O.D.
B.W.L.	1,368 ft. O.D. ⎰	1,700 ft. O.D.	1,630 ft. O.D.
Capacity of Dam	8.5×10^6 yd.³ 6.5×10^6 m³	40×10^6 yd.³ 30.4×10^6 m³	57×10^6 yd.³ 43.5×10^6 m³
Volume of Lake (Acre-feet)	7.1×10^6	20×10^6	62×10^6
Remarks	Dam constructed under water up to 1370 ft. O.D. with 300 yd.³ capacity barges.	As the chord/height ratio of the site is about 1/3, (800/2,350) an arch dam was seriously considered.	Very large reservoir. Grout curtain 350 ft. deep. Large central core with a volume of 12×10^6 yd.³

Arrow Dam

Figure 295a shows a vast deposit of sand and gravel in a moraine overlying grano-diorite on the left bank from 0 ft. to 150 ft. below the surface and on the right bank from 100 ft. to over 500 ft. below the surface.

The sand and gravel were deemed suitable for founding the dam but also for building it. The appearance of grano-diorite on the left bank at the surface was ideal for siting the flood control works. Water tightness of the reservoir is not a problem (*a*) because of the size of the river (Maximum flood recorded 183,000 cusec., 5,180 m³/sec.) and (*b*) although the maximum height of the dam is 54 m, the lowering in water level will not exceed 23 m (76 ft.). As the length of the foundation is in the order of 750 m (2,500 ft.), it is apparent that the hydraulic gradient under the blanket and dam will not exceed 3 or 4 per cent.

Because of the great depth of the river, the scheme would have entailed coffer dams over 100 ft. high if the dam had to have be built in the dry. From nine basic preliminary designs, ranging between a dam in the dry with coffer dams and in the wet by dumping the fill deposited through water, ascertained from small scale tests on the materials in the laboratory and large scale tests in the field, the design of the dam on *Figure 295b* was arrived at.

The tests showed that clay-till could be tipped through the river water without loss of fine material and when loaded it would consolidate.

It was found also that constructional problems were largely overcome by bringing the sloping impervious core as near to the upstream face as possible.

Barges of 300 yd.³ capacity were used for the under water dumping. Above water the dam was raised by conventional methods.

Other dams which have been recently constructed under water are the Jarkvissle rockfill dam in Sweden and a concrete dam at Lake Elvaga, Norway.

It would seem that such experiences as these could be of use for such schemes as the Wash, Morecambe Bay and Solway Firth barrages in Great Britain should they be proceeded with.

Mica Dam

The maximum height being 800 ft. and the crest 2,550 ft. in length, a ratio of 1/3, serious consideration was given to an arch dam. The superficial deposits of sand and gravel had a maximum depth of 150 ft., overlying what appeared to be satisfactory micaschists with intrusions of granite and gneiss. Preliminary estimates indicated little difference in cost between a concrete arch dam and a rockfill dam but further extensive investigation of the micaschist would be necessary. (Cf. Bort dam foundation, page 196).

A rockfill dam was adopted in accordance generally with *Figure 296*. This enormous structure is being constructed with sands and gravels, existing on the site up to 150 ft., which rest on Pre-Cambrian mica-schists and gneisses interbedded with granite into which an impervious core of till (clay) will be taken.

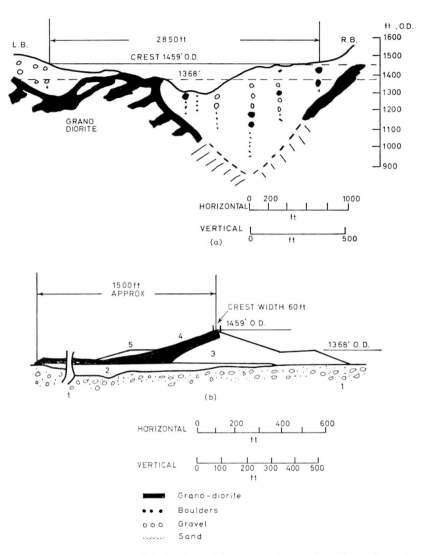

Figure 295. (a) Arrow dam. This dam is founded on moraines in two valleys filled with gravel, sand and some boulders. The filling extends to a depth of 150 ft. in the left and to over 500 ft. in the right valley. This has been proved by several boreholes. (b) The Arrow dam completed in 1969, 190 ft. in height, is composed of sand and gravel with a clay core of till and overburden. The dam was built under water from 1267 ft. O.D. to 1368 ft. O.D. and thence to 1459 ft. O.D by conventional methods

1. River bed sands and gravels.
2. Sand and gravel-fill to replace poor material.
3. Sand and gravel.
4. Sand, till, and overburden for clay core, (shown in black).
5. Wave and scour protection, grano-diorite.

Top material for the rest of the dam will be replaced down to 60 ft below the surface with sound sand and gravel near the till core and sound schists and granites will form the main embanking material compacted with vibrating equipment.

The dam is expected to be finished by 1973.

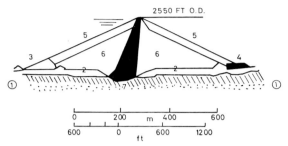

Figure 296. The Mica dam is 800 ft. in height and should be completed in 1973. The clay core will be founded on the Mica-schists and above these strata it will be supported by sands, gravels and fine granites, gneisses and schists.

1. Pre-Cambrian mica-schists with intrusions of granite and gneiss (shaded).
2. Overburden, sands and gravels replacing unsuitable material and layers of fine sand (unshaded).
3. Upstream coffer dam.
4. Downstrdam coffer dam (in black).
5. Compacted rockfill of biotite six schists and fine gravel, gneisses and granites.
6. Compacted sand and gravel.
7. Compacted core of clay and till (in black).

Portage Dam

The dam foundation consists of Cretaceous sedimentary rocks of alternating sandstones and shales with occasional conglomerates and coal seams which dip gently downstream beneath the dam, the strike being almost parallel with the dam (*Figure 297b*).

The permeability of these strata decreases with their depth but for some layers the permeability is unacceptable. The Dunlevy series, chiefly thick sandstones, overlie the Gething series which are predominately shales. Some of the permeable sandstones in the reservoir area outcrop downstream.

A full size grouting experiment was undertaken to investigate an area of the grout curtain 200 m in length and 100 m in depth as shown on *Figure 297*.

This area was tackled in three zones:

0– 20 m depth below surface, very permeable
20– 70 m rather permeable
70–100 m impermeable except for the odd coal measure and conglomerate seams which did not appear to be persistent or appear downstream.

The area was tested with 16,000 m of boreholes with various grouts and pressures, water tests, internal photographs of boreholes, tunnels giving

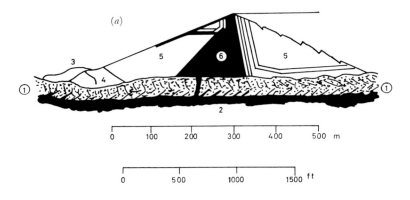

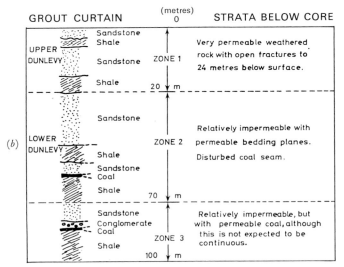

Figure 297. (a) Portage dam, Peace River. This dam is 600 ft. in height, its volume is 57×10^6 yd^3 and the volume of the core is 12×10^6 yd^3. It is founded on approximately 20 m of Cretaceous sandstones underlain by 50 m of semi-permeable material. The lower 30 m are mainly shale.

    1. Cretaceous sandstones and shales.
    2. Grout curtain 100 m deep.
    3. Coffer dam.
    4. Rockfill.
    5. Random silt, sand, gravel.
    6. Impervious core; processed silts and sands.

(b) Upper cretaceous strata in which some 16,000 m of experimental boreholes drilled to a maximum depth of 100 m took on an average 112 kg/m of grout injected up to a pressure of 53 kg/cm^2. It is inferred from these experiments that the overall impermeability will be increased from $10^{-3}/10^{-4}$ to 10^{-5} cm/sec.

visual evidence of grouted areas. It was estimated that the impermeability of the system could be increased from $10^{-3}/10^{-4}$ cm/sec. before grouting, to 10^{-5} cm/sec. after grouting.

BIBLIOGRAPHY

Arrow dam
  GOLDER, H. Q. and BAZETT, D. J. An earth dam built by dumping through water. *I.C.O.L.D.* **IV,** 369 (1967).
  NANCARROW, D. R. Canadian National Committee. Arrow Project. *I.C.O.L.D.* **IV,** 609 (1967).

Mica dam
  NANCARROW, D. R. Canadian National Committee. *I.C.O.L.D.* **IV,** 610 (1967) and **IV,** 22 (1964).

Portage Mountain dam
  BENKO, K. F. Large scale experimental rock grouting for Portage Mountain Dam. *I.C.O.L.D.* **I,** 465 (1964).
  NANCARROW, D. R. Canadian National Committee. Portage Mountain Development. *I.C.O.L.D.* **IV,** 611 (1967).

36

AUSTRALIAN DAMS

An interesting review of dams in Australia is given in the proceedings of the 9th Congress (1967) of the Australian National Committee of Large Dams (Vol. IV, page 545).

Rockfill dams

During the period 1962–66 there were 25 dams completed or under construction and 4 proposed. Those dams over 50 m in height (proposed, under construction and completed) are detailed in the table.

ROCKFILL AND EARTH DAMS

Year of completion	Dam	Max. height ground (m)	Crest length (m)	Foundation	Remarks
—	Talbingo (N.S.W.) (Snowy Mts.)	155	701	Rhyolite	Proposed
—	Cethana (Tasmania)	108	244	Quartzite	Proposed
—	Blowering (N.S.W.)	101	808	Quartzite and Phylite	Under construction
1966	Geehi (N.S.W.)	91·4	265	Granite gneiss and diorite	Completed
—	Wyangala (N.S.W.)	85·6	1,510	Embrechite	Under construction
—	Corin (Victoria)	71·6	282	Sedimentary	Under construction
—	Burrendong (N.S.W.)	67·1	1,110	Gravels	Under construction
—	Jindabyne (N.S.W.) (Snowy Mts.)	67·1	335	Granite	Under construction
—	Kangaroo Creek (South Aust.)	57·9	178	Schist and gneiss	Proposed
1965	Perservance Creek (Queensland)	51·8	197	Granite	Completed

CONCRETE DAMS

Year of completion	Dam	Max. height ground (m)	Crest length (m)	Foundation	Remarks
—	Wuruma (G) (Queensland)	43·9	350	Agglomerate	Under construction
1965	Island Barn (G) (N.S.W.) Snowy Mts.	42·7	146	Granite with weak zones which disintegrate on exposure	Curved to fit the foundation rock
—	Devils Gate (A) (Tasmania)	80·5	119	Massive Cambrian chert of good quality	Double curved arch. Under construction
1962	Myponga (A) (South Aust.)	47·5	226	Dolomite limestone with cavities on S. abutment and phyllitic slates on N. abutment	Arch
1967	Meadowbank (B) (Tasmania)	37·8	244	Alternating beds of Triassic Sandstone and mudstone with potential sliding zones of breccia and fault zones	Buttress
—	Cluny (P) (Tasmania)	30·5	204	Dolerite	Under construction
1949–65	Clark P (part) (Tasmania)	67·4	378	Raised 6·12 m by cables in concrete 1965	Completed

G = Gravity.   A = Arch.   B = Buttress   P = Prestressed

The Warrangamba Dam

Several notable concrete dams were constructed before 1962 e.g. Burrinjuch, Ellcumbene, Hume, Catagunya and Warragamba. The latter controls the water supply of Sydney, and is the subject of a paper by T. B. Nicol (see Bibliography). It is a straight gravity concrete dam 450 ft. in height with a crest 1,150 ft. in length, about 40 miles from the coast, and impounds a reservoir on the River Warragamba.

The capacity of the reservoir is $452,500 \times 10^6$ gallons or 1,665,000 acre-feet. Population to be supplied in 1980, will be 2,630,000. Yield 263 m.g.d. Although the crest/height ratio is about $3\frac{1}{4}$ the Triassic sandstone on which it is founded is 'weak'—having a Young's modulus of 500,000 lb./in.^{2}.

The river gorge is about 500 ft. deep at the dam site, the river bed is 45 ft. O.D., cut into the Narrabeen Group and Hawkesbury Sandstone. The Permian appears in the river bed six miles upstream and the Carboniferous appears over large parts of the gathering ground.

The generalized section of the Triassic rocks in this area is as opposite (page 421).

Figure 298 indicates Chocolate shales at the bottom of the valley about 100 ft. below the base of the Hawkesbury Sandstone. The fault in the centre of the valley is filled with 18 ft. of crushed rock at river bed level—the dip is generally 25 degrees downstream.

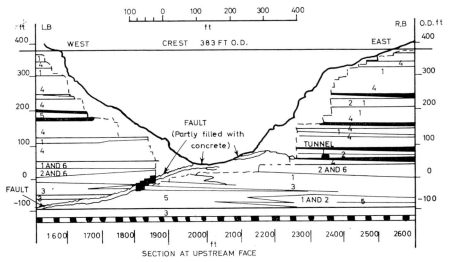

Figure 298. Warragamba dam. The dam is on Triassic Sandstone with intervening shales in a gorge with a chord/height ratio of $3\frac{1}{4}$. A straight concrete gravity dam, 450 ft. in height was adopted for the strata were too jointed, soft, faulted, permeable and with an inadequate modulus of elasticity for an arch dam. A rockfill dam was impracticable because the sandstone of the district was too soft for a dam of such a height and a concrete spillway was necessary in any event. The strata at the dam includes a major fault.

(1) Fine-grained sandstone.
(2) Coarse-grained sandstone.
(3) Very coarse conglomerate.
(4) Current-bedded sandstone.
(5) Shaley laminations.
(6) Shaley breccia.

Shale seams in black; the bottom shale below (3) is known as the Chocolate Shale. The broken line denotes the foundation level of the dam.

GENERALISED SECTIONS OF THE TRIASSIC ROCKS

	ft.	
Wianamatta		
Ashfield Shale	50+	Black shale
Silverdale Sandstone	30–80	Fine sandstone with current-bedding and thin shales
Hawkesbury		
Sandstone	600–700	Fine and coarse sandstone current-bedding shale slump breccias. Thin conglomerate bands and lenticular beds of fine and coarse sandstones with quartz and pebbles
Narrabeen Group	30–100	Mudstones and grey shales, fine and coarse sandstones with quartz conglomerate (20 ft.) and lenticular grey mudstones
	5–20	Chocolate shale
	20–50	Fine to coarse sandstone
	5–10	Chocolate and grey shale
	50–70	Fine to coarse sandstone and conglomerate beds
	300+	Fine sandstone with many shale beds and shale laminae

Reasons for a gravity dam

The gravity dam was adopted because:

(1) There were horizontal seams with vertical joints.
(2) There were numerous shale bands and some clay seams.
(3) There was the faulted zone on the left bank (*Figure* 298).
(4) The Trias sandstones were very permeable. (We know this well in England as their permeability gives one of our major underground water resources.)
(5) The low modulus of elasticity of the sandstone is not suitable to take arch thrusts. This the author enlarged upon, in discussion, showing that there was no precedent even for the alternative of a rockfill dam of such a height with sandstone of adequate quality near the site.
(6) A large concrete spillway was also necessary for carrying 354,000 cusec and,
(7) Concrete dams had been built in the Sydney area before, although they were not so high. (The thin arch dams for Sydney described at the Institution of Civil Engineers in 1909 in a paper by L. A. B. Wade were satisfactory and are still in use.)

Grouting

The grouting of the Triassic sandstones for the dam is extensive:

(*a*) *Intermediate pressure curtain.* When the concrete of the dam had reached a height of about 35 ft. across the valley holes 10 ft. apart, were drilled diagonally under the heel of the dam to reach a depth of 75 ft. below foundation level.

(*b*) *High pressure curtain.* This curtain is to provide a high impermeable zone to reduce leakage and uplift under the dam. Holes were taken to about 150 ft. below the foundation, about a third of the maximum height of the dam. Some holes were deeper and were drilled from galleries after 200 ft. height of concrete had been placed, spacing on left abutment generally 5 ft. on the faulted side, 10 ft. under the spillway in the fine sandstones and 20 ft. in the fine and coarse sandstones and shales of the right embankment.

(*c*) *Wing grouting.* Wing grouting carried out beyond the right and left abutments to reduce leakage around them, is work which went on after the dam was finished. Holes so far have been spaced at 40 ft. and were driven from the surface.

(*d*) *Blanket or contact grouting.* The object of the blanket grouting is

(a) to increase the sheer strength of 2,080 lb./in.^{2} for the fine sideritic or irony sandstone, typical of the river bed, and of 2480 lb./in.^{2} for the coarse alternating siliceous-kaolinitic sandstone of the abutments; and

(b) to increase the compression strength of 3,830 lb./in.^{2} for the fine river bed sandstone and 5,970 lb./in.^{2} for the fine-coarse abutment sandstone.

(*e*) *Marcasite.* Under the central part of the dam a grouted enclosure was put some 400 ft. by 1,200 ft. with holes 130 ft. deep at 10 ft., centres to retain ground-water in the sandstone to prevent oxidation of Marcasite (Fe^{s2}) which could act on concrete as sulphuric acid. As far as could be ascertained the natural underground water level is thought to be retained above the Marcasite by the enclosing rectangular grout curtain. The area enclosed appears to be in shale breccia and coarse conglomerate.

QUANTITIES OF GROUT CONSUMED (1963)

Grout locality	Cement No. of bags	Boreholes (ft)	Bags per ft. of borehole	Purpose
(a)	15,313	27,057	0·57	Minor curtain
(b)	38,055	59,132	0·64	Deep curtain
(c)	48,304	29,524	1·64	Wing curtain
(d)	188,694	1,006,755	0·19	Contact grouting
(e)	45,366	47,502	0·95	Marcasite enclosure
Total	335,732	1,169,970	0·8 mean	

1 Bag = 94 lbs. or 1000 bags = 42 tons

Cethana Dam, Tasmania

An interesting investigation was carried out for the Cethana dam in Tasmania.

The dam proposed, which was to be an arch dam about 350 ft. (106 m) in height, in an attractive quartzite conglomerate valley with a width/height ratio of the order of $240/106 = 2\frac{1}{4}$, seemed to be ideal.

However, an intensive site investigation was undertaken. The site for the dam is on an anticline and the quartzite, particularly, was found to be much faulted. The investigation revealed weak bedding planes, shears and joint systems in the foundation rock and their direction indicated that the dam could slide downstream.

In the laboratory low shear strengths were found in the quartzite and conglomerate and, although the rock itself registered a modulus of 500,000 kg/cm^2, seismic values on site were between 50,000 and 250,000 kg/cm^2.

It was felt, therefore, that a rockfill dam would be a better proposition. This was because of the high cost involved (of an arch dam) in order to remedy the inadequacy of resistance to sliding and deformation possible in both quartzite and conglomerate and particularly the loss of time involved in overcoming these defects.

BIBLIOGRAPHY

Australian National Committee. Australian Large Dams 1962–1965. *I.C.O.L.D.* **IV**, 545 (1967).

NICOL, T. B. The Warragamba Dam. *I.C.E.* **27** 491 (1964).

BROUGHTON, N. O. and HALE, G. E. A. Foundation studies for Cethana arch dam. *I.C.O.L.D.* **I**, 142 (1967).

PART 3
APPENDICES

1. Conversions.	427
2. Relation between storage and yield of reservoirs.	429
3. Assessment of floods.	430
4. Principles of dam design.	431
5. *In situ* tests.	436
6. Seismic methods of testing the condition of rock underground.	438
7. Thixotropy. Clay grout.	440
8. Pore pressures in earth dams.	442
9. Pore pressures in concrete dams.	443
10. Uplift.	444
11. Deformation of earth dams.	445
12. Deformation of concrete dams.	446
13. Deformation of strata (after construction).	448
14. Settlement.	449
15. Approximate cost of dams.	451

APPENDIX 1

CONVERSIONS (APPROXIMATE)

British

1 cusec = 6·23 gal./second = 0·539 million gal./day = $\frac{1}{2}$ million gal./day (approx.)
1 million gal./day = 1·86 cusec
1 acre-foot = 271,300 gal. = 43,560 ft.^{3}
Yield in gal./day = in./yr × area of gathering ground in acres × 62
Storage in gal = storage in in. × area of gathering ground in acres × 22,610
1 gal. = 10 lb.
1 ft.^{3} = 62·3 lb. = 6·23 gal.
p.s.i. = lb./in.^{2}

British to American

1 British gal. = 1·2 U.S. gal.
1 cusec = 7·5 U.S. gal./second
1 acre-foot = 326,000 U.S. gal.

British to Metric

220 gal. = 1 m^3 = 1,000 l. 1 pint = 567/1,000 l. = $\frac{1}{2}$ l. approx.
220 gal./second = 35 cusecs = 1 m^3/second
1 cusec/1,000 acres = 0·0071 m^3/second/km^2
1 cusec/mile^2 = 0·011 m^3/second/km^2
1 gal./day/mile^2 = 1·75 l./day/km^2
1 mile^2 = 640 acres = 2·56 km^2
250 acres = 1 km^2
2$\frac{1}{2}$ acres = 1 hectare = $\frac{1}{100}$ km^2
1,000 acres = 4 km^2
1 ton = 1 tonne (approx.)
1 ton/ft.^{2} = 1 kg/cm^2 (approx.) = 1 atmosphere (14·2 lb./in.^{2})

Grout

1 cwt./ft. = 112 lb./ft. = 167 kg/m
1 cwt./ft.^{2} = 112 lb./ft.^{2} = 550 kg/m^2

1 Lugeon = 1 l./m of borehole/min. at 10 kg/cm^2
       = 0·067 gal./ft./min. at 142·2 lb./inch
       = permeability of 1×10^{-5} cm/sec.

Cement

1 cwt.   = 1·25 ft.^{3}
1 ft.^{3} = 90 lb.
1 m^3 = 1,440 kg
1 bag (U.S.A.)   = 94 lb.
1 barrel (U.S.A.) = 376 lb.
20 bags (Great Britain) = 1 ton
1 quintal (Italy) = 2 cwt = 100 kg

APPENDIX 2

RELATION BETWEEN STORAGE AND YIELD OF RESERVOIRS

For Water Supply to Maintain a Daily Reliable Yield

If the average annual run-off can be assessed, the relation between storage and yield of a reservoir can be obtained from charts based on experience of the behaviour of rivers.

On the Lapworth chart (*Manual of British Water Supply*, Institution of Water Engineers) for various proportions of run-off and storage, the yields are as follows.

Run-off (in./an)	Storage		
	5 in.	15 in.	25 in.
20	10·5	14·0	—
30	15	22·5	—
50	21	34	40
70	25	43	53

Examples

For a 30-in. average annual run-off (perhaps from an average annual rainfall of 50 in. and an evaporation loss of 20 in./yr) from a gathering ground of 1,000 acres, the following relationships between storage of the reservoir to give a reliable daily yield would hold:

(a) Storage 5 in., that is, 5 in. × 1000 × 22610 = 113 million gal.
yield 15 in. (from table), that is, 15 in. × 1000 × 62 = 930000 gal./day.
(b) Storage 15 in., that is, 15 in. × 1000 × 22610 = 340 million gal.
yield 22·5 in. (from table), that is, 22·5 × 1000 × 62 = 1400000 gal./day.
(For conversions see page 427.)

For Hydro-electric Schemes

The capacity of a reservoir is usually calculated to store the flow of the driest year, which is some 50 per cent of the average annual run-off.

APPENDIX 3

ASSESSMENT OF FLOODS

For Great Britain, the Institution of Civil Engineers Floods Committee has compiled records of actual floods for upland areas prior to 1932, and by 1959 many additional records became available chiefly based on records in Somerset.

The amount of flood is proportional to the size of the gathering ground and typical intensities are as follows.

Area of gathering ground (acres)	Approximate upland flood anticipated (cusecs/1000 acres)		Approximate lowland flood anticipated before 1959 (cusecs/1000 acres)
	Before 1932	Before 1959	
2000	1000	4000	300
5000	700	2300	200
10000	500	1800	150
15000	400	1400	125
20000	350	1000	100
50000	250	700	70
100000	200	450	50
250000	140	250	35

The overflows from reservoirs are less, depending on their areas. Thus a reservoir whose area is 5 per cent of the gathering ground, reduces the flow to half the natural flood when there is a depth of water of 3 ft. going over the overflow weir. Or if it is 10 per cent, the flow at the overflow of the dam is only a third of the natural flow of the stream entering the reservoir.

APPENDIX 4

PRINCIPLES OF DAM DESIGN

The following basic principles are intended to enable geologists to appreciate some of the engineering calculations involved in the design of earthen embankments and concrete dams as they affect the strata on which they are placed.

Earthen Embankments

The object of the following calculation is to ascertain:

(1) The steepest slope at which the embankment will remain stable; flat slopes may be more stable but they entail larger embankments and greater cost. In practice slopes vary between 1 to 1 and 1 to 6.

(2) The shear strength required of the material of which the bank is to be made in order to resist slipping.

(3) The shear strength required of the foundation strata under the embankment necessary to resist sliding.

Example

Various profiles or cross-sections at the maximum height of the dam are drawn, one of which is shown in *Figure 299*.

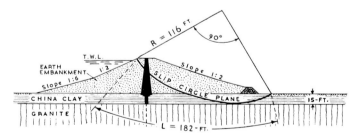

Figure 299

From many trial holes and laboratory tests on the strata it has been ascertained that there are poor strata for a depth of 15 ft. which have a shear stress of only 1·5 lb./in.² on an average.

Referring to the diagram (*Figure 299*), a slip plane, often assumed as a circle, is drawn through the dam and bad ground, along which the embankment is most likely to slip. The circle embraces: (*a*) the top of the dam;

(b) the junction between good and bad material 15 ft. below the surface; and (c) the bad ground beyond the downstream toe. The length of the slip plane (L) which resists the sliding is as short as possible consistent with a circular slip plane in the weakest material.

The radius of the slip plane is 116 ft. and the length $116 \times \dfrac{\pi}{2} = 182$ ft.

If the weight of the embankment material is assumed at 120 lb./ft.^{3} and the thrust is resolved along the slip plane, we get 104160 lb.

Hence the shear stress for each ft. of width to be resisted is:

$$\dfrac{104160}{182} = 572 \text{ lb./ft.}^2 = 4 \text{ lb./in.}^2$$

For safety, the shear strength of the China clay should be half as much again, that is, 6 lb./in.^{2}, and as the bad material in this example is only 1·5 lb./in.^{2} it should be replaced with something better or otherwise treated in order to sustain 6 lb./in.^{2}

The material of the bank should also be able to sustain a shear strength of 6 lb./in.^{2}

If no material can be found in the district for this, the section must be modified, perhaps by a flatter slope, or by a berm, or berms, by a reduction in height, by obtaining better material with which to construct the dam, or by selecting another site or a mixture of several alternative expedients.

Gravity Dams

A gravity dam is so-called because its stability depends upon its weight. Any section, or part of it, would stand by itself and support the weight of water (as shown, for example, by the part of the gravity section which stood by itself following the St. Francis dam disaster).

The basic profile of a gravity dam is a triangle where the width of the base in cross-section must be at least two-thirds of its height.

The broad principles of design are that the base must be wide enough to resist overturning and sliding.

The chief factors, therefore, are: (a) the dimensions of the section, that is, the height of the dam and the width at the crest, whether there is a road on top and whether the upstream face is vertical or battered; (b) the weight of the material of which the dam is made, that is, concrete 120–150 lb./ft.^{3}; (c) the vertical water pressures (Note. Water presses upwards as well as sideways and downwards) tending to lift the dam, due to leakage from the reservoir (or natural springs from the hills which may be laid bare in the excavation of the broad foundation). As it may be considered impracticable to release the water pressure entirely by drains, an allowance of up to 100 per cent of the depth of the water in the reservoir is made which has the effect of increasing the width of the base and hence the size of the dam; (d) the horizontal pressure of water which must not make the dam slide down the valley. These problems are worked out mathematically.

With concrete having a weight of 150 lb./ft.^{3}, the base of a triangular section, and assuming no pressure of water underneath (without any width at the top) is base = height/2·4 ft.

APPENDICES 433

With uplift pressure assuming the full head of water, base = height/1·4 ft. Thus, for different heights of dam the base widths in ft. are as follows.

	Height of dam (ft.)					
	50	100	150	200	300	400
Base width (no uplift)	20·8	41·7	62·5	83·3	125·0	166·7
Base width (full uplift)	35·7	71·4	107·1	120·9	214·3	285·7

Although most gravity dams are built straight, many of them are slightly curved (probably for aesthetic reasons) and although it is not wise, nor has it been the practice, to reduce the thickness or base width on account of any additional strength they may have gained owing to any arching effect, the slight curvature is believed to have substantial advantages. Thus: (a) increased stability due to slight arching effect is now believed to be a fact; (b) there may be stresses due to changes of temperature, that is, the arch effect may give some lateral play with contraction and expansion; (c) conditions are favourable for greater watertightness, that is, pressure by the water against the arch tends to close up cracks.

Buttress and Multiple Arch Dams

The principles of design for the buttresses follow very largely those of the gravity dam except that the loads to be carried by the foundations are 2–3 times greater than that of the gravity dam and the question of uplift can usually be eliminated.

If the strata under each buttress do not have the same bearing strengths, it is permissible for some relative settlement may take place because the heads of the buttresses are cantilevers, and where they join each other expansion joints are usually inserted.

In multiple arch dams where the buttresses support thin arches, there must not be any relative settlement between the buttresses otherwise the arches will crack. There are no expansion joints in an arch and if cracking occurs in the arches it may be possible as an expedient to reconvert them into monoliths by grouting with cement under pressure if the settlement is slight.

Thin Arch Dams

The basic formula on which a theoretical cylindrical arch dam would be designed is as follows.

$t = PR/S$. Where t is the thickness of the arch in ft.
P the water pressure in tons/ft.^{2}
R the radius of the upstream face in ft.
S the permissible compressive strength of concrete in tons/ft.^{2}
   Example for 100 ft. high dam with a radius of 100 ft. and a concrete stress of 10 tons/ft.^{2} considering 1 ft. strip of dam
$P = 100 \times 62·3 \div 2240 = 3$ tons/ft.^{2} (approx.)

R = 100 ft.
S = 10 tons/ft.²
Therefore: t at base $= \dfrac{3 \times 100}{10} = 30$ ft.

The equivalent gravity dam for a dam 100 ft. in height would require a base width of about 75 ft. Hence, an arch dam saves at least 50 per cent of concrete compared with a gravity dam.

In the above example, the radius of 100 ft. is suitable for a narrow valley; for a wide valley it would be too small for the ends of the arch would not reach the sides of the valley. As a dam of gravity section 100 ft. in height would require a base width of 75 ft., the maximum radius (R) for the arch to enable any economy to be made would be, for a 10 ton/ft.² strength of concrete:

$$\dfrac{3 \times R}{10} = 75 \quad \text{or} \quad R = 250 \text{ ft.}$$

For a 20 ton/ft.² concrete the limiting radius for economy would be 500 ft. So that a cylindrical arch having a concrete strength of 20 tons/ft.² would not be economical if the valley is so wide that it has to have a radius of over 500 ft.

This subject was very well explained in a remarkable paper (which has had but scant recognition) by Wade describing the thirteen curved concrete dams constructed in New South Wales, Australia. These dams ranged in height between 25 and 87 ft. and their thicknesses at the base varied between 9 and 24 ft. (In the animated discussion which followed, Sir Alexander Binnie said that in looking at the thin sections without reference to the plans of these dams showing that they were arches, they provided a blood curdling sensation!)

Wade at that time, made the following allowances for the crushing strength of concrete according to the aggregate used. This would affect the limiting radius of curvature to compare with the economy of the gravity sections and affect the thicknesses of the individual dams, as follows.

Permissible allowance of crushing strength of concrete (tons/ft.²)	Aggregate	Maximum radius for economy (ft.)
25	Granite	750
20	Conglomerate Altered slate schist Basalt	500
15	Sandstone	375
12	Sandstone	300
10	Sandstone quartzite	250

For the actual bearing on foundations he says these limiting pressures refer to the concrete of the dam, which, if it is greater than the crushing resistance of the rock at the abutments, it is a simple matter to splay the concrete where it meets the rock in order to reduce the pressure on the rock. He found that igneous rocks needed only washing and coating with cement while sedimentary rocks required the channelling of a shallow trench for the reception of the concrete.

Thick Arch Dams

The calculation necessary to ascertain the thickness for a thin arch ($t = PR/S$) applies to the thinnest sections only, and it has been found that arch dam sections can advantageously be constructed lying somewhere between thin arch and gravity sections. Such thick arch dams have been successfully constructed in valleys having a chord-height ratio of 5 or 6, but the mathematical analysis of these is complex and more reliance on model tests is being placed nowadays. The abutments of these thick arches must be several times more substantial than the foundations of a gravity or buttress dam.

Cupola or Dome Dams

Dome or cupola dams are curved, not only horizontally but also vertically, which results in a further saving of concrete when compared with the plain arch dam, because the thickness of a dome can be made even thinner than the arch. Calculations are complex, especially for unsymmetrical valleys which are sometimes made symmetrical by gravity concrete abutments or plugs in gorges, or both, to receive the thin cupola structures.

Where the chord at the crest of the dam is less than three times the maximum height of the dam, thin arches, and certainly cupola or dome dams, are possible, always providing that the strata at the abutments (and base) of the dam are absolutely above suspicion.

Pre-stressed Concrete Dams

By drilling bore holes into hard rocks, inserting steel cables and subjecting them to a pull of several hundred tons on the top of the dam, the dam can, so to speak, be anchored, and a gravity triangular section can be considerably reduced, say to 40 per cent of the volume of a gravity dam.

To withstand such stresses the foundation rock must again be absolutely sound, not only to bear 2–3 times the weight of a gravity structure, which is now much more concentrated, but also to withstand the pull of the cables.

REFERENCES

JAEGER, C. Hydraulic power plants. *Civil Engineering Reference Book*, **2** 183 Butterworths, London (1961).
WADE, L. A. B. *Min. Proc. I.C.E.*, **178** 1 (1909).

APPENDIX 5

IN SITU TESTS

It has been said that despite the wealth of available data—chiefly microgeological features—of the properties of rock specimens, much is conflicting, uninterpreted or inapplicable to civil engineering problems.

The chief single factor which determines the safe bearing pressure for a dam is probably due to tectonic features of the strata as a whole, *in situ*, such as the extent of bedding, jointing, faults and variability, and macrogeological features. This is why even a broad knowledge of the formation is such a good pointer to the conditions which are likely to be encountered at the chosen site.

Large scale *in situ* experiments, therefore, on the lines of laboratory experiments are in their infancy, but having in mind the full-scale experiments at Génissiat, Bort, Castillon, Beni Bahdel, Drift, Lamaload, and the Lewis Smith dam, herein referred to, and particularly the findings of the Malpasset dam, it may not be out of place to set out a few of the basic analytical methods of the laboratory beginning to be applied on the large scale in the field.

Modulus of Elasticity—Youngs Modulus

Basically, a static modulus of elasticity for rocks is defined as the ratio of 'stress' to 'strain' at crushing where: 'stress' is the compression usually expressed in lb./in.^{2} or kg/cm^2, 'strain' is the inches compressed per inch of material (or cm/cm) just a number.

The modulus, therefore, is the number of lb./in.^{2} at crushing per inch per inch of compression, and is a means of comparing the compressive strength of one rock with another, how much they compress for a given load.

Rocks like granite or basalt would have a high modulus like 10×10^6 or

$$\frac{\text{stress (load per area tested)}}{\text{strain (contraction per length tested)}} = \frac{10}{1/1\,000\,000} \text{ lb./in.}^2$$

whereas a limestone would have a lower modulus such as 6 or 8×10^6, a sandstone 3×10^6, and a shale 2×10^6 lb./in.^{2}

In the laboratory the modulus is ascertained on small specimens in a testing machine, for example, glass 8×10^6, steel 30×10^6.

Dynamic Modulus

Similarly, the dynamic and other moduli are obtained, wet and dry moduli and by vibration.

Seismic Modulus

For large-scale site experiments the velocity of sound through rock gives a measure of the bulk modulus of the rock, and this is what was recommended for ascertaining the cause of failure of the Malpasset dam.

H. Link has shown that the elasticity module of rock derived seismically is higher than that derived from static investigation, I.C.O.L.D. **I,** 833 (1964).

Poisson's Ratio

Another basic method of comparison between one rock and another is the determination of Poisson's ratio.

When a rock is compressed it contracts by so much in the same direction of the load and bulges or extends by so much at right-angles to the load.

For many (hard) rocks it is generally 0·25, that is, the length of compression parallel to the load is 4 times the length of extension or bulging at right-angles to the load.

Sandstones have a value of anything between 0·02 and 0·5, but 0·5 indicates a plastic rock and a Poisson's ratio of 1·0 is a liquid.

Shear Strength

Similarly, the shear strength of clay has been measured *in situ* as for the Lewis Smith dam, where a large shear box in two halves, which for each given vertical load the shear is measured horizontally by sliding one half of the box over the other.

REFERENCE

Judd, W. R. Effect of the elastic properties of rocks on civil engineering design. In *Engineering Geology Case Histories*, No. 3, page 53. Geol. Soc. U.S.A. (1959).

APPENDIX 6

SEISMIC METHODS OF TESTING THE CONDITIONS OF ROCK UNDERGROUND

As mentioned in Appendix 5 the seismic method, by ascertaining the wave velocity of an explosion through a rock, gives valuable indications of the state of the rock underground, i.e. whether it is faulted, dry, saturated, disturbed, altered in texture, or carstic. It is also a valuable method for ascertaining the relative density of concrete foundations and the relative efficiency of grout curtains and contact grouting.

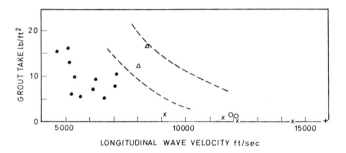

• • • Drained granite at Stithians
△ Saturated granite at Stithians
x Cleaved tuffs and hornfels at Wet Sleddale
○ Denotes cleaved mudstone at Clywedog
+ Denotes quartzose schist at Glen Finglas

Figure 300. Seismic method of testing rocks. By measuring the relative velocity of explosions in a rock a great deal can be learnt about its structure, cavities, faults. Recent work shows also a relationship between the seismic velocity and grout consumed in a grout curtain on various formations. The results of many igneous, sedimentary and metamorphic rocks lie within the curves of the figure

Dr. Knill has correlated the amount of grout consumed with the longitudinal seismic wave *velocity* in part of a grout curtain in Cornish granite at the Stithians Dam near Truro (*Figure 300*).

Under saturated conditions, this granite (softer than Guernsey or Aberdeen granite) which consumed about 15 lb./ft.^{2} of grout screen was found to correspond with a wave velocity of 8,000–9,000 ft./sec. Other velocities were found when the granite was drained and values of other velocities and grout consumptions were found in other formations thus:

Formation	Site	Seismic Velocity ft./sec.		Grout taken lbs./ft.^{2} of screen
Cleaved tuffs and hoinfels	Wet Sleddale	10,000	15,000	0–2
Cleaved mudstone	Clywedog	12,000		2
Quartzose schist	Glen Finglas	9,000	17,000	2–0
Granite saturated	Stithians	8,000	9,000	15
Granite drained	Stithians	4,000	8,000	5–15

Seismic testing is a useful tool especially:

(a) for testing the relative compactness of foundation concrete, grouting and contact grouting of existing dam foundations,

(b) for ascertaining and locating hidden faults which may be in proximity to or unsuspected outside the area which it would ordinarily be considered reasonable to explore.

The author is very grateful to Dr. Knill for reading this note on Seismic methods.

REFERENCE

KNILL J. L. *Conference on in situ investigations in soils and rocks*, London May 1969. Paper No. 8. The application of seismic methods in the prediction of grout taken in rock.

SCALABRINI, M., CARUGO, C. and CARATI, L. Determination in situ of the state of the Frera dam foundation rock by the Sonic Method, *I.C.O.L.D.* **I,** 585 (1964).

APPENDIX 7

THIXOTROPY—CLAY GROUT

An interesting account of the use of bentonite for sinking a cut-off trench without timbering is described by R. S. La Russo in Grouts and Drilling Muds (Butterworths 1963) for the Wanapum dam on the Columbia River, Washington (U.S.A.).

This dam impounds 88 ft. of water. The foundation is on alluvial gravels and sand 40 ft. to 190 ft. in depth resting on basalt. The permeability of the alluvium is estimated to be 1 cm/sec.

The Slurry Trench-Method

The following description of the Slurry Trench-Method is by Mr. La Russo.

'As the overburden is removed from the trench, bentonite slurry is added at a rate such that the trench always remains filled. The slurry is a mixture of high-yield sodium bentonite particles (high water-adsorbing characteristics) and water in proportions of approximately 0·6 lb./U.S. gallon. The mixing is carried out to ensure almost complete hydration of the bentonite particles. The resulting viscous fluid (15–20 cP at 20°C), or, more accurately, colloidal suspension, is slightly heavier than water (67–68 lb./ft.^{3}). As the top of the trench was 3 to 5 ft. above the water-table, there was an excess of fluid pressure of as much as $340 + 5·5 \times$ lb./ft.^{3} at any point along the wall of the trench, where x is the distance (ft.) below the water table. This excess fluid pressure forces the slurry into the void of the material at, and some distance beyond, the walls of the trench. Under the tranquil conditions existing there, the liquid slurry, being a *thixotropic* material, forms a solid soft gel which is relatively impermeable. This gel is the so-called 'filter cake'. The slurry in the trench remains fluid due to agitation caused by the excavation operations and the excavating tool passes readily through the liquid slurry to the bottom of the trench.'

'Apparently, the excess fluid pressure exerted on the walls of the trench, combined with the action of the filter-cake is sufficient to support the walls in their vertical positions. The actions and reactions involved are not fully understood, all classical approaches to the problem indicating that "cave-in" should result. However, quite obviously, instability did not occur for a heavy crane worked within 15 ft. of the edge of the trench, lifting loads of up to 12 tons and swinging them out over the trench without the slightest indication of instability.'

'Once a sufficient length of trench has been excavated backfilling operations can be started. The slurry displaced by the backfill is used in the extension of the trench.'

The maximum depth of 80 ft. was anticipated for the trench, but Mr. La Russo says that he feels that it could have been made deeper. The trench was excavated by 7 yd.³ drag-line excavator $7\frac{1}{2}$ ft. wide, the concrete deposited in the slurry filled trench by 4 yd.³ buckets.

The use of Thixotropic clay was applied in London to the under-pass at Hyde Park Corner and to a building foundation in Horseferry Road, where a long length of trench for a wall foundation was treated in sections of about 40 ft. in length, 20 ft. in depth and 4 ft. wide. Bentonite is much used for grouting and examples of its use in moraines are those at Backwater and Sylvenstein dams.

REFERENCES

LA RUSSO, R. S., *Grouts and Drilling Muds* Butterworth (1963).

BOSWELL, P. G. H. A preliminary examination of the thixotropy of some sedimentary rocks Q.J.G.S. **CIV,** 499 (1949).

APPENDIX 8

PORE PRESSURES IN EARTH DAMS

The measurement of pore pressures when clay material is consolidated is recent in Great Britain and the following example is intended to illustrate the procedure.

The embankment is under construction and is 50 ft. high.
The weight of bank is taken at 124·6 lb./ft.³
Earth pressure on 'tip' = $50 \times 124 \cdot 6 = 6,230$ lb./ft.²
If gauge reading = 80 ft. As the gauge is 20 ft. above the tip, the water pressure in the tip is 100 ft.
Therefore: water pressure in the tip = $62 \cdot 3 \times 100 = 6,230$ lb. which is equal to the earth pressure on tip.

The 'tip' is a porous pot where the pressure of water only inside the tip is measured and not the pressure of the bank on the tip outside. The tips are generally placed downstream of the core and grout curtain.

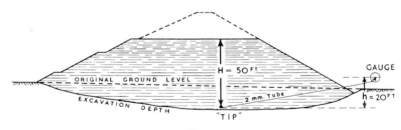

Figure 301

Therefore, the pore water pressure is 100 per cent of embankment loading and is considered to be excessive, and it is suggested that 66 per cent of the bank loading should not be exceeded. This can be done by varying the shape of the dam or the materials from which it is made.

It is suggested for the high earth dams in Switzerland that pore-pressure should be measured weekly during first filling, fortnightly for 2 or 3 years after filling, and thereafter monthly. The suggestions of fortnightly and monthly readings *after the reservoir has been filled* seem unnecessary for most dams, unless they refer to the assessment of pore-pressures *upstream* of the core and to the determination of the ultimate rate of *dissipation* of pressure downstream of the core any special reason or for pure research (see example, page 87).

APPENDIX 9

PORE PRESSURES IN CONCRETE DAMS

It is apparent that just as pore-pressure appears in earth dams, i.e. where the water in the reservoir leaks past (a) the core, (b) the grout curtain, (c) through the strata under the dam which results in, (d) pressing up by and through the dam itself. Then, similarly, it must be theoretically possible for the water in the reservoir to leak through (a) the dam, (b) the grout curtain, (c) the strata under the dam.

Professor A. W. Bishop raised this question (reported in Arch Dams, a review of British research and development I.C.E. 155 (1968)) by asking if pore pressures could be ignored for concrete dams. The answer is 'Probably Yes', but if pressure came against the dam from a ground-water source higher than the top water level of the reservoir as at Cow Green 'No'.

Dr. Knill also suggests that pore-pressures in concrete dams presumably are reduced where there are drainage or access galleries. In any case for concrete dams uplift pressure is of more importance rather than possible water pressure in the concrete of a dam.

APPENDIX 10

UPLIFT

Excessive uplift can arise when water from a full reservoir passes the curtain and trench. This may cause the dam to lift, turn over or slide particularly when the foundation strata are not naturally very permeable or have been made so by excessive contact grouting without being drained or when soil has been deposited against the downstream embankment.

Extreme cases are when a grout curtain and cut off completely stop all leakage; no dangerous uplift pressure under the dam arises unless there is pressure of natural water from the hill-sides to cause uplift). Or, by contrast, when the grout curtain and cut-off do not exist or hardly function and the strata are very permeable no dangerous uplift pressures may occur either. Leakage of course is another story.

If, on the other hand the strata are impermeable or nearly so and there is little leakage from the reservoir, drains may have to be drilled near the downstream side of the core in the strata under the dam to relieve dangerous uplift pressures at the risk of increasing the leakage.

Uplift is generally measured in a series of vertical boreholes penetrating about a metre into rock downstream of the cut-off. Perforated pipes inserted in the strata terminating in plain pipes are led to pressure gauges at a central spot in a gallery where the levels of water can be measured and recorded.

Figure 252a (see page 370) show a gallery downstream of the core for inspection, to house the gauges sufficiently high to allow plant for drilling additional drains to reduce uplift should the need arise.

In this case the need did arise and an excellent paper on uplift is published by F. P. Lacy Jr. and Gary L. Van Schoick of the Tennessee Valley Authority on four dams namely Norris (page 367), Hiwassee (page 371), Douglas (page 372) and Fontana (page 377), (I.C.O.L.D. I, 487 (1967)).

APPENDIX 11

DEFORMATION OF EARTH DAMS

Slips and slumps

Excessive movement in earth dams is probably due to failure to comply with soil-mechanics principles in design and on site to ensure that the shear strength is adequate and that the material in the bank is compacted at its maximum dry density for its optimum moisture content.

However with the object of forecasting slips, and to ascertain what is going on inside a dam, horizontal strains were measured recently in the Balderhead and Bough Beech embankments by means of an electrically stimulated vibrating wire, set between two blocks of concrete, about 6 ft. apart. Any increase of tension caused by movement of the bank is proportional to the increase in the number of vibrations per second of the wire. The wire is in an air-tight container filled with nitrous oxide which protects it from corrosion and any variation in vibration is transmitted through the wire to the recording instruments in the gauge house.

Similarly vertical deformation movements in an earth dam have been measured during construction in the Selset and Backwater dams. The apparatus consists of plates 12 in. square and $\frac{1}{4}$ in. thick through which a vertical tube passes. When an electric induction coil is lowered down the tube the depths of the plates are recorded by the coil and thereby any change of movement can be recorded.

APPENDIX 12

DEFORMATION OF CONCRETE DAMS

Strain Gauges

Strain gauges of the electrically stimulated vibrating wire type are commonly serted in concrete dams in Britain.

The basic principle of the strain gauge is that any change in pressure conditions in the dam changes the frequency of oscillation of the vibrating wire and each gauge may contain several wires oriented in different directions to obtain the duration of the maximum strain. This is comparable, inversely, to the constant frequency of oscillation of a tuning fork attached to a flywheel for measuring its variation in speed during a single revolution. (H. B. Ransom).

Temperature Gauges

Gauges for measuring temperature changes in the concrete of British dams have been installed for many years.

Weathering of Concrete

The Concrete Research Association, is a body which can supply information on slags, fly-ash, sulphurous, gypsiferous and other suspect aggregates requiring special research as to their suitability for resisting acid waters.

Movement of Dam

For detecting movement of a dam, surveying instruments of a high degree of accuracy are available and applied externally give some information on deformation, but they are not easy to use. The amount of deformation of a dam when full and when empty can be gauged much more easily by a Pendulum.

The basic principle of the Pendulum is a wire suspended near the crest inside the dam in still air in a shaft and a plumb-bob in water or oil at the base, with a scale for measuring the deviation from the vertical. This usually measure any tilt or deviation of the crest of the dam above ground.

On filling the reservoir, the tilt would be expected to be downstream if the dam was on an honest foundation.

APPENDICES

The necessity for testing deformation, particularly for high dams is to find out:

(1) The amount of movement of a dam when the reservoir is filled;
(2) Whether the movement returns to zero when the reservoir is emptied, when the movement is called *elastic*;
(3) Whether the movement does not return to zero, when it is called *inelastic*;
(4) Whether the movement continues indicating unstable conditions.
(5) Whether any change in deformation takes place over the years.
(6) The comparison between predicted deformation with actual deformation.
(7) Movement of the strata relative to the dam (see page 448).

APPENDIX 13

DEFORMATION OF STRATA (After Construction)

Inverted Pendulum

After the dam is constructed, effects on the strata can be indicated (by inference) by the inverted pendulum. The basic principle of the inverted pendulum is that the hanging-end of the wire is fixed in concrete at the bottom of a borehole in the rock and the 'bob' end floats about in a concentric trough of water with a scale to give the relative movement or deformation of the strata to the bottom of the dam (or vice versa), before and after filling the reservoir.

The objects of testing deformation of strata with the inverted pendulum, particularly with high dams, with or without deformation of the dam and/or the effect of temperature, is to measure any similar elastic or inelastic effects (outlined in Appendix 12), with the strata.

In short, the accumulation of records of behaviour of different dams on different formations will lead to interesting and useful data for forecasting the likely conditions for any dam on any strata in the future.

APPENDIX 14

SETTLEMENT

Settlement is due mainly to the slow natural consolidation of an embankment during a long period; thus a bank 100 ft. in height could sag 2 ft. in two years after construction and 4 ft. in ten years. If the overflow level were made 5 ft. below top bank level, the resulting freeboard would be only 1 ft.

An occasional surveying check on the levels of the permanent pegs ('Monuments') relative to the level of the spillway is generally all that is required to ensure good behaviour.

As a caution however it is sufficient to relate an actual case where a responsible engineer, well acquainted with an embankment failed to notice that the crest had sunk almost to the level of the overflow weir. His attention was drawn to it casually by a visiting engineer when his comment was 'Oh ah; yes, how extraordinary, I must lower the water immediately'.

Long term natural settlement is generally not of vital significance but for a six-line motorway over the crest of the Scammonden dam (about 250 ft. in height), the expected amount of settlement to be anticipated became important, not because of the dam, but because of the motorway, as settlement would mean constant repairs to the carriageways. The West Riding County Council carried out a full-sized investigation to ascertain the best method for quarrying, depositing and compacting the rock to ensure maximum consolidation and minimum settlement. The embankment for the combined motorway and dam is mainly the Pule Hill Grit (also called Midgley grit), a soft sandstone of the Millstone Grit Series of the Carboniferous formation and very similar to the Pennsylvanian Rocks of the Lewis Smith dam described on page 395.

These extensive experiments costing nearly £200,000 showed that the best compaction occurred:

(1) When a series of charges in large holes in front of a series of small holes in the quarry face proved to give the best fragmentation for subsequent compaction. Small charges behind large ones left the face of the quarry in better shape for the next blast with less overbreak and therefore prevent pieces (on the next blast) from being too large to compact and too expensive to break individually.

(2) When the material was spread in 3 ft. lifts; it was found best to tip it 3 or 4 yds. from the advancing edge of the layer and then to push it over the edge, by a bulldozer, i.e. not tip it over the edge direct from a lorry.

By pushing the material over the edge, the larger pieces went ahead and the voids were filled up with the finer material, whereas with direct tipping over the edge the larger blocks tended together to bridge over voids.

(3) With the particular explosives and best methods of blasting to give the optimum amount of fines for 3 ft. lifts.

(4) When either an $8\frac{1}{2}$ ton or 5-ton vibrating roller made 8 or 12 passes respectively. It is interesting to compare these results with those on page 395 where a 50-ton roller would give a permeability of $2\cdot5 \times 10$ cm/sec.

BIBLIOGRAPHY

WILLIAMS, H. and STOTHARD, J. H. Rock trials for Lancashire–Yorkshire Motorway. *I.C.E.* **36** 607 (1967) and 135 (Sept. 1967).

APPENDIX 15

APPROXIMATE COST OF DAMS

Formulae have been proposed by Mitchell, based on those suggested by Dr. Herbert Lapworth, namely, to give the cost of dams as a first approximation.

Earth Dam

$$\text{Cost} = £ \left\{ \underbrace{\frac{aLH^2}{7 \cdot 5}}_{\text{Dam}} + \underbrace{\frac{bLH}{3}}_{\text{Core}} + \underbrace{\frac{2cLD}{9}}_{\text{Trench}} + \underbrace{4000H + 40000}_{\text{Ancillary works}} \right\}$$

H is the mean height of the dam which is the area of cross-section of the valley from crest to ground level in ft.^{2}, divided by the length of the crest in ft., thus:

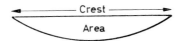

and, depending on the shape of the valley, H is usually between 0·75 and 0·6 times the maximum height of the crest above ground level.

- L = length of crest in ft.
- D = mean depth of cut-off trench in ft.
- a = cost/yd.^{3} of bank.
- b = cost/yd.^{3} of puddle in core wall of bank.
- c = cost/yd.^{3} of cut-off trench filled with puddle or concrete.

In 1970 the following values could be taken in Great Britain:

a = £2.5   b = £6.25   c = £18.75

Concrete Gravity Dam

$$\text{Cost} = £ \left\{ \underbrace{\frac{xLH^2}{72}}_{\text{Dam}} + \underbrace{\frac{xLW^2}{36}}_{\text{Road}} + \underbrace{\frac{yLD_1H}{36}}_{\text{Foundation}} + \underbrace{\frac{2zLD_2}{9}}_{\text{Trench}} + \underbrace{30000}_{\text{Ancillary Works}} \right\}$$

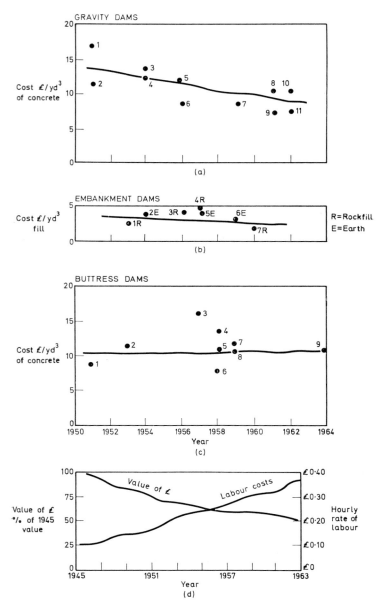

Figure 302. Trend in cost of dams during period 1950–1964. (a) Concrete gravity dams. (b) Earth or rockfill dams. (c) Concrete buttress dams. (d) Unskilled labour hourly rate including insurance, subsistance and other allowances. The numbers on the graphs refer to the table on page 454

APPENDICES

H = mean height of dam above foundation level in ft. (calculated as for earth dam as above).
L = length of crest in ft.
W = width of road, if any, in ft.
D_1 = mean depth of broad foundation below ground in ft.
D_2 = mean depth of cut-off trench below broad foundation in ft.
x = cost/yd.³ of mass concrete.
y = cost/yd.³ excavation
z = cost/yd.³ cut-off trench
In 1970 the following values could be taken in Great Britain:
x = £12.5   y = £1.25   z = £15

Trend in Cost of Dams

Since 1945 it has been difficult to make precise estimates of costs owing to the rise in cost of labour, the rise in cost of materials and the lowering in the cost of constructural processes for dams, e.g. excavation, muck-shifting and certain manufacturing processes. Many dams in the period 1950–1964 were built in Scotland where the physical and geological conditions may have some degree of similarity, and Mr. Fulton, and Mr. Dickerson in discussion, gave some interesting costs:

(1) Showing a fall in cost of concrete and earth dams and a slight rise in buttress dams (*Figure 302a, b, c*).

(2) Showing a rise in the cost of labour and a fall in the value of the £ (*Figure 302d*).

Some 50 dams have been constructed within 15 years in the north of Scotland, we have a useful guide to the trend in cost of dams. Even so the diagrams show a scatter and the buttress dam No. 3 has to be ignored because of 'difficult foundations'.

Trend in cost of dams 1950–1964

Some 50 dams have been built in this period in the North of Scotland and the estimated costs are for the dams alone, i.e. elimination of ancillary works such as power-stations, fish lifts, and where other items have been included in the same contract. What is believed to be a reasonable estimate of the cost of the dam at the date when it was finished is divided by the volume of the dam and the following curves are arrived at:

Figure 302a. The cost per cubic yard of a gravity dam shows a definite downward trend from £14 to £8 per cubic yard.
Figure 302b. The embankment dam tends to get slightly cheaper.
Figure 302c. The buttress dam tends to get slightly dearer.
Figure 302d. The value of the £ based on the per cent of its value in 1945; for comparison the unskilled hourly rate of labour including insurance, subsistence and other allowances is added.

The names of the dams on the diagrams are as follows:

	Figure 302a Concrete gravity dams Cost £/yd³			Figure 302b Earthfill or Rockfill Cost £/yd³			Figure 302c Concrete buttress dams Cost £/yd³	
Ref.	Dam	Volume yd³	Ref.	Dam	Volume yd³	Ref.	Dam	Volume yd³
1	Benevean	64,500	1	Col	44,817 (R)	1	Sloy	208,000
2	Mullardoch	286,000	2	Shira (lower)	70,390 (E)	2	Tarsan	56,000
3	Luichart	48,000	3	Quoich	391,600 (R)	3	Errochty	248,000
4	Shira (Lower)	16,830	4	Fannich	81,000 (R)	4	Lubreoch*	159,300
5	Loyne	66,000	5	Vaich	435,000 (E)	5	Lawers	118,700
6	Cluanie	232,000	6	Orrin	218,000 (E)	6	Lednock	128,000
7	Orrin	233,000	7	Breaclaich	275,000 (R)	7	Shira (Main)	268,600
8	Glashan	47,300				8	Giorra	92,450
9	Bhlaraidh	13,500				9	Cruachan	120,000
10	Loichel	20,000						
11	Nant	37,000					*Difficult foundations	
	1950–1964			1950–1964			1950–1964	

REFERENCES

DICKERSON, L. H. and MORTON, L. R. Review of some recent dams built for Scottish conditions. *I.C.O.L.D.* **III,** 321 (1961).

FULTON, A. A. and DICKERSON, L. H. I.C.E. in the discussion of their Paper 'Design and constructural features of hydro-electric dams built in Scotland since 1945'. *I.C.E.* **29** 713 (1964) and **33** 474 (1965) (Discussion).

MITDHELL, P. B. Reservoir site investigation and economics. *I.W.E.*, **5,** 445 (1951).

PIPPARD, A. J. S. British National Committee of International Commission on Large Dams. *I.C.O.L.D.*, **IV,** 121 (1964).

INDEX

For ease of reference, where dams are listed under countries, the type of dam is denoted by initial letters as follows: (G) *gravity (concrete or masonry);* (B) *buttress;* (MA) *multiple arch;* (A) *concrete arch;* (E) *earth;* (R) *rockfill;* (P) *prestressed whole or in part.*

Abutments
  artificial concrete blocks
    Esch-sur-Sure Luxembourg, 283
    Gmeund, Austria, 147
    Kops, Austria, 147
  collapse
    Les Cheurfas dam, 313, 314
    Malpasset dam, 59–62, 167
    St. Francis dam, 56
  damage
    bomb damage, Clywedog, 104
    earthquake, Blackbrook, 53
    earthquake, Pontéba, 50
  strata strengthened
    Ancipa dam, Miocene/Eocene, 270
    Barcis, Cretaceous limestone, 264
    Bort, Gneiss/mica schist, 193–200
    Castillon, Jurassic limestone, 186–189
    Chaudanne, Jurassic limestone, 189–190
    Fontana, Lower Cambrian, 374, 376
    Les Cheurfas, Miocene limestone, 311, 313
    Ridgegate, Millstone grit, 27
Achensee lake (Austria), 143, 148, 150
Achten, A., 162
Adam Beck Power Station, 409–411
Aigas dam, 120
Aigle dam (France), 167, 190, 191
Albigna dam (Swiss), 295, 298
Ailleret, P., 28, 102, 181
Ain Skrouna spring, reverse action of, 334
Ainono dam (Japan), 358
Aiken, P. L., 130
Akiba dam (Japan), 359
Alabama, 12, 391
Algerian dams, 311–346
  Bakhadda, (R) Miocene 20m on Jurassic, 311
  Beni-Bahdel, (P, MA) U. Jurassic, grits and marls, 5, 10, 330–335
  bibliography, 345–346
  Bou Hanifia, (R) Miocene marl and grits, weak, 335–341

Algerian dams *continued*
  Col Nord, (MA) Mid. Jurassic, marls and thin grits, 330–335
  D'Erraguéne, (P, B, MA) Lower Chalk schists, 311
  Digue de la Route (MA), L. Jurassic, clays and limestones, 330–333
  D'Iril Emda, (R) Cretaceous (weak), 311
  Fergoug, (Perrégaux) (P, G) Lower Miocene, 328–330, 339
  Fodda, (G) U. Lias limestone, 5, 326–328
  Foum el Gherza, (A) Cretaceous limestone, 312
  Foum el Gueiss, (R) Oligocene 312
  Ghrib, (R) Miocene, grit and marl, (weak), 12, 38, 320–325
  Hamiz, (G) Eocene and Permo-Triassic grits, 312
  Ksob, (MA) Eocene, 313
  Les Cheurfas, (P, G) Miocene limestone, 313–320
  map of, 312
  Perrégaux, (P, G) Lower Miocene (case history), 328–330, 339
  Pontéba, (G) Miocene, earthquake damage, 50
  Sarno, (E) Pliocene/Oligocene, sandy marl, 341–345
  Zardézas, (G) Miocene clay and conglomerates, 313
Algerian reservoir enclosing dams
  Beni Bahdel dam, 331
  Col Nord dam, 330
  Digue de la Route dam, 330–335
Allt-na-Lairige dam, 10, 88–90
Almansa dam (Spain), 286
Alno dykes, 289
Alps, 225, 228–231, 236, 250
Alternative types of dam suggested, Piave di Cadore dam, 237
Alto Càvado dam (Portugal), 284
Alto Rabagão dam (Portugal), 10, 284
Alvito dam (Portugal), 284
Alwen dam, 10, 76, 100

455

INDEX

Amphibolites (Gneiss, mica schists), 4
Ancipa dam (Italy), 270, 271
Andesite, see specific countries
André Blondel dam (France), 214
Anticlines, *see specific dams and valleys*
Arch dams (*see also under specific countries*)
   Algerian, 311–346
   Austrian, 142–161
   British, 75–141
   French, 166–224
   Italian, 236–277
   Japanese, 358–362
   Portuguese, 284, 285
   Spanish, 286–288
   Switzerland, 291–306
   Turkey, 307–310
   U.S.A. 363–400
Argal dam, 3, 76, 89–92
Arguis dam (Spain), 286
Arimine dam (Japan), 358
Arrow dam (Canada), 413–415
Artic flora, 70
Ashdown Sand, nature of, 13, 19, 78–82
   source of embanking material, 78–82
Ashop valley, 63
Aswan dam (Egypt), 11
Aswan high dam (Egypt), 11
Asymmetrical dome dam, example, 262
Atkinson, Alan, 97
Aussois dam (Swiss), 25
Austin dam (U.S.A.), 58
Australian dams
   bibliography, 423
   Cataract, (G) Coal Measures, 42
   Cethana, (R) Quartzite, 423
   Warrangamba, (G) Trias sandstone, 420–423
Austrian dams, 142–161
   Achensee Lake, (E) Trias limestones, 143, 148, 150
   Bächental (Dome) Trias dolomite, 148
   bibliography, 161
   Biel, (R) Moraine, 145
   Burg, (G) Mica-schist, 150
   Danube development, 13, 143, 159–161
   Diessbach, (R) U. Trias limestone, 154
   Dobra, (A) Gneiss and mica-schist, 158
   Drossen, (A) various schistase rocks, 143, 152
   Durlassboden, (R) Schist and moraine, 148
   Freiach, (E) Trias, Lias limestones and shales, 157
   Gepatch, (R) Gneiss and talus, 70, 148–149
   Gmuend (Gerlos), (A) Quartz schist, 70, 143–146
   Grossraming, Trias sandstone and marl, 143, 156
   Jochenstein (B) Granite, 13, 143, 159–161
   Kaprun dams, 143–150

Austrian dams *continued*
   Kops, (G, A) Gneiss, 146, 147
   Limberg, (A) Calc. mica-schist, 143, 150–1
   Lunersee, (G) U. Trias dolomite, 143, 145
   map of, 143
   Margaritze, (R, G) Calc. mica schists, 143, 152–154
   Möll, (A) Calc. mica schist, 143, 152
   Mooser, (R, G) Calc. schist and other schists, 143, 152–154
   Mühldorfersee, (G) Gneiss, 157
   Mühlrading, Miocene, 143, 156
   Ottenstein, (A) Miocene/Granite, 156, 159
   Rosenau, (G) Cretaceous, 143, 156
   Rotgüldensee, (R) Gneiss, 156
   Salza, (A) U. Trias Dolomite, 156, 158
   Silvretta dams, (G, R) Moraines mica-schist, 145
   Spullersee, (G) Lias Marls, 144
   Staning, Miocene, 143, 156
   Ternberg, Trias sandstone and marl, 143, 156
   Vermunt, (G) Gneiss, 143, 145
   Ybbs-Persenbeug, 143, 159
Austrian reservoirs
   Möll, 155
   Moosenboden, 154
   Wasserfall, 150
Austro-German development, 159–161

Bachental (Austria), 148
Backwater dam, 7, 110, 113
Baitings dam, 6
Bakhadda dam (Algeria), 311
Baldwin, A. B., 63–67
Banks, J. A., 72, 123, 125
Barbier, R, 223, 224
Barcis dam (Italy), 9, 143, 262–266
Barnhart Island Power dam (St. Lawrence Seaway), 40, 402
Barrages (*see specific names and sites*)
Barrage, tidal, estuary, 220–223
Basalt pitching, 234
Basle, Rhine development, 213
Basse Alps, 186–190
Bassenthwaite lake, 97
Bate, Stanley R., 29
Bates, John S., 50
Bean Nachram dam, 120
Beauharnois dam (St. Lawrence Seaway), 13, 401, 402
Behaviour, *see* Deformation, Deviation
Belgian dams, 162–165
   Butgenbach, 162
   La Gileppe (G), Devonian, 162–163
   map of, 167
   Robertsville, 162
   Vesdre, (G) Cambrian, 164–165

INDEX

Bendelow, L., 63–67
Beni-Bahdel dam and reservoir (Algeria), 5, 10, 330–335
Bennett, A. L., 48
Bentonite, 440
Bergamo laboratories, 236
Bergeforsen dam (Sweden), 289
Berlin, Stettin Canal, leakage from, 28
Bibliography and References
  Algerian, 345, 346
  Australian, 423
  Austrian, 161
  British, 22, 31, 43, 49, 55, 62, 72, 140, 141
  Canadian, 418
  French, 223–224
  German, 233
  Irak, 350
  Italy, 276–277
  Japanese, 362
  Jugoslavian, 279
  Pakistan, 357
  Portuguese, 285
  Rhodesian, 353
  Spanish, 288
  Swedish, 290
  Swiss, 306
  Turkish, 310
  U.S.A. and St. Lawrence Seaway, 411, 412
Biel dam (Austria), 145
Bilberry dam failure, 58, 63
Binnie, G. M., 347, 354–357
Binnie, W. J. E., 64
Birmingham
  Alabama, Lewis Smith dam near, 391
  England, water supply of, 102, 103
Blackbrook dam, Loughborough, earthquake damage at, 51–55
Blatherwycke dam, 44
Blight, J. H., 21, 48
Bomb damage of Clywedog dam, 102–105
Bomb damage of Möhne dam, 54, 55
Bomba dam (Italy), 268
Boone dam (U.S.A.), 18, 380–382
Borrowdale rocks, Haweswater dam, 96
Bort dam (France), 10, 167, 193–200
  crush zone between mica schist and gneiss under, 193–198
  large scale tests on rocks at, 198, 199
Bort reservoir
  leakage from, 23, 28, 200
Boscathno reservoir leakage, 48
Bottoms dam slip, 44, 45
Bou Hanifia dam (Algeria), 335, 341
Boucã dam (Portugal), 284
Bough Beech, 7, 89
Boulder Clay
  foundation for dams on, 7, 95, 100, 235
  material for construction with, 105, 107, 109, 227

Boulder Clay *continued*
  pore pressure discovery, 107–109
  vertical drains in, 95, 105
  (*See also under specific countries for dams built on*)
Boulder dam, 4
Boulders, embedded in clay, 95
Bouzey dam failure (France), 58
Box Canyon dam (U.S.A.), 363
Braunau dam (Germany), 143
Breachlaich rockfill dam, 72, 106, 138, 139
Bristol, water supply of, 7, 13
British dams, 75–141
  Aigas, (G) (L.B.) Old Red Sandstone. (R.B.) Conglomerate on Moine schists, 120
  Allt-na-Lairige, (P, G) Moraine on schists, 11, 106, 123–125
  Alwen, (E, G) Silurian, (G) on Denbighshire grit (E) on Boulder clay, 18, 76, 100
  Argal, (G) Granite, 3, 76, 89, 92
  Backwater, (E) Moraine and recent deposits, 7, 110–113
  Beannachran, (G) Mica-schist, 120
  bibliography, 140, 141
  Blackbrook, (G) Pre-Cambrian, 51–55
  Blatherwych, (E) Lias, 44
  Boscathno, (E) Granite, 48
  Bottoms, (E) Boulder Clay, 44, 45
  Bough Beech, (E) Weald Clay, 89
  Breachlaich, (R) Metamorphic, 106, 138, 139
  Broomhead, (E) Millstone Grit, 63–67
  Caban Côch, (G) Ordovician, slates, 11, 76, 102
  Chiliostair, (A) Granite gneiss Pre-Cambrian, 121
  Cluanie, (G), 106
  Clunie, (G), 106
  Clywedog, (B) Ordovician mudstones, 102–105
  Craig Côch, (G) Ordovician slates and conglomerates, 11, 76, 101, 102
  Cruachan, (B) Diorite Phyllite, 135
  Daer, (E) Silurian, 106, 109
  Derwent, (G) 'Yoredales' sandstone and shale, 6, 11, 76, 92, 93
  Drift, (G) Granite and China Clay, 19, 21
  Dunalastair, 13, 106
  English, 75–99
  Errochty, (B) Quartzite and granulite, 6, 11, 106, 130, 131
  Eyebrook, (E) Lower Lias Clay and limestones, 5, 13, 19, 76, 82
  Garry, 106, 136, 137
  Gaur, 13, 106
  Giorra, (B) Granulite, 130, 134
  Haweswater, (B) Andesite/Rhyolite, 26, 76, 95–99

British dams *continued*
  Howden, (G) 'Yoredales' sandstone and shale, 6, 11, 76, 92
  Kilmorak, (G) Conglomerate, Sandstone Lenses, 120
  King's Mill, (E) Coal Measures, 40
  Knockenden, (E) Moraine Boulder Clay, 106–109
  Ladybower, (E) Carboniferous Grits and shales, 6, 13, 76, 93
  Laggan, (G) 106, 114
  Lake District, (E) Silurian, Borrowdales, Ordovician, 76, 97
  Lamaload, (B) Yoredales grit and shales, 6, 392
  Langsett, (E) Millstone Grit, 13
  Lednock, (B) Epidiorite, 11, 106, 130–132
  Llanvaches, (E) Limestone, clogged by clay, 20
  Lluest Wen, (E) Coal Measures, 40–42, 49, 76
  Lochan-na-Lairige, Phyllites, quartzite, 106, 130, 133
  Loch Katrine, 106
  Loichel, (G) Schist, 120
  Loyne, (G), 106
  Lubreoch, (B) Quartz mica-schist, 130, 133
  maps of, 76, 106
  Monar, (A) Granulite Pre-Cambrian, 119–121
  More Hall, (E) Millstone Grit, grits and shales, 63
  Moriston, 106
  Muirfoot, (E) Moraine, Boulder Clay, 37, 107, 108
  Orrin, (G) (E) Granulite, 122, 123
  Pen-y-Gareg, (G) Ordovician slates and conglomerate, 11, 76, 101
  Pitlochry, (G) Moraine quartzite, 4, 13, 106, 114–118
  Powdermill, (E) Ashdown sand, 13, 19, 76, 78–80
  Quoich, (R) Metamorphic Schists, 12, 106, 136, 137
  Rannoch, 106
  Ridgegate, (E) Yoredales grits and shales, 27
  Scottish, 105–139
  Selset, (E) Boulder Clay, 76, 89, 95
  Shira (Glen)
    Lower Sron, (E) Phyllites; (G) Quartz Mica-schists, 4, 11, 25, 106
    Upper Sron, (B) Phyllite; Quartz (Dalradian), 125, 130
  Sloy, (B) mica-schist, fissured, 4, 11, 106, 130, 132
  Sutton Bingham, (E) Cornbrash, Forest Marble, 5, 13, 19, 32, 76, 82–89
  Stithians, (A, G) Granite, 6

British dams *continued*
  Treig, (E), 106, 136
  Trentabank, (E) Yoredales grits and shales, 7, 76, 93–95
  Usk, (E) Old Red Sandstone, clay, silt, 37, 76, 105
  Vyrnwy (Masonry), Silurian, Moraine, 11, 26, 76, 100
  Weir Wood, (E) Ashdown Sand, 13, 19, 76, 80–82
  Welsh, 100–105
  Wentworth (*see* Llanvaches), 20
Broomhead, 62–67
Brown, J. Gutherie, 85
Brushwood foundation, 235
Bulges
  definition of term, 75
  examples, 79–85
Bürg dam (Austria), 150
Burguillo dam (Spain), 287
Butenbach dam (Belgium), 162
Buttermere lake, 97
Buttress dams (*see specific dams under countries*)

Caban Côch dam, 11, 76, 102
Cables, pre-stressed, 92, 124, 144, 189, 319, 329
Cabril dam (Portugal), 284
Cachi dam (Portugal), 285
Calcareous tufa, 5, 52, 327, 332
Caldew Head, buried valley at, 23, 25, 28
Camarasa dam (Spain), 5, 36
Cambambe dam (Portugal), 285
Camber of crest, *see* Settlement of dams
Cambers in strata, definition of term, 75
Cambrian, dams founded on, see *under specific countries*
Campilhas dam (Portugal), 284
Canadian dams (Arrow, Mica, Portage), 413–418
  St. Lawrence seaway, 400–405
Canal (embankment) schemes
  Donzère Mondragon, 159–163, 166
  Montélimar, 163–167
  Rhine, 211, 213
  Rhône, 211, 230
  Stettin, 28
  Suez, 14
Canicada dam (Portugal), 284
Capillary attraction, cause of leakage, 48
Carboniferous Coal Measures, coal mining under dams in, 40–43
Carboniferous Limestone, trough fault leakage at Dol-y-Gaer, in, 28, 29
Carboniferous Millstone Grits, similarity of strata in Cheshire and Alabama, 12, 392
Casoli dam (Italy), 266–268
Castello-do-Bode dam (Portugal), 4, 284
Castiletto dam (Switzerland), 295, 296

INDEX

Castillon dam (France), 186–189
Castro dam (Spain), 286
Cataract dam (Australia), 42
Cavdarh Isar dam, (Turkey), 307
Caverns in limestone at Génissiat dam, 205
Cecita dam (Italy), 269
Cement grout, see Grouting
Cenajo dam (Spain), 287
Chambon dam (France), 166–170
Chastang dam (France), 167, 190, 191
Chatelot dam (Switzerland), 295
Chaudanne dam (France), 5, 10, 167, 189
Cheddar reservoir, 7, 14
Cheoah dam (U.S.A.), 364–366
Chew dam, 13
Chickamauga dam (U.S.A.), 5
Chilhowee dam (U.S.A.), 377–380
China, Shin Mung dam, 50
China Clay
    difficulties caused by, 3, 19–21, 92
    origin of, 92
Chingford dam settlement, 7, 38
Chord-height ratio, 8–11, (see also specific dams)
Claerwen dam site, 11, 76, 102
Clark, J. F. F., 350
Clay (see specfic dams under countries)
Clegg, Colin, 64
Clercq, J. de M., 162
Cluanie dam, 106
Clywedog dam, 102–105, 439
Clunie dam, 106
Coal Measures, dams founded on (see specific dams under countries)
Coastal barrage, 221–224
Cockrane, N. J., 102
Cogoti dam, Chile, 50
Col Nore dam (Algeria), 330–335
Compaction of dam embankments
    Garrison, 395, 396
    Gavins Point, 396
    Lewis Smith, 394, 395
    Oahe, 397
    Rosshaupten, 228
    Tuttle Creek, 396, 397
Compagnie Nationale du Rhône, 212
Composite dams
    Alwen, 18, 76, 100
    Boone, 18, 380–382
    Chilhowee, 377–380
    reasons for, 18
Compressive strength of rocks
    China Clay, 20
    conglomerate, 56
    gneiss, 198, 199
    limestones, 396, 397
    mica schist, 198, 199
    phyllite, 374
    quartzite, 374, 385
    sandstone, 388

Compressive strengh of rocks *continued*
    schists, 371
    shales, 396, 397
Comrie, Scotland, epicentre of earthquakes, 50
Concrete block dams (Italian)
    Platani and Pian Palu, 273
    Pozzillo, 271–273
Concrete dams (see specific dams under countries)
Concrete versus puddle in trench, 19
Conglomerate foundations (see specific dams under countries)
    failure of, at St. Francis dam, 56–58
    mistaken identity at Les Cheurfas dam of, 313
Coniston Water (lake), 97
Contra dam (Switzerland), 305
Conversions (Appendix 1), 427, 428
Cornbrash limestone, dam founded on, 82, 89
Cost of dams 451–454
Couesque dam, 3
Cow Green Reservoir, 68–70
Cover, G. W., 49
Coyne, M., 16, 192
Craig Côch dam, 11, 76, 101, 102
Cretaceous limestone foundations (see specific dams under countries)
Cronin, H. F., 47, 48
Crummock Lake, 26, 97
Crushing strength of rocks (see Compressive strength of rocks)
Cubak dam (Turkey), 308
Cumberland
    Caldew Head, proposed dam in, 23, 25, 28
    details of lakes in, 97
    formation of lakes in, 26
Cupola dams (see specific arch dams and chord-height ratios)
    geological conditions for, 8, 16

D'Erraguéne dam (Algeria), 311
Daer dam, 106, 109
Dale Dyke dam disaster, 58, 63
Dalradian Metamorphic Complex, situation of dams in, 125, 130
Dams (see also specific names and sites)
    'landscaping' of, 328
    plains, on, 13, 14, 160–161, 165
    principles of design, 431–435
Danube development, 13, 143, 159–161
Dargeau, M., statement on geology at Malpasset dam, 59–61
Deacon, G. F., 100
Decelle, A., 186, 190
Deformation (Deviation), 293, 295, 296
    after filling, 298, 299
    of concrete dams, 446

Deformation (Deviation) *continued*
 of earth dams, 445
 of strata, 448
 seasonal, 120
Delattre, P., 200, 215, 216
Demerkapru dam (Turkey), 309
Denizot, Professor, 219
Derwent dam, 6, 11, 76, 92, 93
Deterioration (*see also* Earthquakes, Leakage and Slips)
 concrete, loss of, 292
 grout curtain, 35
 leakage through absence of grout curtain, 35
 leakage through weak concrete, 33
 of pipes, 44, 45
 pore pressure, increase of
  Douglas, 372
  Fontana, 377
  Hiwassee, 371, 372
  Norris, 366
Derwentwater (lake), 97
Devonian formation, *see specific dams under countries*
Dickerson, L. H., 120, 122, 135, 453
Diessbach dam (Austria), 154
Differential settlement of the Vega de Tera dam, 59
Digue de la Route dam (Algeria), 11, 330–333
D'Iril Emda dam (Algeria), 311
Disasters, dam, 56–62 (*see also* Failures of dams)
 Vaiont reservoir, 8, 9, 143, 247–251
Diseworth, near Loughborough, epicentre of earthquake, at, 51
Dobra dam (Austria), 158
Dogger, dolomitized limestone in Jurassic at Vaiont, 243
Dokan dam, 28, 347–350
Dolgarrog dam disaster, 58, 59
Dollar, Dr., 51, 55
Dolomite limestones, *see* Italian dams, 236–277
Dol-y-gaer dam, 28, 29
Dome dams, surface features for (*see specific arch dams and chord-height ratios*)
Donzère Mondragon scheme, 212–217
 canals, 217
 diversion barrage (Donzère), 213
 effect on underground water levels, 215–217
 leakage losses, 220
Dordogne River, 8–10, 191–193
Dorias dam (Spain), 286
Dossett, J. H., 45, 47, 93
Douglas dam (U.S.A.), 372
Dowdeswell dam, 39
Drac River, 5, 27, 167, 176–185
Drift dam, 19–21
Drift deposits (*see under specific dams*)

Drossen dam (Austria), 143–152
Dunalastair dam, 13, 106
Durance Valley, 7, 9
'Duckbill' overflow, 335
Durlassboden (Austria), 148
Dyke, Zuider Zee, 235
Dykes sills, 69, 121

Earth or rockfill dams, *see also under specific countries*
 Algerian, 311–346
 Austrian, 142–161
 Australian, 419–423
 British, 75–141
 Canadian, 413–418
 French, 166–224
 German, 225–233
 Holland, Zuider Zee, 234
 Italian, 236–276
 Japanese, 358–362
 Pakistan (Mangla), 354–357
 Portuguese, 284
 Spanish, 286–288
 Swedish, 289, 290
 Switzerland, 291–306
 Turkey, 307–310
 United States of America, 363–400
Earthquakes
 Algerian dam damage from, 50, 51
 British dam damage from, 50–55
 California (an earthquake state), 399
 Chilian dam damage from, 50
 Chinese dam precautions against, 50
 Japanese dam precautions against, 359
 Scottish epicentre of, 50
 Turkish dam precautions against, 308–309
El Molinar dam (Spain), 287
Embankments (*see also* Earth dams)
 London M.W.B. reservoirs, 17
 Rhône diversions, canals, 215, 220
 Zuider Zee, 234
Elche dam (Spain), 286
Eleuterio reservoir (Italy), 273–276
Emali dam (Turkey), 308
Emergency emptying of reservoirs, 48
En el Rio Cubillas dam (Spain), 287
English dams, notes on, 75–78 (*see also* British dams)
Ennerdale Lake, 97
Eocene, *see under countries for* dams founded on
Epidiorite, *see under countries for dams founded on*
Epigenic valleys
 definition of, 23, 25
 examples and sites of dams in or near
  Aussois (Switzerland), 25
  Bort (France), 206

INDEX

Epigenic valleys *continued*
  examples and sites of dams *continued*
    Caldew Head (England), 23–25
    Génissiat (France), 204, 210
    Kaprun area (Austria), 154, 155
    Pitlochry (Scotland), 116
    St. Pierre-Cognet (France), 182, 183
    Sautet (France), 179
    Sron (Scotland), 25, 126
    Tignes (France), 173
    Valserine (France), 208, 210
Erhmann, M., 186
Errochty dam, 6, 11, 106, 130, 131
Escales dam (Spain), 287
Esch-sur-Sure dam (Luxembourg), 283
Etheridge, R., 20
Etna, dams near, 271
Ewden valley, 38, 63–67
Eyebrook dam, 5, 13, 19, 76, 82

Failures of dams
  Bilberry, upward pressure, 58
  Bouzey, scour under dam, 58
  Dale Dyke, undermining by water, 58
  Dolgarrog, foundation not deep enough, 58
  Dol-y-Gaer, trough fault, leakage, 28, 29
  Fergoug, flood, 328, 329
  Les Cheurfas, mistaken conglomerate, 58, 314
  Malpasset, weak foundation of left bank, 59–62
  Pérregaux, flood, 328, 329
  Pontéba, earthquake, 50, 51
  St. Francis, conglomerate faulty, 56, 57
  Vega de Tera, sliding, 59
  Woodhead, upward pressure, 58
Fedaia reservoir dams (Italy), 11, 27, 143, 257–261
Fergoug dam (Algeria), 328–330, 339
Flooding
  destruction due to,
    Fergoug or Perregaux, dam, at, 328
    Salzburg Bridge, at, 228
  prevention of, Sylvenstein dam, at, 228–230
  provision for
    Aigle, 'ski' jump, 191
    Barcis, 'morning glory', 262
    Beni Bahdel, at, 335
    Bort, 'ski' jump, 194
    Bou Hanifia, overflow channel and stilling pool, 340–341
    Digue de la Route, 'duckbills', 333
    Génissiat, 'ski' jump, 207–208
    Ghrib, overflow channel, 321
    Marèges, 'ski' jump, 192
    Pontesei Maè, 'morning glory', 255
    Sarno, 'Marguerite', 344

Flow data
  Arrow dam (Canada), 414
  assessment of (Appendix 3), 430
  Barcis, at, 266
  Beni Bahdel, at, 235
  Bou Hanifia, at, 339
  Chilhowee, at, 378
  Donzère (Rhône), at, 213
  Ghrib, at, 321
  Iroquois (St. Lawrence), at, 401, 404
  Pontesei, Maè at, 253–255
  St. Pierre-Cognet, at, 185
Fluoresein test for leakage, Fedaia reservoir, at, 27, 261
Flysch
  Barcis dam, at, 267
  Valle di Cadore dam, at, 239
Fodda dam (Algeria), 5, 326–328
Fontana dam (U.S.A.), 366, 372–377
Forest Marble, Jurassic, foundation for dam, 13, 16, 83
Forks dam, 4
Foundations *(see specific dams under countries)*
Foum el Gherza dam (Algeria), 312
Foum el Gueiss dam (Algeria), 312
France *(see French dams)*
Fréjus (France) Malpasset dam disaster, 59
French dams, 166–224
  Aigle (G) Gneiss (solid), 167, 190, 191
  André Blondel Power dam, (G) Cretaceous /Pliocene, 214
  bibliography, 223, 224
  Bort, (A) Gneiss/mica-schist, clay between, 4, 10, 167, 193–200
  Bouzey, Trias sandstone, 58
  Castillon, (A) Upper Lias limestone fissured, 5, 9, 10, 167, 186–189
  Chambon, (G) Gneiss, Trias, Lias under drift, 166–170
  Chastang, (G) Granulite (solid), 167, 191
  Chaudanne, (A) U. Lias limestone, 5, 10, 167, 189, 190
  Donzère Rhône diversion barrage, (G) Cretaceous/Pliocene, limestone/clay, 213, 214
  Génissiat, (G) Cretaceous/Eocene, 5, 25, 167, 200–208
  Henri Poincaré Power Dam, (E) Sandstone, 219
  Malpasset, (A) Gneiss with dolerite, 59–62, 167
  map of, 167
  Marèges, (A) Granite (Solid), 3, 8, 16, 167, 192, 193
  Monteynard, (G) Lias limestone, 185
  Rance Barrage, (G, E) Gneiss, 220–223
  Rochemaure Rhône diversion barrage, (G) Gargasian Marl, 217, 218
  Roche qui Boit dam, 192

French dams *continued*
  St. Pierre-Cognet, (A) Upper Lias limestone, 5, 167, 178–185
  Sautet, (A) Lias, alternating clays and limestones, 5, 9, 32, 167, 174–177
  Serre Ponçon, (E) Alluvium, 7
  Tignes, (A) Triassic Quartzite, 167, 170–174
  Zola, (MA), 192
French reservoirs
  Bort, Moraine, 200
  Génissiat, Cretaceous Quaternary, some Jurassic, 200, 209–211
  St. Pierre-Cognet, Upper Lias, alluvium, 178–182
  Sautet, Lower Lias, alluvium, 25, 28, 68, 167, 174–177
Fuller's earth, 33
Fulton, A. A., 114, 453, 454
Freiach dam (Austria), 157
Funil dam (Brazil), 285
Füssen, Rosshaupten dam, near, 28, 143, 225–228
Futase dam (Japan), 360

Gabbro (*see under countries for dams founded on*)
Gallina dam (Italy), 5, 9, 143, 241–243
Galloway Power Company, curved dams, 107
Garrison dam (U.S.A.), 395, 396
Garry dam (Garry and Moriston Hydro-Electric Scheme), 106, 136, 137
Gaur dam, 13, 106
Gavins Point dam (U.S.A), 396
Geddes, W. G. N., 109
Generalisimo dam (Spain), 287
Génissiat dam (France), 5, 25, 167, 200–208
Génissiat Reservoir, 209–211
Geological conditions for types of dam, 15–18
Geophysical surveys, adverse effect of variable strata, 22
Gepatch dam (Austria), 70, 148, 149
Gerlos dam (Austria), 63, 143, 146
German dams, 225–233
  bibliography, 233
  Braunau, 143
  Jochenstein, granite, 13, 143, 159–161
  map of, 143
  Möhne (G), 54, 55
  Passau (Kachlet) Granite, 143, 159–161
  Rosshaupten, (G) Oligocene; alternating sands and marls, 143, 225–228
  Sylvenstein, (E) Glacial deposits on dolomite, 143, 228–233
Ghrib dam (Algeria), 12, 38, 320–325
Gignoux, M., 223, 224
Giorra dam, 130, 134
Giudici (Vaiont Valley research), 247, 248
Glacial deposits, material, moraines, *see specific dams built on*

Glaciated valleys
  dams in, 23, 25
  lakes in, 26
  reservoirs in, 25, 26
Glaciers, action and effect of, 23, 24
Glasgow, Loch Drunkie, 107
Glen Shira dams, Sron Mor, 25, 106, 125–130
Glendevon dam, 35
Gmuend (Gerlos) dam (Austria), 70, 143–146
Gneiss (*see specific countries for dams founded on*)
Goguel, Prof. (Malpasset dam), 60
Gore, Jnr., C. E., 391
Gorge 'plugs' for dam foundations
  Barcis, 262
  Piave di Cadore, 237
  Sautet, 176
  Vaiont, 246
  Val Gallina, 242
  Valle di Cadore, 240
Gorges, types of dams in, 8
Göscheneralp dam (Switzerland), 295, 299, 300
Gourley, H. J. F., 30
Grande Dixence dam (Swiss), 295, 303
Granite formation (*see specific countries for dams founded on*)
Granulite (*see specific countries for dams founded on*)
Gravel (*see specific countries for dams founded on*)
Gravity dams (*see under specific countries*)
Great Neath fault, 29
Grenoble, dams near, 25, 28, 166, 174, 178
Greywacke
  Great Smoky district, in, 369
  Japanese dams on, 358–362
Grits (*see under specific countries for dams founded on*)
Grossraming barrage (Austria), 143
Grout
  early development, 30, 31
  modern development, 32, 33
Grout Manchette method of precise injection, 112
Grouting
  Argal, Granite, 91, 92
  Backwater, Moraine on schistose grits, 112
  Barcis, Cretaceous limestone, 266
  Boone, Ordovician and Cambrian (Knox Dolomite), 362
  Bou Hanifia, Miocene marl, 337, 339
  Casoli, Miocene limestone, 266–268
  Castillon, Upper Jurassic limestone, 187–189
  Chambon, Gneiss, 170
  Chiliostair, Granite gneiss, 122
  Daer, Boulder clay on Silurian limestone, 109
  Dokan, Dolomite limestone, 349

INDEX

Grouting *continued*
  Fedaia, Tuff, 261
  Fodda, Lias limestone, 327
  Fontana, Cambrian quartzite and phyllite, 374
  Genissiat, Cretaceous limestone, 207, 208
  Ghrib, Miocene grits and marls, 321
  Grande Dixence, Presinites, phyllites, 303
  Haweswater, Andesite, 99
  Hiwassee, Pre-Cambrian quartzite and schists, 371
  Isola, Gneiss, 300, 301
  Japanese, various, 358–360
  Kariba, Gneiss, 352
  Ladybower, Carboniferous grits and shales, 93
  Lednoch, Epidiorite, 132
  Les Cheurfas, Miocene limestone, 316–318
  Mauvoisin, Calcareous schists, 299
  Monar, Pegmatites, 120, 121
  Orrin, Granulite, 122
  Piave di Cadore, Upper Trias dolomite, 239
  Pitlochry, Moraine on Quartzite, 115, 116
  Pontesei (Maè), Upper Trias limestone, 252, 253
  Pozzillo, Miocene Sandstone, 271
  St. Pierre-Cognet, Upper Lias limestone, 184, 185
  Sarno, Pliocene conglomerate, 342, 343
  Sautet, Lias limestone, 175
  Sutton Bingham, Forest marble (mudstone) and Cornbrash limestone, 85, 87
  Sylvenstein, Moraine on upper Trias dolomite, 231–233
  Val Gallina, Upper Trias dolomite, 243
  Valle di Cadore, Middle Trias dolomite, 241
  Warragamba, Trias sandstone, 422
  Weir Wood, Ashdown sand, 81, 82
  Zervreila, Gneiss, 297, 298
Gruner, Edward, 62, 306
Guild, E., 20
Guilhofrei dam (Portugal), 284
Gulls, definition, 75
Gulneri Gorge (Dokan), 248, 249

Hamiz dam (Algeria), 312
Hatori dam (Japan), 361
Haweswater dam, 26, 76, 95–99
Hawksley, T., 20, 64
Hazards of sites, 19–22
Henri Poincaré Power dam (France), 219
Henry, M., 216, 219
Hill, G. H., 20
Hill, H. P., 30
Hiwassee dam (U.S.A.), 368–372
Holland, Zuider Zee Dyke, 234, 235

Hollingworth, S. E., 24, 75
Hollins reservoir site, 67
Holmfirth, early example of grouting at, 31, 63
Hoover dam, 4
Howden dam, 6, 11, 76, 92
Hunter and Keefe, 55
Hydraulic fill dams, 13, 17 (*see also* Earth dams)
Hydro-electric schemes on the Rhône, 200–208, 213, 214, 217–219

Igneous dykes, 11, 121
Igneous rocks (*see under specific countries for dams founded on*)
Ikari dam (Japan), 358
Ikowa dam (Japan), 359
Ishibuchi dam (Japan), 361
Irak, Dokan dam, 5, 347–350
Iroquois control dam (St. Lawrence Seaway), 13, 403, 404
Irrigation improvement on Rhône, 13
Isère, French Department, 166–185
Island Barn dam, trees on, 46, 48
Isola dam (Switzerland), 295, 300–303
Isotopes
  Fedaia dam, 257–261
  Rosshaupten dam, 28, 225, 228
  test for leakage, 27
Italian dams, 236–277
  Ancipa (Sicily), (B) Eocene, Soft sandstones, 270
  Barcis, (A) Upper Trias limestone, 143, 262–266
  bibliography, 277
  Bomba, (R) Alluvium on scagliosa; Plio-miocene clay, 268
  Casoli, (B) Miocene limestones, brescias, shales, 266–268
  Cecite, (A) Granite-Diorite, 269
  Fedaia, (B, G) Middle Trias dolomite; (E) Moraine, 11, 27, 143, 257–261
  Gallina, *see* Val Gallina
  Maè (Pontesei), (A) Upper Trias limestone, 143, 251–257
  map of, 143
  Piave di Cadore, (A) Upper Trias dolomite, 143, 237–239
  Pozzillo, (G) (Blocks) Miocene, 271–273
  Rossella, (E) Clay, Alluvium and compact clay, 273–276
  Scanzano, (E) Clay, Alluvium and compact clay, 273–276
  Vaiont, (A) Mid and Upper Jurassic limestone, 8, 143, 243–246
  Val Gallina, Upper Trias dolomite, 5, 143, 241–243
  Valle di Cadore, (A) Middle Trias dolomite, 143, 239–241

INDEX

Italian reservoirs
  Eleuterio, Cretaceous, 273–277
  Fedaia, 273–276
  Vaiont, Eocene, Cretaceous, Jurassic, 143, 247–251

Jacob, Professor, 219
Japanese dams, 358–362
Järkvissle dam (Sweden), 290
Jeanpierre, M., 198
Jochenstein barrage, 13, 143, 159–161
Joints, types of, 371
Jones, Lancaster, P. F. F., 35, 347
Jugoslavian dams, 278
Jurassic formation (see under specific countries for dams founded on)

Kachlet barrage, 143, 159–161
Kale, Saim, 310
Kamishiiba dam (Japan), 360
Kansas (U.S.A.), 396
Kaolization of Cornish granite, 92
Kaprun dams (near Gross Glockner), (Austria), 143, 150–155
Kariba dam, 351–353
Karroo formation, 351
Karstic limestone, 200–208
Keban dam (Turkey), 309
Keefe, Hunter and, 55
Kennard, J., 95
Kennard, M., 69
'Kentledge' (loading) tests, 20
Keuper Marl (see specific countries for dams founded on)
Kilmorak dam, 120
Kimmons, G. H., 390
King's Mill dam, 40
Knill, Dr. John, 69, 439
Knockenden dam, 106–109
Knox Dolomite, 380, 381
Kobayashi, Tai, 358
Kops dam (Austria), 144
Krauss, Dipl.-Ing. Joseph, 228
Ksob dam (Algeria), 313
Kurobegawa dam (Japan), 360

La Coruna dam (Spain), 286
La Gileppe dam (Belgium), 162, 163, 167
Ladybower dam, 6, 13, 76, 93
Laggan dam, 106, 114
Lake District, 76, 97
Lake Eigiau (Dolgarrog dam), 58
Lake Geneva, 211
Lake Ontario, 403
Lake Sanabria (Vega de Tera dam), 59
Lamaload dam, 6, 392
'Landscaping' of reservoirs, 328

Landslides
  Maè reservoir, 71, 255, 257
  Vaiont reservoir, 71, 73, 247, 251
Langsett dam, 6, 13
Lapworth, H., 34, 49, 64, 79
Large-scale tests (see Tests)
La Sarra dam (Spain), 286
Las Conchas dam (Spain), 286
Las Picades dam (Spain), 286
Lavaud-Gelade dam, 3
Leakage
  calculation of, prior to construction of St. Pierre-Cognet dam, 179–182
  canal, Donzère Mondragon, 220
  detection by isotopes, 27, 226, 261
  from reservoirs (because of pre-glacial valleys), 28
    Bort, 200
    St. Pierre-Cognet, 179–182
    Sautet, 28, 177, 178, 179
  through capillary attraction, 48
  under dams, 27–29
    Broomhead, Millstone grit, 28
    Castilletto, Serpentine, 298
    Castillon, Limestone, 188
    Cheoah, Quartzite, 366
    Digue de la Route (Beni-Bahdel reservoir), Jurassic limestone, 334
    Dokan, Cretaceous limestone, 349
    Dol-y-Gaer, Carboniferous limestone, (in trough fault), 29
    Glendevon, Andesite, 35
    Isola, Gneiss (increase), 302, 303
    Les Cheurfas, Miocene limestone, 315–318
    May dam, Alluvium, 309
    Nantahala, Quartzite, 384
    Sarno, Conglomerate, 343
    Schrah dam, Cretaceous, 292
    Tignes, Quartzite, 174
Lednoch dam, 11, 106, 130, 132
Leggett, R. F., 43
Leonardi, Prof. Piero, 249, 250
Les Cheurfas dam (Algeria), 38, 313–320
Lewis Smith dam (U.S.A.), 12, 391–395
Lewisian gneiss, 121
Lewiston Power-House, 411
Lias (see under countries for dams founded on)
Limberg dam (Austria), 143, 150, 151
Lime water, pumped into ground, 289, 290
Limestones (see also specific countries for dams founded on)
  effect on dam construction, 5
Llanvaches dam, condemned site successful for, 20
Lloyd Roberts, W. M., 47
Lluest Wen dam, 40–42, 49
Loading tests, full scale (see Tests)
Lochaber scheme, Laggan and Treig dams, 86, 114, 136

464

INDEX

Lochan na Lairige dam, 106, 130, 133
Loch Katrine, 106
Loch Quoich dam, 106, 136, 137
Loch Sloy dam, 4, 11, 106, 130, 132
Loch Treig dam, 106, 136
Loire valley, Rieutord dam, 4
Lockport Dolomite, 406, 409, 410
Loichel dam, 120
London Clay
  earthen embankments on, 77
  effect on dam construction, 6
Long Sault dam (St. Lawrence Seaway), 403
Lorenz, Dr. Ing. W., 228
Loriga, C., 247
Los Peares dam (Spain), 286
Lower Sron Mor dams, 4, 11, 106, 125, 130
Lower Lias (*see under countries for dams founded on*)
Lower Trias (*see under countries for dams founded on*)
Lower Carboniferous (*see under countries for dams founded on*)
Loweswater (Lake), 97
Loyne dam, 106
Lubreoch dam, 130, 133
Luestwen dam, 40, 42, 49, 76
Lugeon, M., 175, 209
Lunersee dam (Austria), 36, 143, 145
Luxembourg (Esch-sur-Sure dam), 283

Maè dam (Italy), 9, 71, 143, 251–257
Makio dam (Japan), 360
Malpasset (Fréjus dam) (France), 59–62, 167
Malvaglia dam (Switzerland), 295
Manchette method of precise grouting, 112
Mantovani, M. C., 247
Mattmark dam (Switzerland), 305, 306
Maps (sites of dams)
  Algeria, 312
  Austria, 143
  Belgium, 167
  England, 76
  France, 167
  Germany, 143
  Italy, 143
  Jugoslavia, 143
  Luxembourg, 167
  St. Lawrence Seaway, 400
  Scotland, 106
  Turkey, 308
  U.S.A. Tennessee Valley (part), 389
  Wales, 76
Maranhão dam (Portugal), 284
Mangla dam (Pakistan), 354–357
Marcello, Dott. Ing. C., 273
Mardale Church, 99
Maregès dam (France), 3, 8, 167, 192, 193
Margaritze dam, 143, 153–154
'Marguerite' overflow, 344–345

Marr, J. E., 26
Martel, M., 203
Maruyama dam (Japan), 358
Massena dam (U.S.A.), 404
Mauvoisin dam (Switzerland), 295, 298
Mer de Glace, 26
Mesozoic strata (*see under countries for dams founded on*)
Metamorphic rocks (*see under countries for dams founded on*)
Metropolitan Water Board
  Chingford dam, 38
  Island Barn reservoir, 47, 48
  long earthen embankments, 17
Miboro dam (Japan), 360
Mica dam (Canada), 413, 414, 416
Mica schist (*see under countries for dams founded on*)
Mill, H. M., 97
Millstone Grit (*see* Carboniferous Millstone Grits *and under countries*)
  pre-glacial gorge in, near Yarrow reservoir, 28
Mining and dams, 40–43
Miocene (*see under countries for dams founded on*)
Missouri Basin earth dams
  compaction data, 395–397
Models
  Bou Hanifia stilling pool, 341
  Ghrib overflow channel, 324
  reliability of, 9
Möhne dam (German), 54, 55
Moine schist plane, 119
Moiry dam (Switzerland), 295
Molasse (Oligocene formation), 225
Möll dam (Austria), 143, 153
Möll reservoir, 152–154
Monar dam, 9, 119–121
Mondragon (*see* Donzère Mondragon) 212–220
Montargil (Portugal), 284
Moor Hall dam, 63
Montélimar scheme, 212, 217–220
  canal, 219, 220
  diversion barrage (Rochemaure), 218
  power dam (Henri Poincaré), 219
Monteynard dam (France), 185
Mooser dam (Austria), 143, 152–154
Moraines (*see specific dams on*)
Moriston dam, 106
Morning glory spillways for earth dams, 255, 262, 268, 388
Movement in dams
  concrete, *see under* Deformation
  earth curious forward movement, 390
Mühldorfersee dam (Austria), 157
Mühlrading dam (Austria), 143
Muirfoot dam, 39, 107, 108
Multiple arch dams
  Algerian, 330–335

INDEX

Multiple arch dams *continued*
  fundamental requirements of site, 11, 16
Munich, flood abatement for, 228

Nanpantan, geophysical survey, 22
Nantahala dam (U.S.A.), 12, 382–384
Narrow valleys, types of dam in, 9, 10
Narugo dam (Japan), 360
Navigation improvement
  Danube, 159–161
  Rhine, 13, 14
  Rhône, 13, 200, 220
Niagara Falls
  improvement of flow by control dam, 14, 405–407
  pumped storage hydro-electric scheme, 14, 407–411
Nicholson, W., 64
Norris dam (U.S.A), 366, 367
North Carolina, 364, 372, 382
North Dakota, 395
North of Scotland Hydro-Electric Board, 115–139
Nozori dam (Japan), 361
Nukabira dam (Japan), 359

Oahe dam (U.S.A.), 397
Oberaar dam (Switzerland), 295
Odomari dam (Japan), 359
Ogochi dam (Japan), 358
Old dams, 44–49, 286, 307, 308
Old Red Sandstone
  South Wales dams founded on, 100
  trough fault in marl, 29
Oliana dam (Spain), 287
Oligocene (*see under countries for dams founded on*)
Oliver, G. C. S., 45, 82
Omorigawa dam (Japan), 359
Ordovician formation (*see under specific countries for dams founded on*)
Orleansville, earthquake epicentre, 50
Oroville dam (U.S.A.), 398–400
Orrin dam, 122
Orukaya dam (Turkey), 307
Ottenstein dam (Austria), 156
Ova Spin dam (Switzerland), 305
Overflow channels, special treatment at
  Bou Hanifia dam, 340, 341
  Ghrib dam, 323, 324
Overflow weirs
  types of bellmouth spillways
    'duckbills', 335
    'marguerite', 344
    'morning glory', 255, 268, 388
    movable, 262
Oxfordian (*see specific countries for dams founded on*) (*see also Jurassic*)

Pakistan, Mangla dam, 354–357
Palaeozoic rocks (*see specific countries for dams founded on*)
Palo, G., 368, 384, 390
Paradela dam (Portugal), 284
Passau, Kachlet barrage (Germany), 143, 159–161
Paton, J., 125
Paton, T. A. L., 351
Pego de Altar dam (Portugal), 284
Pendred, B. W., 59
Pendulum measurements (*see* Deformation)
Pennines
  Dale Dyke disaster, 58
  grouting of reservoirs in, 32, 33
  multiplicity of dams in, 17
  Scammonden motorway dam, 6
Pennsylvania Series, Alabama, U.S.A., 12, 392
Pentwyn dam, 29
Pen-y-Gareg dam, 11, 76, 101
Percolation (Permeability), *see* leakage
Permian (*see specific countries for dams founded on*)
Pernegg dam (Austria), 143
Perrégaux dam (Algeria), 328–330, 339
Perrott, W. E., 35
Peterson, E. R., 400
Phyllites (*see specific countries for dams on*)
Piave di Cadore dam (Italy), 10, 143, 237–239
Picote dam (Portugal), 284
Pitlochry dam, 4, 11, 106, 114–118
Plains, dams sited in, 13, 214, 219
Pliocene (*see specific countries for dams on*)
Plugged gorges
  Barcis, 262
  Piave di Cadore, 237
  Vaiont, 246
  Val Gallina, 242
  Valle di Cadore, 240
Pneumatolysis, action on Cornish granite, 92
Pontéba dam (Algerian), 50, 51
Pontesei dam (Italy), 71, 143, 251, 257
Pontsticill reservoir (Taf Fechan), 28
Pore pressure, uplift
  concrete dams (Appendix 9), 443
  Douglas dam, 372
  earth dams (Appendix 8), 442
  Fontana dam, 377
  Hiwassee dam, 371, 372
  Isola dam, 301
  Knockenden dam, 107–109
  Norris dam, 366
  Rosshaupten dam, 228
  Sutton Bingham, 87–89
  uplift under dams (Appendix 10), 444
  Usk dam, 105
Portage dam (Canada), 4, 13, 416–418
Portuguese dams, 284, 285
Poudingues, Sarno dam, 341–345
Powder-Mill dam, 13, 19, 76, 78–80

Pozzillo dam (Italy), 271, 273
Pracana dam (Portugal), 284
Pre-Cambrian (*see specific countries for dams founded on*)
Pre-glacial valleys
  Caldew (England), 23, 24
  Dordogne (France), 167, 200
  Drac (France), 25, 167, 177, 185
  Glen Shira (Scotland), 106, 125, 129
  Isère (France), 25, 167, 170, 171
  Rhône (France), 167, 204
  Tummel (Scotland), 106, 116
Pre-stressed concrete
  Allt na larige dam, 124
  raised dams, 92, 144, 329
  reinforcing strata, 189, 319
Prosperina dam (Spain), 286
Puenta Alta dam (Spain), 286
Punt dal Gall (Switzerland), 305

Quartz, and Quartz porphery, Quartzite (*see under specific countries for dams founded on*)
Quoich dam, 106, 136, 137

Raffety, S. R., 71
Raising dams, 92, 144, 156, 319, 329
Rance Barrage (France), 220–223
Rannoch dam, 106
Ratherichs boden dam (Switzerland), 105, 295
Redhill source of Bentonite, 33
Reinforcing strata, 187, 189, 195, 196, 257, 270, 319 (*see also* Grouting)
Repairs to dams, 27, 44, 45
Reservoir geology, 63–72
  Ashop Valley, 63
  Cow Green reservoir, 69–70
  Ewden Valley, 63–67
  Hollins reservoir, 67
  Pontesei Valley, 71
  Sautet reservoir, 68
  Scammonden Valley, 67
  Scarborough Cliff, 67
  slides in Austria, 70
  Vaiont Valley, 72, 247–251
Reservoirs (Safety Act), 62
Reservoirs (*see specific reservoirs and dams*)
Respomuso dam (Spain), 287
Reyran River, 59
Rhine development, 213
Rhodesia, Kariba dam, 351–353
Rhône valley
  above Génissiat dan, 209–211
  below Génissiat dam,
    canalization of, 211–220
    diversion, purpose of, 212–213
    geology of diversion works, 200–208
    leakage loss in canal, 220

Rhône valley *continued*
  below Génissiat dam *continued*
    pre-glacial valley in, 204, 205
    underground water interference with, 215–217
Ricobayo dam (Spain), 286
Rivers (*see specific rivers*)
Robert Moses barrage (Barnhart Island dam, U.S. side), 401
Robert Saunders barrage (Barnhart Island dam, Candian side), 401
Roberts, C. M., 120
Roberts, G. Ewart, 40
Robertsville dam (Belgium), 162
Rochemaure diversion barrage, (France), 217, 218
Roche qui boit dam (France), 192
Rock-bar at Vyrnwy dam, 26, 100
Rocks (*see under specific countries for dams founded on*)
Rodriguez dam (Mexico), 364
Roots of trees, 48
Rosenau dam (Austria), 143
Rossella dam (Italy), 273–276
Rossens dam (Switzerland), 295
Rosshaupten dam (Germany), 28, 143, 225–228
Rio di Bosco Nero stream, 255
Rötguldensee (Austria), 155
Rothe, M., 50
Russi, D. H., 247

Sabden shale, olim Yoredales, 93
St. Francis dam, California, 56–58
St. Lawrence River development, 13, 14, 400–404
St. Pierre-Cognet dam (France), 5, 9, 33, 167, 183–185
St. Pierre-Cognet reservoir (France), 167, 172–182
  confirmation of predicted loss, 182
Sakuma dam (Japan), 358
Salamonde dam (Portugal), 284
Salime dam (Spain), 286
Salza dam (Austria), 158
Salzach River, 143, 150, 228
Sambuco dam (Switzerland), 295, 296
San Esteban dam (Spain), 286
San Juan dam (Spain), 287
Sand, Sandstone, (*see under specific countries for dams founded on*)
Santa Eulalia dam (Brazil), 285
Santa Luzia dam (Portugal), 284
Sarno dam (Algeria), 8, 341–345
Sautet dam (France), 5, 9, 32, 167, 174–177
Sautet reservoir (France), 25, 28, 68, 167, 177–178
  leakage at, 177–178
Sazanamigawa dam (Japan), 360

INDEX

Scaglia (Upper Cretaceous), 243
Scammonden dam, 6, 449
Scanzano dam (Italy), 273–276
Scarborough slip, 67
Schists and Mica schists (*see specific countries for dams founded on*)
Schräh dam (Switzerland), 291, 292
Schwabeck dam, 143
Scottish dams, notes on, 105–107
Sedimentary rocks, 11–13, 17
Seepage (*see* Leakage)
Seeuferegg dam (Switzerland), 3
Seismic tests, 438, 439
Selli, R., 249–251
Selset dam, 76, 89, 95
Semenza, Dott. Ing. Carlo, 16, 71, 72, 237, 239, 257
Semenza, Prof. Eduardo, 71, 247–250
Sericite, 61
Serre Ponçon dam, location of (France), 7, 167
Settlement of dams, 37–39, 449, 450
  concrete (Bort), 199–200
  earth (castiletto), 296
  earth (Dowdeswell), 39
  earth (Göscheneralp), 299
  rockfill (Bou Hanifia), 337
  rockfill (South Holston), 390
  rockfill (Watauga), 386
Shear (*see* Sliding of dams)
Sheepstor dam, deep trench in China Clay, 17
Sheet piling, as a grout screen, 364
Shin Mung dam, 50
Shira (Glen) dams, 4, 11, 25, 106, 125–130
Shira Wata dam (India), 35
Silent Valley dam, 20
Silicate sodium (*see* Grouting)
Silting in reservoirs
  gravel embankment, stabilization, 217
  Jurassic, Fodda reservoir, at, 328
  Lias, Sautet reservoir, 178
Silurian (*see also under specific countries*)
  Niagara Falls, at, 405–406
Silves dam (Portugal), 284
Silvretta dam (Austria), 145
Sipsey Fork dam, 391, 395
Site investigation (Code of Practice), 19
Site maps (*see* Maps)
Siting of dams in glaciated valleys, 23–26
Skerries, 22
Ski-jumps
  Aigle dam (Gneiss), 191
  Bort dam (Gneiss), 194
  Marèges (Granite), 192
Skiddaw, Ordovician, proposed Caldew Head dam, 23–25
Sliding of dams or embankments
  Chingford, London Clay, 38
  Muirfoot, Boulder Clay, 37

Sliding of dams or embankments *continued*
  St. Francis, schists, 56
Slips of strata
  Ashop Valley, Sabden Shales, alim Yoredale, 63
  Bort Abutment, Schists and clays, 194
  Chaudanne abutment, Upper lias limestone, 189, 190
  Ewden Valley Waldershelf Slip, Millstone Grit, 38, 63–67
  Ghrib over channel, Miocene marls, 35, 323
  Gmuend, Quartz schists, 70, 71
  Maè, Pontesei, Trias limestone, 257
  Sautet reservoir (potential lias limestones and clays), 178
  Scarborough, Coastal clays, 67
  Vaiont reservoir, Cretaceous, 71, 72, 247, 251
Sloy dam, 4, 11, 106, 130, 132
Smith, C. C., 30
Sodium silicate (*see* Grouting)
Soft material (*see specific dams built on*)
South Dakota, 396, 397
South Holston dam (U.S.A.), 384, 390
Sowers, Professor G. F., 391
Spanish dams, 286–288
Spillways (*see* Overflow weirs and channels)
Spitallamm dam (Switzerland), 3, 292–295
Spüllersee dams (Austria), 144
Sron Mor dams (*see also* Shira dams), 4, 11, 25, 106, 125–130
'S'-shaped dams (on plan)
  Biel, 145, 146
  Fedaia, 258, 260
  Mangla, 355
  Pozzillo, 272
Staines reservoirs, 7
Staning dam (Austria), 143
Stewarts and Lloyds, 45, 82
Stithians dam, 438, 439
Storage and yield of reservoirs, relation between, 429
Subsidence, 37, 38 (*see also* Settlement and slips)
Suez Canal, 14
Sufers dam (Switzerland), 295
Sunnen, T., Dr. Theo., 285
Suorva dam (Sweden), 289
Surface features, determination of types of dams, 8–14
Suspicious old dams, 44–49
Suter, M., 186
Sutton Bingham dam, 2, 13, 19, 32, 76, 82–89
Swedish dams, 289–290
Swiss dams, 291–306
  Albigna, (G) Granite 295, 298
  bibliography, 305, 306
  Castiletto, (E) Serpentine and Serpentine slide, 295, 296

U.S.A. dams *continued*
  Lewis Smith, (R) (Olim Sipsey Fork) Lower Coal Measures, 391–395
  Nantahala, (R) Cambrian quartzite, 12, 382–384
  Norris, (G) Cambrian dolomite, 366, 367
  Oahe, (R) Tertiary sands, clay, 397
  Oroville, (R) Gravel, gold mining tailings, 398–400
  Rodriquez (Mexico), (A) Alluvium, 364
  South Holston, (R) Ordovician sandstone, 384–390
  Tims Ford, (R) Limestones/shales, 390–391
  Trinity, (R) Basalt, andesite, 397–398
  Tuttle Creek, (R) Permian shale, 396, 397
  Watauga, (R) Cambrian quartzite, 12, 384–390
U.S.A. and Canada
  Niagara Falls flow control dam, 405–407
  Niagara Falls pumped storage schemes, 407–411
  St. Lawrence Seaway Development, 400–405
    Beauharnois Dam, (G) Sandstones, 401
    Barnhart Dam, (G) Ordovician, 402
    Iroquois Dam, (G) Ordovician, 404–405
    Long Sault Dam, (G) Ordovician, 403
    Massena Dam, (G) Ordovician, 404
Underground water
  dumping clay for dam foundation on, 414
  interference with, 216, 217
  limits, top water level, 68–70
Uplift (Appendices 8, 9, 10), 442, 443, 444 (*see also* Pore pressure)
Upper Lias (*see under specific countries for dams founded on*)
Ural, O. M., 308
Usk dam, 37, 76, 105

Vaiont dam (Italy), 8, 143, 243–246
Vaiont reservoir (Italy), 8, 9, 143, 247–251
Vale de Gaio dam (Portugal), 284
Val Gallina dam (Italy), 5, 9, 143, 241–243
Valle di Cadore dam (Italy), 143, 239–241
Valserine, 25, 208
Vega de Tera dam (Spain), 59
Venda Nova dam (Portugal), 284
Vermunt dam (Austria), 143, 145
Verney, A. F. C., 48
Verrou (glacial hillock) definition, 25
  Sron dams, 126–128

Verrou (glacial hillock) *continued*
  Kaprun dams, 154–155
Vertical strata, dams, on
  Casoli, 268
  Hamiz, 312
  Pozzillo, 271
  Rosshaupten, 225–228
  Spüllersee, 144
Vesdre dam (Belgium), 164, 165
Vilar dam (Portugal), 284
Villalcampo dam (Spain), 286
Volcanic tuffs, 4
Vyrnwy dam, 11, 26, 76, 100

Wade, L. A. B., 43
Wadhurst Clay, 78
Wafa, T. A., 12
Waggital dam (Switzerland), 5
Wakefield, Baitings dam, 6
Wasserfall reservoir (Austria), 150
Wast-Water Lake, 97
Watauga dam (U.S.A.), 12, 384–390
Weald, 13
Weathering, effect on dam construction, 5
Weir Wood dam, 13, 19, 76, 80–82, 292
Welsh dams, 76, 100
Wentworth dam (alias Llanvanches dam), 20
Whitehaven, water supply for (Ennerdale lake), 26
Wide valleys, types of dams in, 10–14
Windermere Lake, 97
Wing trench grouting, 97
Woodhead embankment, 58, 63
Workington, water supply for (Crummock Lake), 26

Yarrow reservoir, 28
Ybbs-Persenbeug dam (Austria), 143, 159
Yoredale rocks (sabden shales), dams founded on, 73
Young, A., 42
Yorkshire, Pennine reservoirs, 32

Zardézas dam (Algeria), 313
Zerbine dam (Italy), 4
Zervreila dam (Switzerland), 295–298
Zeuzier dam (Switzerland), 295
Zola dam (France), 192
Zuider Zee dyke, 14, 234, 235

INDEX

Swiss dams *continued*
  Chatelot, (A) Jurrassic limestone, 295
  Contra, (A) Gneiss, 305
  Göscheneralp, (E) Alluvium/Granite, 295, 299, 300
  Grande Dixence, (G) Schists slates, 295, 303
  Isola, (A) Gneiss, clay, 295, 300–303
  Malvaglia, (A) Rock, 295
  Mattmark, (E) Moraine, 303, 305
  Mauvoisin, (A) Calcareous schists, 295, 298, 299
  Moiry, (A) Metamorphic, 295
  Oberaar, (G) Gneiss/schists, 295
  Ova Spin, (A) Dolomite, 305
  Punt dal Gall, (A) Dolomite, 305
  Ratherichsboden, (G) Granite (compact), 295
  Rossens, (A) Miocene, 295
  Sambuco, (A) Gneiss, 295, 296
  Schrah, (G) Cretaceous, 291, 292
  Seeuferegg, (G), 3
  Spitallamm, (G) Granite (Sound), 292–295
  Sufers, (A) Gneiss, 295
  Waggital, Alluvium, 5
  Zervreila, (A) Gneiss, 295–298
  Zeuzier, (A) Jurassic limestone, 295
Sylvenstein dam, 7, 143, 228, 233

Taf Fechan dam, 29
Talybont reservoir, 100
Tase dam (Japan), 359
Taylor, G. E., 95
Tennessee Valley dams, 4, 5, 12, 18, 366, 373, 377, 380–384, 390
Ternberg barrage (Austria), 143
Terzaghi, 28, 48
Tests (Appendix 5), 436–437
  bearing and settlement
    Allt-na-Lairige dam, 123
    Beni-Bahdel dam, 332, 333
    Bort dam, 199, 200
    Drift dam, 19, 20, 21
  compaction
    Garrison dam, 395, 396
    Gavins Point dam, 396
    Lewis Smith dam, 394, 395
    Oahe dam, 397
    Scammonden, 449, 450
    Tuttle Creek dam, 397, 398
  permeability, *see* Leakage
  settlement
    Bort dam, at, 198
    South Holston dam, at, 385
    Watauga dam, at, 383
  shear, Lewis Smith dam, at, 395

Tests *continued*
  sliding, Beni-Bahdel dam, at, 333
Thick arch dams
  fundamental requirements of site, 16
  reliability of, 10
Thin arch dams, fundamental requirements of site, 16
Thirlmere Lake, 26, 97
Thixotropic clay-grout, 440–441
Thompson, R. W. S., 72
Tiddeman, R. H., 20
Tignes dam (France), 10, 25, 28, 167, 170–174
Tims Ford dam (U.S.A), 390, 391
Tonoyama dam (Japan), 360
Topographical conditions and dams, 15–21
Tranco de Beas dam (Spain), 287
Trees on embankment, 46–49
Treiber, Dr. Ing. Fredrich, 225
Treig dam, 106, 136
Trentabank dam, 7, 76, 93–95
Trevison, R., 249–251
Trias (*see specific countries for dams founded on*)
Trinity dam (U.S.A.), 397, 398
Tufa (calcareous)
  Beni-Bahdel dam, 332
  seepage through dam, 52, 53
Tuff (volcanic)
  Fedaia dam, at, 261
  Japanese dam, at, 361
Turkish dams, 307–310
  ancient dams, 307–308
  bibliography, 310
  earthquake-resisting dams, 308–310
  map of, 307
Tuttle Creek dam (U.S.A.), 396, 397
Types of dams, topographical and geological conditions for, 15–18
Types of rocks, as they affect dam construction, 3–7

Ullswater Lake, 97
U.S.A. dams, 363–400
  bibliography, 411–412
  Boone, (G, E) Ordovician knox dolomite, limestone and shales, 18, 380, 382
  Box Canyon, (A) Alluvium, sand, 363
  Cheoh, (G) Cambrian quartzite, 364–366
  Chilhowee, (G, E) Cambrian sandstone and shales, 377–380
  Douglas, (G) Knox dolomite Ordovician, 372
  Fontana, (G) Cambrian quartzite, 366, 372–377
  Garrison, (E) Tertiary sand and clay, 395, 396
  Govins Point (E) Niobrara chalk, 396
  Hiwassee, (G) Pre-Cambrian quartzite, 368–372